Available Energy and Environmental Economics

Available Energy and Environmental Economics

Editors

Junpeng Zhu
Xinlong Xu

MDPI • Basel • Beijing • Wuhan • Barcelona • Belgrade • Manchester • Tokyo • Cluj • Tianjin

Editors
Junpeng Zhu
Central China
Normal University
China

Xinlong Xu
Hunan Normal University
China

Editorial Office
MDPI
St. Alban-Anlage 66
4052 Basel, Switzerland

This is a reprint of articles from the Special Issue published online in the open access journal *Energies* (ISSN 1996-1073) (available at: https://www.mdpi.com/journal/energies/special_issues/Available_Energy_Environmental_Economics).

For citation purposes, cite each article independently as indicated on the article page online and as indicated below:

LastName, A.A.; LastName, B.B.; LastName, C.C. Article Title. *Journal Name* **Year**, *Volume Number*, Page Range.

ISBN 978-3-0365-7280-2 (Hbk)
ISBN 978-3-0365-7281-9 (PDF)

Contents

energies

Article

Can the Relationship between Atmospheric Environmental Quality and Urban Industrial Structure Adjustment Achieve Green and Sustainable Development in China? A Case of Taiyuan City

Xi Zhang [1], Jiayu Zheng [2,*] and Ligang Wang [2,*]

1 School of Economics, Beijing Technology and Business University, Beijing 100048, China; zhangxi@btbu.edu.cn
2 School of Economics and Management, University of Chinese Academy of Sciences, Beijing 100190, China
* Correspondence: zhengjiayuzjy@126.com (J.Z.); tyzfwlg@163.com (L.W.)

Abstract: Atmospheric environmental quality affects the high quality and sustainable development of the economy. The optimisation and upgrading of the industrial system are important to improve the operation efficiency of the economy and society. Firstly, this paper constructs the theoretical analysis framework of coupling and coordination between the atmospheric environment system and the industrial system and analyses the internal mechanism of the interaction and coordinated development of the two systems. Then, it puts forward the combination of the coupling coordination model and the VAR model (Vector autoregressive model) and presents the analysis and evaluation method of the relationship between them from the two perspectives of "static" and "dynamic". Finally, the empirical study is conducted in Taiyuan, a resource-based city in China. The results show that: (1) The two systems in Taiyuan have an obvious interaction and develop in the direction of benign coupling. (2) The impact of the two systems on each other is mainly in the medium and long term and dominated by the role of the atmospheric environment system on the industrial system. This study provides a theoretical framework and evaluation methods for evaluating and analysing the relationship between the urban atmospheric environment system and the industrial system in China, and then provides suggestions for policymaking.

Keywords: atmospheric environmental quality; industrial structure; Taiyuan city; coupling coordination model; VAR model

Citation: Zhang, X.; Zheng, J.; Wang, L. Can the Relationship between Atmospheric Environmental Quality and Urban Industrial Structure Adjustment Achieve Green and Sustainable Development in China? A Case of Taiyuan City. *Energies* **2022**, *15*, 3402. https://doi.org/10.3390/en15093402

Academic Editors: Junpeng Zhu, Xinlong Xu and Luigi Aldieri

Received: 14 April 2022
Accepted: 3 May 2022
Published: 6 May 2022

1. Introduction

Atmospheric environmental quality is the key index to measure economic quality and sustainable development. After experiencing rapid industrialisation, China is also facing problems, such as environmental pollution restricting economic development, and endangering health. As early as 1982, China promulgated the standards and specifications for evaluating atmospheric environmental quality, which were revised and improved in 1996, 2000, and 2012. Existing studies have conducted rich research on atmospheric environmental quality evaluation methods. The main methods and models include the RBF (Radial Basis Function) network analysis method [1,2], the fuzzy mathematics method [3], the set pair analysis method [4], the grey clustering method [5], the AHP model (Analytic Hierarchy Process) [6,7], the DPSIR model (Drive-Pressure-State-Impact-Response model) [8,9], etc. Lu et al. [10] studied and analysed that under the new air quality evaluation standards, the atmospheric environment of more than 70% of prefecture-level cities in China is overloaded. The improvement of environmental carrying capacity is very urgent. Dong et al. [11] analysed China's atmospheric environmental quality from 2015 to 2019 according to the data of monitoring stations and found that China's air quality index and the concentration of six types of pollutants were significantly improved during the study sample period. The study pointed out that PM2.5 (particulate matter) is the most important pollutant affecting China's air quality. As people pay more and more attention to environmental issues, the Chinese government has adopted a positive action plan to promote the governance of the atmospheric environment. In 2013, the action plan for preventing and controlling air pollution was issued, which made specific action guidelines for "winning the Blue-Sky Protection Campaign". In 2018, the three-year action plan for winning the Blue-Sky Protection Campaign was further released to deepen the governance of end problems. The plan aims to improve air quality in key areas through coordinated control of multiple pollutants. Based on the quasi-natural experimental method, Wu and Yin [12] found that the implementation of the action plan can improve the atmospheric quality through the impact on the industrial structure, especially in resource-based cities. The study emphasises that compared with the development of the advanced industrial structure, the rationalisation of the industrial structure plays a more obvious role in improving air quality. Only when the adjustment of industrial structure is compatible with the overall characteristics of the region can the policy really play a role.

Industrial structure adjustment is an important way to optimise economic and social factors. William Petty first put forward the theory of industrial structure, while Clark revealed the evolution law of industrial structure. How to measure the change in industrial structure and whether the adjustment is reasonable is a more concerning issue in the research of industrial structure adjustment. Fu [13] took the proportion of three industries as the corresponding weight of each industry and aggregated it to build an advanced index of industrial structure. Gan et al. [14] took the ratio of the tertiary industry to the secondary industry to measure the upgrading of the industrial structure. However, relevant studies believe that it is too one-sided to measure the rationality of the industrial structure from the change of the proportion of three industries. Liu [15] believes that the rationality evaluation system of industrial adjustment should have the functions of judgment, selection, control, guidance, and early warning, and should follow the principles of scientificity, comprehensiveness, independence, feasibility, and stability. Based on the above functions and principles, 5 primary indicators, 13 secondary indicators, and 37 tertiary indicators are selected to build an evaluation system for industrial structure rationalisation. In empirical research, most choices are mainly measured by constructing the ratio of the output value of various industries to labour productivity [16–18].

How to coordinate the relationship between the upgrading of the industrial structure and the improvement of atmospheric environmental quality is a very important issue for China in the green transition period. Many scholars have also carried out research on this issue. Zhang et al. [19] used the SDA method (structural decomposition analysis method) to evaluate China's economic development during the Eleventh Five Year Plan period. They pointed out that if China does not change its economic structure and development model, it may not be able to fulfil its commitment to reduce the emission of industrial air pollutants. Zhang et al. [20] found that the adjustment of the industrial structure had a positive impact on reducing carbon emissions based on the econometric model. Among them, the reduction of the proportion of energy-intensive secondary industry and the optimisation of energy structure have a particularly obvious effect on improving atmospheric environmental quality [21,22]. Ding et al. [23] have tried to analyse the relationship between industrial structure and the atmospheric environment through an integrated and systematic method. They constructed the PSR-LQI (pressure–state–response and level–quality–innovation model) index evaluation system and presented the corresponding evaluation results. Zheng et al. [24], based on the threshold model and the empirical analysis of China's provincial panel data, found that the impact of the industrial structure on pollutant NOx and PM2.5 is divided into three stages, while the impact mechanism of pollutant SO_2 is two stages. The adjustment of the industrial structure can improve the impact of economic development on air pollution. Zhou et al. [25] pointed out that the strictness of environmental supervision helps to optimise the industrial structure and then improve the quality of the atmospheric environment. Relevant studies also show that the impact of the industrial structure on atmospheric environmental quality is long-term and increases with time [26].

What is the relationship between atmospheric environmental quality and industrial structure adjustment? How important is the adjustment of the industrial structure to improve atmospheric environmental quality? What difficulties are faced in the process of improving atmospheric environmental quality through industrial structure adjustment? Previous studies have fully analysed the single system of atmospheric environmental or industrial systems and used the econometric model to empirically test and evaluate the relationship between them. Many studies have also mentioned the long-term nature, dynamic interaction, and segmentation of the relationship between the two systems [24–26]. However, the existing studies mainly analyse the relationship between the two systems from the perspective of a single system or comparative static and have not been able to analyse the dynamic process of the interaction between the two systems and present an intuitive evaluation of the coupling and coordination between the two systems. The development of the atmospheric environmental system has experienced a strategic change from the total amount control of single pollutants to the quality improvement of multi-pollutant collaborative control. To maintain the sustainable development of its own system, it needs to maintain coordination with other economic and social subsystems through continuous optimisation. On the other hand, after transforming the industrial structure level from low to high level, the industrial system needs to consider the rationalisation within the system. After reaching coordination and balance within the system, it will develop in the direction of rationalisation among systems. The development between the two systems can be roughly divided into three stages: the low-level coupling and coordination stage, the mutual running in and coordination stage, and the high-level coupling and coordination stage. In the first stage, the atmospheric environment system has constraints on the industrial system, and the industrial system cooperates with the realisation of the governance goal of the atmospheric environment system. In the second stage, the regulation of the atmospheric environment system on the industrial system is gradually optimised. The optimisation and improvement of the industrial system will promote and guide the improvement of the atmospheric environment system. In the third stage, the two systems maintain benign coupling and need coordination, and jointly develop in the direction of sustainability. For the leapfrogging of different stages and the drive of the system itself, the systems will also affect each other. The strategic improvement of the atmospheric environment system will enhance the optimisation requirements of the industrial system, and the improvement of the matching and adaptability of the industrial system will also enhance its impact on the atmospheric environment system. Of course, due to the complexity of the urban development model, the industrial system and atmospheric environment system may not only promote each other's leap forward, but also regress. Therefore, this paper constructs the analysis framework of coupling and coordination between the atmospheric environment system and the industrial system. As shown in Figure 1, it analyses the internal and external evolution processes of the two systems in the interaction process from the theoretical level, and then uses the coupling and coordination model and the VAR model for empirical analysis.

There are three main contributions of this paper. Firstly, it constructs the theoretical analysis framework of coupling and coordination between the atmospheric environment system and the industrial system and clarifies the internal mechanism of the interaction and coordinated development of the two systems. Secondly, the coupling coordination model and the VAR model are combined to test the relationship between them from the perspectives of "static" and "dynamic", which not only provides an overall intuitive evaluation of the relationship between them, but also analyses the dynamic process of interaction. Thirdly, as a resource-based city in China, Taiyuan is typical and representative. This paper selects the data of Taiyuan for the empirical test, provides an analytical framework and comprehensive evaluation method for evaluating the situation of the two systems of the city, and provides support for government departments to formulate policies and evaluate the effectiveness and rationality of urban policies.

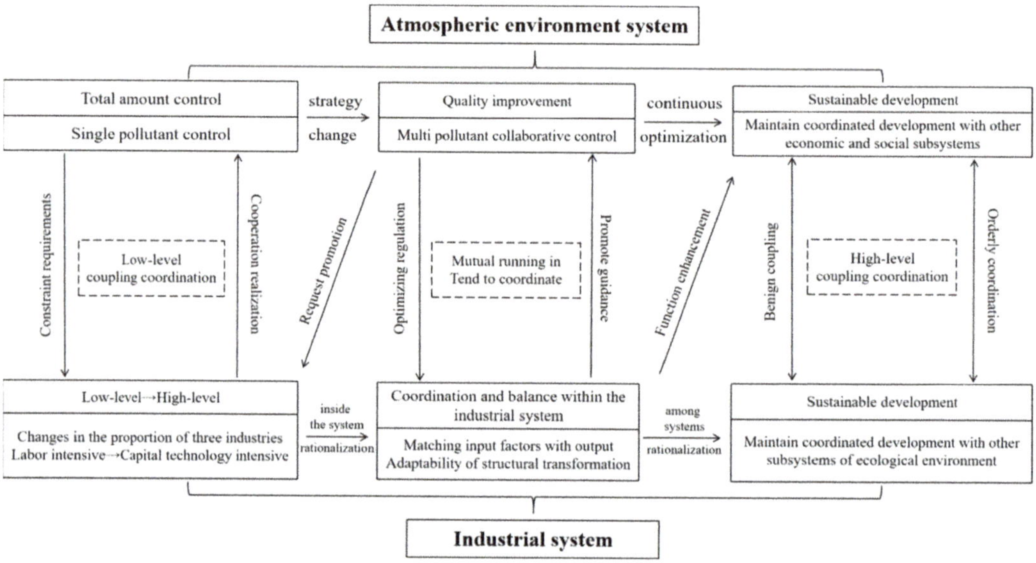

Figure 1. Analysis framework of coupling and coordination between the atmospheric environment system and the industrial system.

2. Study Area

Taiyuan is the capital city of Shanxi Province in China (as shown in Figures 2 and 3). The terrain is surrounded by mountains in the east, west, and north. The central and southern part is the Fenhe River valley plain. The whole terrain is high in the north and low in the south, in the shape of a dustpan. Therefore, the average ground wind speed is small, the static wind frequency is high, and the precipitation is scarce, which is unfavourable to the city's diffusion of air pollutants.

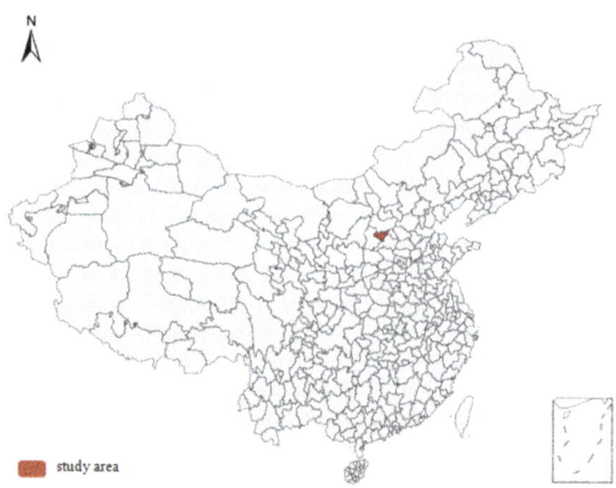

Figure 2. Location of the study area in China.

Figure 3. Location of the study area in Shanxi Province.

As a typical resource-based city, Taiyuan has a special industrial structure. The industry is mainly heavy. Coal mining and combustion greatly impact the quality of the atmospheric environment. Taiyuan has also actively adjusted and upgraded its industrial structure for many years. As shown in Figure 4, the proportion of primary industry in Taiyuan is relatively small and in a downward trend, and the proportion of secondary industry and tertiary industry shows a "K" trend. After 2009, the tertiary industry became the leading industry in Taiyuan, basically maintaining a stable industrial pattern of "tertiary, secondary, and primary industry", reflecting the industrial structure characteristics of urbanisation. However, from 2013 to 2020, the air quality ranking of Taiyuan was always in the lower position of the environmental air quality ranking of national key cities. The poor atmospheric environmental quality has become an important factor restricting the development of Taiyuan.

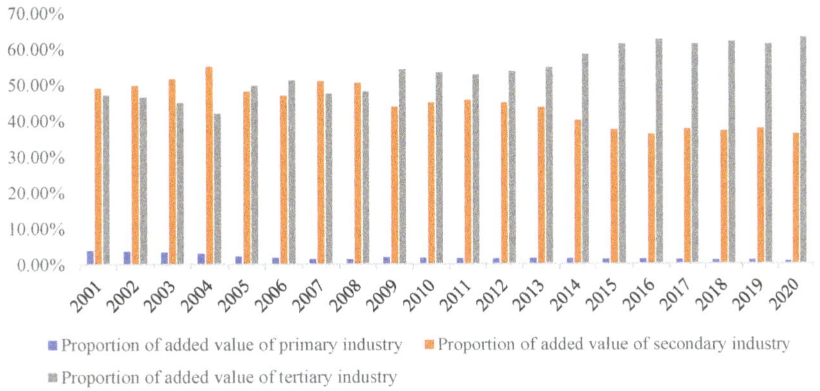

Figure 4. Changes in the proportion of three industries in Taiyuan over the years.

3. Methods and Data

The coupling coordination model provides an intuitive measurement and evaluation of whether the system elements can interact and the overall coordinated development. The VAR model can measure the size and continuous influence of the interaction of elements between systems in the process of dynamic development. Referring to the practice of Liao et al. [27], this paper combines the two models to comprehensively analyse the coupling and internal mechanisms between atmospheric environmental quality and urban industrial structure adjustment in Taiyuan from the perspectives of "static" and "dynamic". This section will specifically introduce the above models and the required data.

3.1. Coupled Coordination Model

Coupling originates from the physical concept, which refers to the phenomenon that two or more systems and their elements affect each other and finally achieve synergy through interaction. The coupling degree in the model reflects the degree of mutual dependence and restriction between systems. The coupling coordination degree reflects the degree of benign coupling between systems in their interaction. The more obvious the trend of orderly development between systems, the greater the index.

For the comprehensive system composed of the atmospheric environment subsystem and the urban industry subsystem, subsystem i has n indicators, which are $x_1, x_2, \cdots, x_n$. In order to eliminate the influence of dimensionality, the indicators need to be standardised before using these indicators to construct the comprehensive index of subsystem i. Referring to the existing research methods, this paper used the range method to process the data. When using this method, it needs to distinguish and process according to the positive and negative contributions of the index to the system. The greater the x_{ij} value, the better the system function. It is called the positive index, and Formula (1) is used for standardisation. The smaller x_{ij} value indicates the better system function, which is called the negative index, and Formula (2) is used for standardisation. The specific formulae are as follows:

$$\text{Positive index}: \ d_{ij} = (x_{ij} - x_{ijmin})/(x_{ijmax} - x_{ijmin}) \tag{1}$$

$$\text{Negative index}: \ d_{ij} = (x_{ijmax} - x_{ij})/(x_{ijmax} - x_{ijmin}) \tag{2}$$

where d_{ij} is the standardised value of system i index j, $0 \leq d_{ij} \leq 1$. x_{ijmax} is the maximum value of system i index j, x_{ijmin} is the minimum value of system i index j, and x_{ij} is the value of system i index j.

The comprehensive index of the atmospheric environment subsystem and the industrial subsystem is based on the weighted synthesis of the contribution of all indicators in each system to the subsystem, and its calculation formula is:

$$U_i = \sum_{j=1}^{n} w_{ij} \cdot d_{ij} \tag{3}$$

where w_{ij} is the weight of index j of system i, where $w_{ij} \geq 0$ and $\sum w_{ij} = 1$. The weight corresponding to each index reflects the ability of the index to provide comprehensive information about the subsystem, so it needs to be determined according to the amount of information contained in the index. The principal component analysis method can effectively deal with the influence of information repetition and interaction between different indicators. Therefore, this method is also used in this paper. Through principal component analysis, the corresponding variance value of each index is obtained as the weight of the corresponding value index, and then the comprehensive index of each system and the coupling degree, C, between systems are calculated.

The calculation formula of coupling degree is as follows:

$$C = \left[\frac{\prod_i^m U_i}{\left(\frac{1}{m}\sum_i^m U_i\right)^m} \right]^{\frac{1}{m}} \tag{4}$$

where m is the number of subsystems. Since the number of subsystems of the atmospheric environment and the urban industrial structure integrated system constructed in this paper is 2, $m = 2$. Therefore, the above coupling calculation formula is simplified as follows:

$$C = \sqrt{\frac{U_1 U_2}{(\frac{U_1+U_2}{2})^2}} = \frac{2\sqrt{U_1 U_2}}{U_1 + U_2} \tag{5}$$

where U_1 and U_2 are the comprehensive indexes of the atmospheric environment and the industrial subsystem, respectively. The value of C is between 0 and 1. The greater the value of C, the closer the relationship between systems. Referring to the median segmentation method adopted by most studies, it can be divided into the following four stages, listed in Table 1.

Table 1. Median segmentation of coupling degree.

C	Corresponding Stage
$0 < C \leq 0.3$	Low-level coupling
$0.3 < C \leq 0.5$	Antagonistic stage
$0.5 < C \leq 0.8$	Running in stage
$0.8 < C \leq 1$	High-level coupling

However, the coupling index only reflects the function degree of the two systems. In order to account for the respective development of the two systems, it is necessary to further study whether the two systems are coordinated. Coordination is used to measure the degree of harmony of each subsystem. There is a close relationship between the industrial and the atmospheric environment subsystems. On the one hand, the atmospheric environmental carrying capacity requires the optimisation and upgrading of the industrial structure, and the atmospheric environmental quality will react to the process of industrial structure adjustment. On the other hand, the adjustment of the industrial structure not only needs to realise the healthy and sustainable development of the industrial system, but also needs to match with the carrying capacity of the atmospheric environment. The coupling coordination degree can better evaluate the coordination degree of the interaction coupling between industrial structure adjustment and atmospheric environmental quality improvement. The calculation formula is:

$$T = \alpha U_1 + \beta U_2 \tag{6}$$

$$D = \sqrt{C \times T} \tag{7}$$

where D is the coupling coordination degree, C is the coupling degree, and T is the comprehensive coordination index of the industry and the atmospheric environment, reflecting the overall synergistic effect or contribution of industrial structure adjustment and atmospheric environment quality improvement. α and β are the weights of the two systems. This paper believes that the improvement of the atmospheric environment is as important as the optimisation and upgrading of the industrial structure, so $\alpha = \beta = 0.5$.

When the D value is larger, it shows that the coordinated development of the two systems is better. According to the judgment method of existing research [28], the coordination can be divided into ten stages according to the coupling coordination degree, D, as shown in Table 2.

Table 2. Coupling coordination level segmentation.

D	Coordination Level
[0, 0.1)	Extremely uncoordinated
[0.1, 0.2)	Seriously uncoordinated
[0.2, 0.3)	Moderately uncoordinated
[0.3, 0.4)	Slightly uncoordinated
[0.4, 0.5)	On the verge of uncoordinated
[0.5, 0.6)	Barely coordinated
[0.6, 0.7)	Slightly coordinated
[0.7, 0.8)	Moderately coordinated
[0.8, 0.9)	Well-coordinated
[0.9, 1]	Quality coordinated

Considering the availability of data and indicators' comprehensive information-carrying capacity, this paper constructs a comprehensive evaluation index system, as shown in Table 3. Relevant data can be obtained from the official website of the Taiyuan Bureau of Statistics (http://stats.taiyuan.gov.cn, accessed on 5 March 2022).

Table 3. Comprehensive evaluation index system.

Coupling Systems	Indicators	Unit	Indicators Direction
Atmospheric environment system	Industrial sulphur dioxide emissions	10,000 tonnes	−
	Industrial smoke (powder) dust emission	10,000 tonnes	−
	Days with air quality above grade II	day	+
Industrial system	Rationalisation of industrial structure (ISR)	/	−
	Advanced industrial structure (ISH)	/	+
	Comprehensive utilisation rate of general industrial solid waste	%	+
	Total investment in fixed assets	100 million yuan	+

For the evaluation indicators of the atmospheric environment system, China mainly evaluates the atmospheric environment by monitoring the concentration of six types of pollutants (including PM10, PM2.5, O3, NO_2, SO_2, and CO) and the air quality index (AQI) in practice. However, because some data are difficult to obtain, this paper selects three indicators: industrial sulphur dioxide emission, industrial smoke (powder) dust emission, and days with air quality above grade II.

As for the evaluation indicators of the industrial structure system, the upgrading of the industrial structure aims to measure the degree of the evolution of the industrial structure from low-level to high-level. This paper constructs the following indicators based on the idea of Fu [13]:

$$ISH_t = (q_{2t} + q_{3t}) \times \frac{q_{3t}}{q_{2t}} \tag{8}$$

where q_{2t} and q_{3t}, respectively, represent the proportion of the secondary and tertiary industries in Taiyuan's GDP in the t period.

The rationalisation of the industrial structure is to consider the coordination between different industries, the adaptability between input factors and output, and the adaptability of structural transformation ability. The existing studies mostly use the structural deviation degree improved by Gan et al. [14] based on the Theil index. This method was also used in this paper. The specific calculation formula is:

$$ISR_t = \sum_{j=1}^{3} (q_{jt}) ln \frac{Y_{jt}/Y_t}{L_{jt}/L_t} \tag{9}$$

where q_{it} indicates the proportion of the j industry in the region's GDP during the t period, Y_{jt} refers to the gross domestic product of the j industry in Taiyuan during the t period, and L_{jt} refers to the number of employees in the j industry in Taiyuan during the t period. Y_t and L_t, respectively, represent the regional GDP (Gross domestic product) and total employment of Taiyuan in the t period. The closer ISR is to 0, indicating the closer the industrial structure is to the equilibrium state, the more reasonable the industrial structure is.

Referring to the index construction of the industrial rationalisation system by Liu [15], the comprehensive utilisation rate of general industrial solid waste reflects the improvement of factor utilisation within the industry. The total investment in fixed assets is a strategic plan reflecting the industry's long-term development. Therefore, this paper selected the above four indicators as the evaluation indicators of the industrial system.

3.2. VAR Model

The coupling coordination model can conduct an intuitive evaluation of the coupling coordination of the two systems from a comparative static point of view, and the VAR model is an econometric model used to estimate the dynamic relationship of joint endogenous variables. The model is established according to the statistical characteristics of the data without setting any constraints in advance. Each variable in the system is regarded as endogenous, and the lag term of all variables is included in the constructed function model. The VAR model is mainly used to analyse the response of interconnected time-series systems under the dynamic impact of system variables. The analysis of the model is mainly to observe the impulse response function and variance decomposition of the system. The former refers to the system's response to a random impact of one of the variables and how long this response will last. The latter is an important method to judge the dynamic correlation between economic series variables. In essence, it decomposes the prediction mean square error of the system into the contribution of the shocks of various variables in the system. Through the Granger causality test, we can analyse the causal effect of variables in time. This paper used the VAR model to analyse the dynamic action of the atmospheric environment system and the industrial system in Taiyuan. The specific model is constructed as follows:

$$\begin{cases} y_{1t} = \alpha_{10} + \gamma_{11}y_{1,t-1} + \ldots + \gamma_{1p}y_{1,t-p} + \beta_{11}y_{2,t-1} + \ldots + \beta_{1p}\beta_{2,t-p} + \varepsilon_{1t} \\ y_{2t} = \alpha_{20} + \gamma_{21}y_{1,t-1} + \ldots + \gamma_{2p}y_{1,t-p} + \beta_{21}y_{2,t-1} + \ldots + \beta_{2p}\beta_{2,t-p} + \varepsilon_{2t} \end{cases} \quad (10)$$

where y_1 and y_2 are the comprehensive indexes of the atmospheric environment system and the industrial structure system, respectively. p represents the lag order, t represents the time, γ and β represent the regression coefficient, α represents the intercept term, and ε represents the residual term.

4. Results and Discussion

4.1. Descriptive Statistics

The original data of various indicators of the Taiyuan atmospheric environment system and the industrial system can be obtained from the official website of the Taiyuan Bureau of Statistics. Among them, the number of days with air quality above grade II, the upgrading of the industrial structure, and the total investment in fixed assets include the data for 20 years, from 2001 to 2020. Industrial sulphur dioxide emission, industrial smoke (powder) dust emission, and the comprehensive utilisation rate of general industrial solid waste cover 15 years of data, from 2003 to 2017. The data of industrial structure rationalisation include the data for 16 years, from 2003 to 2018, as shown in Table 4. Considering the amount of information and availability of data, this paper finally selected the data from 2003 to 2017 for empirical analysis. Due to the different dimensions of the above data, the range method was used to standardise the above data. According to the positives and negatives of the index, we used Formulas (1) and (2) to deal with it, respectively. The results are shown in Table 5.

Table 4. Descriptive analysis of each index.

System	Indicators	Observed Value	Mean	SD	Min	Max
Atmospheric environment system	Days with air quality above grade II	20	228.8	59.10	120	324
	Industrial sulphur dioxide emissions	15	100,576	49,399	9759	183,656
	Industrial smoke (powder) dust emission	15	44,052	15,140	17,086	72,171
Industrial system	Advanced industrial structure (ISH)	20	1.232	0.325	0.739	1.726
	Rationalisation of industrial structure (ISR)	16	0.024	0.01	0.009	0.039
	Comprehensive utilisation rate of general industrial solid waste	15	0.488	0.0519	0.422	0.560
	Total investment in fixed assets	20	978.0	610.2	122.7	2028

Table 5. Standardised treatment results.

Year	Days with Air Quality above Grade II	Industrial Sulphur Dioxide Emissions	Industrial Smoke (Powder) Dust Emission	Advanced Industrial Structure (ISH)	Rationalisation of Industrial Structure (ISR)	Comprehensive Utilisation Rate of General Industrial Solid Waste	Total Investment in Fixed Assets
2003	0.117	0.011	0.000	0.106	0.139	0.059	0.000
2004	0.383	0.000	0.102	0.000	0.185	0.059	0.072
2005	0.512	0.191	0.285	0.279	0.788	0.177	0.128
2006	0.611	0.316	0.420	0.340	0.761	0.131	0.163
2007	0.660	0.443	0.469	0.184	0.975	0.000	0.204
2008	0.864	0.481	0.573	0.206	1.000	0.380	0.273
2009	0.827	0.536	0.609	0.492	0.608	0.462	0.317
2010	0.877	0.514	0.678	0.440	0.688	0.730	0.390
2011	0.901	0.434	0.508	0.410	0.812	0.784	0.450
2012	1.000	0.471	0.546	0.453	0.000	0.839	0.612
2013	0.000	0.545	0.638	0.513	0.169	0.892	0.804
2014	0.216	0.575	0.231	0.722	0.308	0.946	0.846
2015	0.420	0.684	0.684	0.910	0.250	1.000	0.999
2016	0.432	0.966	0.913	1.000	0.173	0.660	1.000
2017	0.080	1.000	1.000	0.896	0.298	0.039	0.417

4.2. Evaluation Results of Coupling Coordination Model

Since the indexes selected in this paper are inter-related and the information covered is overlapped, this paper adopted the principal component analysis method to extract the information contributed by each index to each subsystem, determine the corresponding weight coefficient of each index according to the component matrix and corresponding variance (see Table 6), and then calculate the comprehensive index of the atmospheric environment system and the industrial system according to Formula (3). The comprehensive index of atmospheric environmental and industrial systems, coupling degree, coupling coordination degree, comprehensive coordination index, and corresponding coupling coordination stages of the two systems are shown in Table 7 and Figure 5.

Table 6. Corresponding weight of each index.

System	Indicators	Weight Coefficient
Atmospheric environment system	Days with air quality above grade II	0.278
	Industrial sulphur dioxide emissions	0.333
	Industrial smoke (powder) dust emission	0.389
Industrial system	Advanced industrial structure (ISH)	0.328
	Rationalisation of industrial structure (ISR)	−0.054
	Comprehensive utilisation rate of general industrial solid waste	0.352
	Total investment in fixed assets	0.374

Table 7. Coupling and coordination of the atmospheric environment system and the industrial system in Taiyuan.

Year	Atmospheric Environment System U1	Industrial System U2	Coupling Degree, *C*	Comprehensive Coordination Index, T	Coupling Coordination Degree, *D*	Coupling Stage	Coordination Stage
2003	0.036	0.048	0.991	0.205	0.451	High level	Verge of uncoordinated
2004	0.146	0.037	0.806	0.303	0.494	High level	Verge of uncoordinated
2005	0.317	0.159	0.944	0.488	0.679	High level	Slightly coordinated
2006	0.439	0.177	0.906	0.555	0.709	High level	Moderately coordinated
2007	0.514	0.084	0.694	0.547	0.616	Running in	Slightly coordinated
2008	0.623	0.249	0.904	0.660	0.773	High level	Moderately coordinated
2009	0.645	0.410	0.975	0.726	0.841	High level	Well-coordinated
2010	0.679	0.510	0.990	0.771	0.874	High level	Well-coordinated
2011	0.592	0.535	0.999	0.751	0.866	High level	Well-coordinated
2012	0.647	0.673	1.000	0.812	0.901	High level	Quality coordinated
2013	0.430	0.774	0.958	0.776	0.862	High level	Well-coordinated
2014	0.342	0.869	0.900	0.778	0.837	High level	Well-coordinated
2015	0.611	1.011	0.969	0.900	0.934	High level	Quality coordinated
2016	0.797	0.925	0.997	0.928	0.962	High level	Quality coordinated
2017	0.744	0.448	0.969	0.772	0.865	High level	Well-coordinated

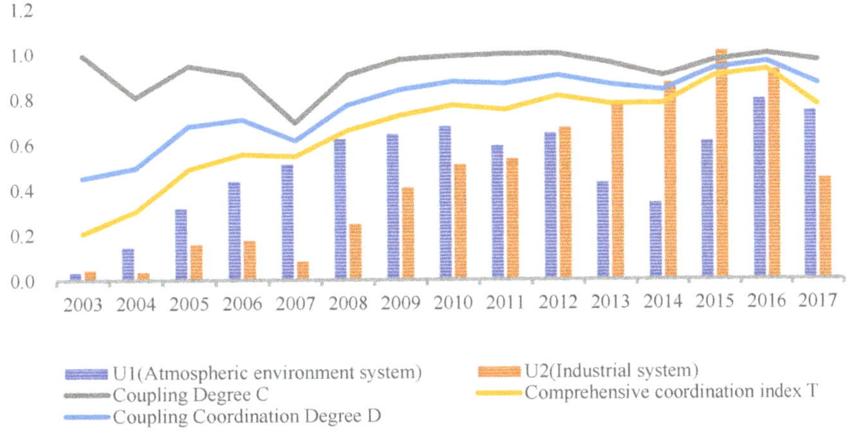

Figure 5. Coupling and coordination trends of the atmospheric environment system and the industrial system in Taiyuan from 2003 to 2017.

It can be seen from the chart that the atmospheric environment of Taiyuan is systematically rising. From 2003 to 2010, it maintained a rapid upward trend, and the comprehensive index increased from 0.036 to 0.679, indicating the continuous improvement of atmospheric quality. However, the composite index fell in shock, and fell to 0.342 in 2014. This is mainly because China revised the ambient air quality standard (GB 3095-2012) in 2012 and adopted more stringent standards. After four years of adjustment and adaptation to the new standard, the development of the atmospheric environment system in Taiyuan showed an upward trend again, and the comprehensive index reached 0.797.

In terms of the industrial system, the comprehensive index of the industrial system basically maintained an upward trend from 2003 to 2015, rising from 0.048 to 1.011. Taiyuan's industrial structure has experienced the transformation from "secondary, tertiary, and primary" to "tertiary, secondary, and primary". The advanced index of the industrial structure continued to rise. However, the rationalisation index fluctuated, which reflects that the matching degree of elements in Taiyuan's industrial system and the internal optimisation of the three industries are important factors affecting the development of the system. After 2015, the decline of the comprehensive index of the industrial system also reflects this situation to a certain extent. After 2017, Taiyuan also began to publish relevant statistical data of strategic emerging industries and high-tech industries to refine and analyse the development quality within the industry.

The coupling degree, C, of the two systems fluctuated greatly before 2009 and became stable and close to 1 after 2009, indicating that the interaction between the atmospheric environment system and the industrial system is obvious. The coupling and coordination dispatching, D, maintained an upward trend, indicating that the two systems were developing in the direction of benign coupling. The coordination of the two systems has experienced the process of "uncoordinated—primary coordination—moderately coordinated—well-coordinated—quality coordination", indicating that the development between the systems tends to be gradually synchronised. However, it is also noted that the coupling coordination degree of the two systems slightly fluctuated, which reflects that the two systems in Taiyuan are currently in the transitional stage of "mutual running in and tend to coordination—high-level coupling coordination" in the above theoretical analysis framework.

4.3. Results of the VAR Model

The coupling and coordination model mainly measured and evaluated the annual coupling and coordination of the atmospheric environment and industrial systems from a static perspective. In this section, the VAR model was used to analyse the interaction force and continuous influence between the two systems. The specific steps and results were analysed as follows.

4.3.1. Decision Lag Order

To estimate the VAR model, we first needed to determine the lag order, p, of the model according to the information criterion. According to the results in Table 8, the lag order of this model is 4.

Table 8. Decision lag order.

Lag	LL	LR	df	p	FPE	AIC	HQIC	SBIC
0	5.754				0.0017	−0.6825	−0.7280	−0.6101
1	14.611	17.714	4	0.001	0.0007	−1.5656	−1.7024	−1.3486
2	21.750	14.279	4	0.006	0.0005	−2.1364	−2.3644	−1.7747
3	29.907	16.314	4	0.003	0.0003	−2.8923	−3.2115	−2.3859
4	38.754	17.693 *	4	0.001	0.0003 *	−3.7735 *	−4.1839 *	−3.1224 *

Note: * indicates the lag order selected by the corresponding information criterion.

4.3.2. Stationary Test

In order to ensure the stability of the model, it was also necessary to test the stationarity of the model. As shown in Figure 6, all unit roots were in the unit circle, indicating that the model is stable.

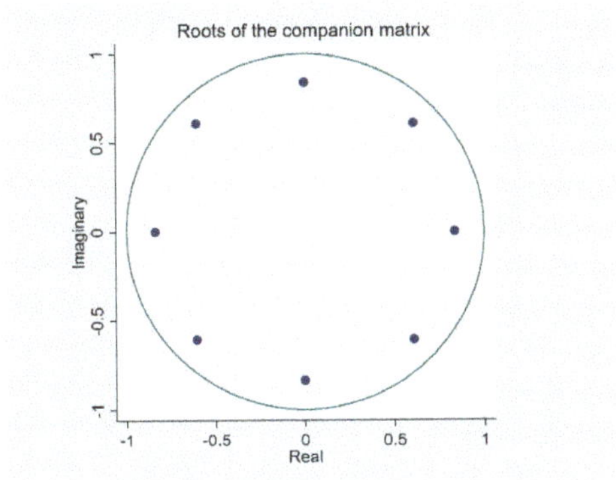

Figure 6. Stationary test.

4.3.3. Granger Causality Test

The Granger test can analyse the causal relationship and action direction of variables in time. It can be seen from Table 9 that the comprehensive index of the atmospheric environment system is the Granger cause of the comprehensive index of the industrial system, and the assumption that the comprehensive index of the industrial system is not the Granger cause of the comprehensive index of the atmospheric environment system cannot be rejected. This is consistent with the optimisation requirements of the industrial system for the strategic improvement of the atmospheric environment system mentioned in the theoretical analysis framework of this paper. From the results of coupling and coordination, it can be seen that the development synchronisation of the two systems in Taiyuan needs to be strengthened. Therefore, Taiyuan should continue to optimise its industrial system, so as to enhance its role in the atmospheric environment system.

Table 9. Granger causality test results.

Hypothesis	chi2	df	df_r	Prob > chi2	Conclusion
U2 is not the Granger cause of U1	0.961	1	8	0.3558	Cannot reject
U1 is not the Granger cause of U2	25.872	1	8	0.0009	Reject

4.3.4. Impulse Response

Since the VAR model contains many parameters, it cannot directly explain the economic meaning of parameters, so it is mainly analysed through the impulse response. Figure 7 shows the impulse response of the Taiyuan atmospheric environment system comprehensive index U1 and the industrial system comprehensive index U2. The horizontal axis represents the lag order of impact (unit: year), and the vertical axis represents the response value of relevant variables. It can be seen from the figure that when the comprehensive index U1 of the atmospheric environment system was used as the pulse variable, the comprehensive index U2 of the industrial system fluctuated significantly during phases 3–5, which shows that the optimisation of the atmospheric environment system will have a positive impact on the industrial system in the medium and long term. When the industrial system comprehensive index U2 was used as the impulse response variable, the atmospheric environment system comprehensive index U1 had positive benefits in phases 3–5, but the benefits were small.

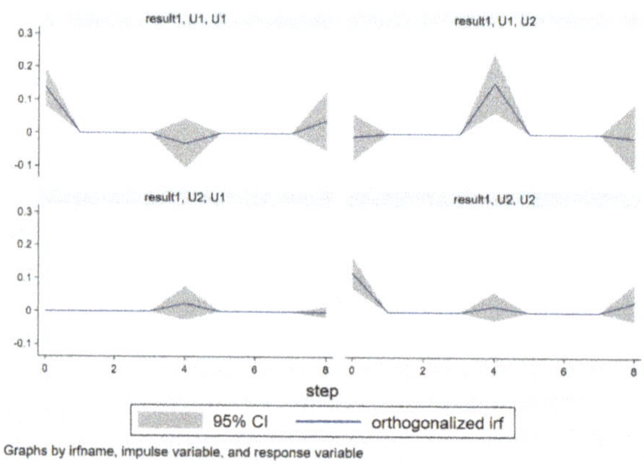

Figure 7. Impulse response.

4.3.5. Variance Decomposition

Variance decomposition can explain the contribution of each variable shock in the system. It can be seen from Figure 8 that the change of the atmospheric environment system mainly came from its own change impact, which is consistent with the practice in China. The governance and improvement of the atmospheric environment system are mainly regulated by administrative means. In the short term, the change of the industrial system is attributed to itself, but in the long term, it is jointly affected by its own change and the impact of the change of the atmospheric environment system. In phases 1–4, the contribution rate of the impact of the industrial system itself to its variance was close to 1, but it dropped to 37% in the subsequent phases. In comparison, the explanation degree of the change of the atmospheric environment system U1 to the industrial system U2 reached 63%.

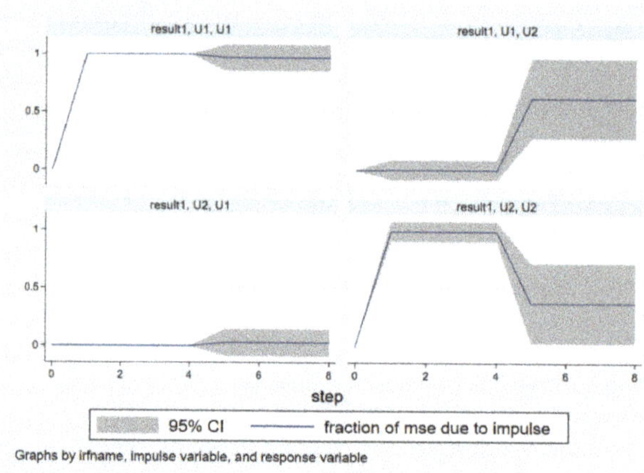

Figure 8. Variance decomposition.

5. Conclusions

The quality of the atmospheric environment affects the high-quality and sustainable development of the economy, and the optimisation and upgrading of the industrial system is an important way to improve the efficiency of economic and social operations. Analysing the relationship and interaction between the two systems is significant in promoting the coordinated development of the environment and economy. By constructing the theoretical analysis framework of coupling and coordination between the atmospheric environment system and the industrial system, this paper clarified the two systems' internal mechanisms of the interaction and coordinated development. Then, it put forward the combination of the coupling coordination model and the VAR model and provided the analysis and evaluation method of the relationship between them from the two perspectives of "static" and "dynamic".

The atmospheric environmental quality is an important factor affecting the development of Taiyuan. For many years, Taiyuan has also been committed to adjusting the industrial structure to make the industrial development adapt to the local environmental carrying capacity. Therefore, based on the theoretical analysis framework and the proposed evaluation method, this paper empirically analysed the data of the typical resource-based city Taiyuan from 2003 to 2017. The results show that: (1) Based on the results of the coupling coordination model, the interaction between the two systems in Taiyuan was obvious and developed towards benign coupling. However, there were still fluctuations in some years, reflecting that the two systems are still in the transition stage of "mutual running in and tend to coordination—high-level coupling coordination". (2) According to the results of the VAR model, the impact of the two systems on each other was mainly in the medium and long term and dominated by the effect of the atmospheric environment system on the industrial system.

Based on the above analysis results, the following policy implications can be obtained. Firstly, Taiyuan should continue to maintain the coordination and sustainability of atmospheric environmental governance and industrial structure adjustment. It should promote and maintain the synchronous and orderly development between the two systems. Secondly, it should be optimised from within the industrial system to improve the scientificity and effectiveness of industrial policies, so as to enhance its internal and external benefits, and then promote the sustainable development of the two systems in Taiyuan and make the two systems enter a high-level coupling and coordinated development stage. Thirdly, when formulating and evaluating urban environmental and industrial policies, central government departments should fully consider the particularity of different urban development stages and the long-term effect of policies so as to avoid "one size fits all" and short-sighted behaviour.

This paper presented an analysis framework and long-term evaluation method of the relationship between the urban atmospheric environment system and the industrial system, which provides theoretical and methodological guidance for policymakers to evaluate relevant policies. However, because some data cannot be obtained through public channels, only the evaluation results of Taiyuan from 2003 to 2017 were presented. Relevant policy departments can conduct the internal evaluation with reference to the methods proposed in this paper and take measures according to the latest situation of the two systems. On the other hand, this paper's evaluation and analysis methods were mainly studied from the macrosystem level. In the future, we can further analyse the response of micro-subjects (such as enterprises) to different types of environmental regulation and industrial structure adjustment policies in different stages, find the action mechanism at the micro-level, and then provide more targeted policy suggestions. In addition, in recent years, the COVID-19 pandemic has had a significant impact on all economies in the world, and the global health crisis has become an external factor that cannot be ignored. The impact of this factor on the atmospheric environment industry coupling system proposed in this paper will be worthy of more extensive and in-depth research in the future.

Author Contributions: Conceptualisation, X.Z. and L.W.; methodology, J.Z.; software, J.Z.; validation, X.Z., J.Z. and L.W.; formal analysis, X.Z., J.Z. and L.W.; investigation, J.Z.; resources, J.Z. and L.W.; data curation, J.Z. and L.W.; writing—original draft preparation, J.Z. and L.W.; writing—review and editing, X.Z., J.Z. and L.W.; visualisation, J.Z.; supervision, X.Z.; project administration, X.Z.; funding acquisition, X.Z. All authors have read and agreed to the published version of the manuscript.

Funding: This research was funded by the National Natural Science Foundation of China, grant number 72003009.

Institutional Review Board Statement: Not applicable.

Informed Consent Statement: Not applicable.

Data Availability Statement: The sources of relevant data acquisition have been described in the text.

Acknowledgments: The authors dedicate their earnest gratitude to the editor and the anonymous reviewers for their active comments, which improved the quality of the manuscript.

Conflicts of Interest: The authors declare no conflict of interest.

References

1. Cao, A.; Tian, L.; Cai, C.; Zhang, S. The research and prediction of atmospheric pollution based on wavelet and RBF neural network. *J. Syst. Simul.* **2006**, *18*, 1411–1413.
2. Nouri, A.; Lak, M.G.; Valizadeh, M. Prediction of PM2.5 Concentrations Using Principal Component Analysis and Artificial Neural Network Techniques: A Case Study: Urmia, Iran. *Environ. Eng. Sci.* **2021**, *38*, 89–98. [CrossRef]
3. Yang, W.; Li, L. Fuzzy mathematics-based comprehensive evaluation of atmospheric environmental quality in Huaian. *Clim. Environ. Res.* **2009**, *14*, 451–454.
4. Xing, Y.; Wang, X.; Hang, D. Set pair analysis for regional atmospheric environmental risk assessment. *China Env. Ment. Sci.* **2016**, *36*, 634–640.
5. Feng, S.; Song, Z.; Wang, H. Evaluation of atmospheric environment quality for resource type cities based on Grey cluster method. *Earth Environ.* **2015**, *43*, 345–349.
6. Jiang, C.Y.; Zhang, X.; Hu, L.; Wang, Q.; Zhang, W.J. Application of AHP Method in Evaluation of Atmospheric Environment Comprehensive Quality in Xi'an. *Adv. Mater. Res.* **2012**, *347–353*, 2054–2057. [CrossRef]
7. Lv, W.; Ji, S. Atmospheric environmental quality assessment method based on analytic hierarchy process. *Discret. Contin. Dyn. Syst. S* **2019**, *12*, 941–955. [CrossRef]
8. Zebardast, L.; Salehi, E.; Afrasiabi, H. Application of DPSIR framework for integrated environmental assessment of urban areas: A case study of Tehran. *Int. J. Environ. Res.* **2015**, *9*, 445–456.
9. Cao, Y.; Bian, Y. Improving the ecological environmental performance to achieve carbon neutrality: The application of DPSIR-Improved matter-element extension cloud model. *J. Environ. Manag.* **2021**, *293*, 112887. [CrossRef]
10. Lu, Y.; Jiang, H.; Liu, N.; Cheng, X.; Ye, W. The carrying capacity assessment of the national atmospheric environment based on the new air quality standards. *Environ. Monit. China* **2017**, *33*, 65–72.
11. Dong, J.; Chen, X.; Cai, X.; Xu, Q.; Guan, Y.; Li, T.; Liu, S.; Chen, F. Analysis of the temporal and spatial variation of atmospheric quality from 2015 to 2019 based on China atmospheric environment monitoring station. *J. Geo Inf. Sci.* **2020**, *22*, 1983–1995.
12. Wu, Z.; Yin, Y. Has the air pollution prevention and control action completed the "blue sky protection campaign" of resource-based cities? *Ind. Econ. Res.* **2022**, *1*, 43–56. (In Chinese)
13. Fu, L. An empirical study on the relationship between the upgrading of China's industrial structure and economic growth. *Stat. Res.* **2010**, *27*, 79–81. (In Chinese)
14. Gan, C.; Zheng, R.; Yu, D. The impact of China's industrial structure change on economic growth and fluctuation. *Econ. Res. J.* **2011**, *46*, 4–16+31. (In Chinese)
15. Liu, S. Study on the construction of evaluation index system for the rationalisation of industrial structure. *Sci. Technol. Manag. Res.* **2011**, *31*, 66–69. (In Chinese)
16. Zhou, X.; Zhang, J.; Li, J. Industrial structural transformation and carbon dioxide emissions in China. *Energy Policy* **2013**, *57*, 43–51. [CrossRef]
17. Zhu, B.; Zhang, M.; Zhou, Y.; Wang, P.; Sheng, J.; He, K.; Wei, Y.; Xie, R. Exploring the effect of industrial structure adjustment on inter-provincial green development efficiency in China: A novel integrated approach. *Energy Policy* **2019**, *134*, 110946. [CrossRef]
18. Wan, L.; Zheng, Q.; Wu, J.; Wei, Z.; Wang, S. How does the ecological compensation mechanism adjust the industrial structure? Evidence from China. *J. Environ. Manag.* **2021**, *301*, 113839. [CrossRef]
19. Zhang, W.; Wang, J.; Zhang, B.; Bi, J.; Jiang, H. Can China Comply with Its 12th Five-Year Plan on Industrial Emissions Control: A Structural Decomposition Analysis. *Environ. Sci. Technol.* **2015**, *49*, 4816–4824. [CrossRef]
20. Zhang, J.; Jiang, H.; Liu, G.; Zeng, W. A study on the contribution of industrial restructuring to reduction of carbon emissions in China during the five Five-Year Plan periods. *J. Clean. Prod.* **2018**, *176*, 629–635. [CrossRef]
21. Cheng, X.; Fan, L.; Wang, J. Can energy structure optimisation, industrial structure changes, technological improvements, and central and local governance effectively reduce atmospheric pollution in the Beijing-Tianjin-Hebei area in China? *Sustainability* **2018**, *10*, 6443. [CrossRef]
22. Wang, S.; Li, C.; Zhou, H. Impact of China's economic growth and energy consumption structure on atmospheric pollutants: Based on a panel threshold model. *J. Clean. Prod.* **2019**, *236*, 117694. [CrossRef]

23. Ding, L.; Chen, K.; Hua, Y.; Dong, H.; Wu, A. Investigating the relationship between the industrial structure and atmospheric environment by an integrated system: A case study of Zhejiang, China. *Sustainability* **2020**, *12*, 12783. [CrossRef]
24. Zheng, Y.; Peng, J.; Xiao, J.; Su, P.; Li, S. Industrial structure transformation and provincial heterogeneity characteristics evolution of air pollution: Evidence of a threshold effect from China. *Atmos. Pollut. Res.* **2020**, *11*, 598–609. [CrossRef]
25. Zhou, Q.; Zhong, S.; Shi, T.; Zhang, X. Environmental regulation and haze pollution: Neighbor-companion or neighbor-beggar? *Energy Policy* **2021**, *151*, 112183. [CrossRef]
26. Wang, W.; Han, Y. An empirical analysis of the impact of the evolution of industrial structure on the ecological environment in Anhui province. *Stat. Decis.* **2017**, *8*, 142–145. (In Chinese)
27. Liao, K.-C.; Yue, M.-Y.; Sun, S.-W.; Xue, H.-B.; Liu, W.; Tsai, S.-B.; Wang, J.-T. An Evaluation of Coupling Coordination between Tourism and Finance. *Sustainability* **2018**, *10*, 2320. [CrossRef]
28. Li, Y.; Li, Y.; Zhou, Y.; Shi, Y.; Zhu, X. Investigation of a coupling model of coordination between urbanisation and the environment. *J. Environ. Manag.* **2012**, *98*, 127–133. [CrossRef]

energies

MDPI

Article

The Impacts of Resource Endowment, and Environmental Regulations on Sustainability—Empirical Evidence Based on Data from Renewable Energy Enterprises

Hongyi Zhang, Hsing Hung Chen, Kunseng Lao and Zhengyu Ren *

School of Business, Macau University of Science and Technology, Taipa 999078, Macau; 1709853nbm30001@student.must.edu.mo (H.Z.); hhchen@must.edu.mo (H.H.C.); 1709853abm30001@student.must.edu.mo (K.L.)
* Correspondence: 2109853ebf20001@student.must.edu.mo

Abstract: In today's socio-economic context where environmental protection and sustainable development are equally important, how renewable energy enterprises can achieve sustainable development has become a topic of academic interest in recent years. This paper investigates the link between sustainable growth (*SG*) of renewable energy firms, resource endowment (*RE*), and environmental regulatory (*ERs*) issues through a fixed-effects model and a GMM model. Through empirical analysis, it was found that economical environmental regulations have the greatest positive impact on sustainable growth, followed by legal environmental regulations and supervised environmental regulations. Resource endowment is positively related to sustainable growth for non-state-owned renewable energy enterprises, but the negative impact on sustainable growth reflects the effect the of "resource curse". In addition, resource endowment has a negative moderating effect on environmental regulations and sustainable growth. Thus, the most significant effect is on the relationship between economical environmental regulations and sustainable growth, followed by legal environmental regulations and supervised environmental regulations. Therefore, the flexible and concurrent application of multiple environmental policies is an important way to ensure effective regulations and promote sustainable business growth.

Keywords: renewable energy business; fixed-effects model; GMM model; sustainable growth (*SG*)

Citation: Zhang, H.; Chen, H.H.; Lao, K.; Ren, Z. The Impacts of Resource Endowment, and Environmental Regulations on Sustainability—Empirical Evidence Based on Data from Renewable Energy Enterprises. *Energies* **2022**, *15*, 4678. https://doi.org/10.3390/en15134678

Academic Editor: Edmundas Kazimieras Zavadskas

Received: 11 April 2022
Accepted: 9 June 2022
Published: 26 June 2022

1. Introduction

Energy is crucial to a country's economy and people's livelihoods since it is the material underpinning for human survival and development [1]. The majority of people have realized that excessive energy consumption and pollution from the production and use of chemical products will cause an environmental crisis, despite the rapid expansion of the renewable energy industry.

The need to strike a balance between energy corporations' long-term expansion and environmental protection has become a pressing concern [2].

According to Ren et al. (2018), the environmental regulations discussed in this paper are related to those mandatory regulations issued by the government [3]. Environmental regulation (*ER*) is one of the policies initiated by the government to control and protect environmental resources [4]. Environmental regulation's role has received a lot of academic attention. Telle and Larsson (2007) suggested that *ER* does not reduce industrial productivity [5], but Xie et al. (2017) argued that environmental regulation can boost enterprises' industrial production competitiveness [6]. Wang, Y. et al. (2022) consider the diversity of the regulatory role of environmental regulations on the energy sector [7]. *ERs* have a favorable impact on sustainable growth up to a certain amount, but beyond that point, environmental regulation is detrimental to sustainable growth. Later, it was further suggested that environmental regulation has a facilitating effect on firms' performance and sustainable growth [8]. In addition, *ERs* have a facilitating and then inhibiting effect on the green economy [9]. Further, a combination of policy subsidies and carbon taxes is an effective way to develop low-carbon environmental protection.

Sustainability is the ability to improve living standards within ecological tolerances. Elkington (1994) proposed the concept of a sustainability floor based on social, economic, and environmental perspectives, which is now widely recognized by the academic community [10]. Das et al. (2020) argue that the concept of achieving growth without compromising the prospects of the next generation is increasingly becoming a core concept in business philosophy [11]. Salzmann et al. (2005) and Engert et al. (2016) argue that the regulatory effect of *ERs* on the sustainability of a business or economy varies in effect in different situations [12,13]. There are three main relationships between *ERs* and *SG*: first, environmental regulation has a positive contribution to sustainable growth [14]; second, Curtis and Lee (2019) argue that environmental regulation has a reverse inhibitory effect on sustainable growth [15]. Third, Curtis and Lee (2019) argue that the link between environmental regulation and sustainable growth is considered to be a "hump-shaped" relationship that varies over time [15].

Without resources, no business can expand sustainably, and resource endowment is the most important aspect in promoting long-term success. Zhai and An (2020) discovered that human capital, financial capabilities, technical innovation, and government conduct have a significant positive impact on *SG*, based on survey data from 500 Chinese manufacturing enterprises in 2017 [16]. The factors of education, expertise, and the availability of local entrepreneurial capital, fluctuate along the stages of the entrepreneurial process [17]. It has been argued that resource endowment has a long-term positive effect; specifically, the opportunity cost since resource endowment has the impact of inhibition in the early stage, and when it waits for the later order, resource endowment starts to show its positive effect, thus favoring long-term growth. Wang, S et al. (2022) argue that countries and regions that are rich in natural resources tend to have poorer green economic growth [18]. Government subsidies are the main driver for the long-term growth of renewable energy firms (Yang et al., 2019) [19], and they have been an important policy tool to nurture and promote the renewable energy industry in China (Song et al., 2020) [20]. Peng and Liu (2018) argue that government subsidies have a moderating effect on firm development, with negative and then positive effects developing over time [4]. When businesses receive government subsidies, it signifies that the government has accepted their legal status, which allows them to obtain more resources [21]. According to Lu et al. (2020), finance and subsidy impacts can raise the export size, impacting the long-term viability of firms' export expansion, and social capital plays an important role in fostering long-term growth, since high social capital firms are subjected to more lenient non-price lending requirements, resulting in lower bond interest rates [22].

With its large population and vast land area, China needs to import and use large amounts of energy in its development. The energy structure of China is dominated by traditional energy sources of fossil fuels, and the massive use of fossil fuels will certainly lead to a slew of significant environmental issues, including energy scarcity and pollution. Wu, L. et al. (2021) point out that energy endowment is a major factor in the growth of carbon emissions [23]. People's tolerance for environmental pollution decreases as their money and living standards rise, and the strength of environmental regulation steadily rises, opening up prospects for renewable energy development. Wang, Q et al.'s (2022) research found that renewable energy gives a significant boost to the economy [24]. Renewable energy is crucial for modifying the energy structure, lowering greenhouse gas emissions, and fostering long-term growth. Renewable energy businesses have grown quickly in recent years as high-tech industries throughout the world. Increased R&D expenditure is required to strengthen technical innovation and promote long-term growth in order to expand quickly and profitably. Furthermore, a major portion of China's renewable energy businesses are state-owned companies (SOEs), which have greater resources than private businesses. As a result, the research object for this study is renewable energy enterprises.

Do environmental regulations inhibit or promote sustainable company growth? How do different types of environmental regulatory regimes affect the sustainable growth of firms? With the increasing emphasis on resource endowment by firms, how does environmental regulation affect resource endowment and further contribute to firms' sustainable growth? To address these questions, this paper will provide insights into the impact of environmental regulation and resource endowment on firms' sustainable growth from the perspective of Chinese renewable energy firms and provide constructive suggestions for the government to develop more accurate environmental and energy policies.

The following are the main contributions: First, this is the first article that divides environmental regulation into three levels, including economic environmental regulation, legal environmental regulation, and supervisory environmental regulation. Second, for the first time, the impact of environmental regulation on the sustainable development of renewable energy firms is studied and specifically analyzed from the perspective of micro data of firms. Third, this paper is the first to study the relationship between environmental regulation and sustainable growth using resource endowment as a moderating variable. Fourth, this paper analyzes the ownership structure heterogeneity of the impact of resource endowment on sustainable growth, which helps to provide targeted policy recommendations for improving the sustainable growth of renewable energy firms with different ownership structures. Fifth, considering the accuracy and comprehensiveness of variable calculation, this paper uses a weighted algorithm to calculate environmental regulations and principal component analysis to calculate resource endowments.

2. Literature Review

2.1. Impact of Environmental Regulations (ERs) on Sustainable Growth (SG)

Environmental regulations (*ERs*) are a collection of features for government environmental policies aiming at reducing businesses' influence on the natural environment and providing an atmosphere conducive to environmental innovation [25]. According to López-Gamero et al. (2010), *ERs* are a collection of environmental behaviors that are either mandatory or discretionary and are disseminated directly or indirectly by economic organizations or governments [26]. Pargal and Wheeler (1996) established the notion of informal *ERs*, arguing that in poor nations where institutional regulation is weak or non-existent, many communities have struck emission reduction agreements with local firms [27]. Informal regulation is the term for this occurrence. For command and control and market-based rules, Li and Ramanathan (2018) find a positive non-linear relationship between Environmental regulations (*ERs*) and Sustainable growth (*SG*). [28]. The informal *ERs* represented by environmentally related technology and education levels, according to Wang and Shao (2019), have a favorable and substantial influence on *SG* [29]. *ERs* have a statistically significant and positive connection with *SG*, according to Javeed et al. (2020a) [14]. Higher *ERs* intensity might drive manufacturing, resulting in a more concentrated economy with lower CO_2 emissions, hence promoting *SG* [30]. Firms can increase staff quality at or beyond the *ER* threshold level, according to Song et al. (2018), for additional gains in *SG* [31]. Labor cost and *ERs*, according to Zheng et al. (2019), have a synergistic influence on company growth and structural adjustment [32]. Zhao et al. (2018) suggested that if appropriate *ERs* are used, then in a short period of time, *ERs* and financial returns can produce a win–win situation [33]. The effect of *ERs* on the link between technical innovation and *SG* is theoretically good, but not substantial, showing that there is still an "implementation gap" [34].

According to Ramanathan et al. (2017), depending on their resources and expertise, firms that adopt a more dynamic approach to reacting to *ERs* innovatively and taking a proactive approach to managing their environmental performance are generally better able to reap the *SG* [35]. Regulatory and supervisory actions based on actual market conditions, according to Xie et al. (2017), have a non-linear connection and can be favorably associated with "green" production [6]. Dasgupta et al. (2001) discovered a substantial positive correlation between the frequency of *ER* agency inspections and the *SG* [36]. According to

Liu et al. (2018), *ERs* have a net negative effect on energy usage, which is advantageous for reducing energy pressures [4]. In China, the energy-saving impact of *ERs* is dynamic, with complicated outcomes arising from the "Green Paradox" and "compliance cost." It has also been found that *ERs* help stimulate technological progress in manufacturing, which indirectly saves energy [37].

Based on Liu et al. (2018), this paper breaks down environmental regulations into three aspects: economic, legal, and supervision [4], and makes the following hypotheses about their relationship with sustainable growth (*SG*), respectively.

Hypothesis 1a (H1a). *Economic environmental regulation is conducive to sustainable growth.*

Hypothesis 1b (H1b). *Legal environmental regulation is conducive to sustainable growth.*

Hypothesis 1c (H1c). *Supervised environmental regulation is conducive to sustainable growth.*

2.2. The Impact of Resource Endowment (RE) on Sustainable Growth (SG)

Resource endowment (*RE*), also known as factor endowment, relates to a country's ownership of numerous production components such as labor, money, land, technology, and management. The concept of the "resource curse" was coined by Auty (1993), where the dependence on natural resources and its potentially detrimental relationship with economic growth is referred to as a "curse". His research found that the world's natural resource-rich countries were unable to use their environmental wealth to improve their economies, and he introduced the concept of the "resource curse", and as a result, their economies grew at a slower rate than those without natural resources [38]. The evidence that *RE* negatively impacts *SG* remains compelling, especially in Chinese cities that produce fossil energy, and the direct influence of *ERs* on economic development exhibits an "N" curve connection, according to survey data [39]. However, there was some dissent to this widely held belief; Hilmawan and Clark (2019) found no evidence of a "resource curse" using yearly fixed effects and first-order difference regression analysis [40]. It is worth noting that even the most ardent proponents of the "resource curse" are not arguing that states with abundant natural resources would be better off without them [39].

Basic and diverse resources are the two types of enterprise resources [41]. Human resources, financial resources, material resources, technical resources, information resources, and other basic resources are required for company technological innovation operations. The heterogeneity of heterogeneous resources is expressed in the variability of the unique use value [42]. Enterprise culture, which transforms basic resources into diverse resources while encouraging technical innovation abilities, ensures the survival and development of businesses. Energy companies' *RE* is unique, and their financial *RE* mostly consists of government subsidies and financing limits. The value contained in social interactions between individuals or groups is referred to as social capital, which may help spread knowledge, communicate information, and share resources, lower transaction costs, and enhance financial performance. The "relationship finance hypothesis," presented by Chakravarty and Scott (1999), holds that social capital plays a crucial function in enhancing a company's ability to raise funds [43].

Government subsidies are the most visible kind of social capital in the energy sector. The value of the government subsidy is derived from the company's financial statements' remarks. The value of government subsidy items is calculated using the amount from direct subsidies, tax refunds, and other things [44]. The subsidies granted by the government can help companies with customer service shortage of funds and are an important source of cash for companies [45]. Hu (2001) found no evidence of a link between government subsidies and increased productivity in subsidized firms in Chinese industries [46]. According to Yang et al. (2019), government subsidy policies have a positive moderating effect on investment in the renewable energy sector in China [47]. The contribution of government subsidies to renewable energy investment increases significantly when energy consumption intensity is high, but bank credit is more restrictive, and the degree of economic growth is below the threshold. Both cash subsidies and tax incentives can encourage renewable energy investment, with tax incentives having a greater impact. Overall, government subsidies are the main driver of renewable energy firms.

Another expression of *RE* in energy businesses is a financial limitation. Energy companies face three types of financial constraints: loan financing, equity financing, and internal financing. Short-term liabilities, according to Cutillas and Sánchez (2014), can prevent businesses from making unproductive investments [48]. Enterprises' expansion initiatives are directly hampered by financial restrictions [37]. The most significant impediment to the development of SMEs is the absence of funding channels; high financing costs and lack of professional advice are the main obstacles to external financing [49]. Access to funding is a crucial growth restriction for SMEs, according to Beck and Demirguc-Kunt (2006), and financial and legal institutions play a key role in alleviating this limitation [50]. Ferris et al. (2017) found that social capital lowered the cost of equity borrowing using data from 1999 to 2012 [51]. Information asymmetry and the agency problem are reduced as a result of social relationships, lowering the cost of equity. Hypothesis 2 is offered based on the preceding discussion:

Hypothesis 2 (H2). *A positive relationship exists between resource endowment and long-term growth.*

2.3. The Role of Resource Endowment (RE) in Mediating the Connection between Environmental Regulations (ERs) and Sustainable Growth (SG)

According to Yang and Song (2019), the link between *ERs* and the "resource curse" is inverted U-shaped, and the "resource curse" can only be broken when the *ERs'* intensity passes the turning point [19]. In complete samples, *ERs* can also break the "resource curse" issue indirectly by increasing green technical innovation, reducing resource reliance, and speeding up enterprise development. Reducing financing constraints and implementing government subsidies are one of the main sources of *RE* for energy firms. Financial and tax assistance are the key way for *SG* of firms in nations with excellent renewable energy development [52,53]. *RE* has the potential to not only assist the long-term growth of energy businesses, but also to achieve the government's environmental policy objectives. Environmental management and debt finance, according to Xu and Chen (2020), have a good association with firm sustainability [54]. When businesses have less social capital, such as government subsidies and limited funding, their growth is constrained to some extent. The impact of *ERs* on business *SG* will be readily apparent at this time. The better the *RE*, the less influenced by Ers it is, reducing the impact of Ers on the *SG* of businesses and promoting their long-term development.

In addition, this paper reviews in the literature review the findings of the effect of *ERs* on factors such as CO_2 emissions, firm structure, and *SG* in different research contexts in different literature. Based on the results of existing studies, the hypotheses to be tested in this paper are presented:

Hypothesis 3a (H3a). *With the increased resource endowment, the impact of environmental regulation on sustainable growth will be weakened.*

Hypothesis 3a (H3b). *With the decreased resource endowment, the impact of environmental regulation on sustainable growth will be strengthened.*

3. Research Methodology

3.1. Modeling

The experimental analysis in this paper is based on the following theoretical framework, as shown in Figure 1.

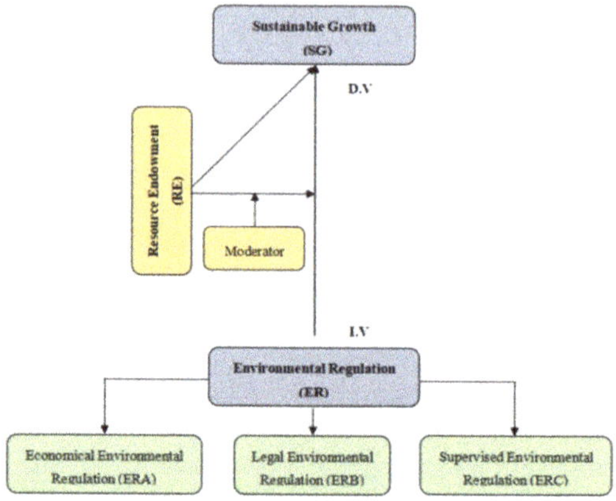

Figure 1. Conceptual framework.

First, the paper tests the effects of environmental regulations and resource endowments on sustainable growth using the following model:

$$SG_{it} = \alpha_0 + \alpha_1 ER_{it} + \alpha_2 RE_{it} + \alpha_3 Ctrl_{it} + \theta_{it} \quad (1)$$

$$Ctrl_{it} = AT_{it} + HHI_{it} + FDI_{it} + FS_{it} + LEV_{it} + PPE_{it} \quad (2)$$

To measure the impact of environmental regulation on sustainable growth under the moderating effect of resource endowments, Equations (1) and (2) are further extended in this paper as follows.

$$SG_{it} = \beta_0 + \beta_1 RE_{it} + \beta_2 Ctrl_{\ it} + \mu_{it} \quad (3)$$

$$SG_{it} = \gamma_0 + \gamma_1 ER_{it} + \gamma_2 RE_{it} + \gamma_3 ER_{it} \times RE_{it} + \gamma_4 Ctrl_{it} + \varepsilon_{it} \quad (4)$$

where Equation (3) measures the effect of resource endowment per se on sustainable growth, and Equation (4) measures the effect of resource endowment on the relationship between environmental regulation and sustainable growth. Where i and t stand for listed businesses and time periods, respectively. The dependent variable SG stands for long-term growth. Environmental regulation is represented by ERs, resource endowment is represented by RE, and the moderating influence of resource endowment on environmental regulation and sustainable growth is represented by $ER \times RE$. We have also added some important control variables represented by $Ctrl$, such as asset turnover (AT), the Herfindahl–Hirschman Index (HHI), foreign direct investment (FDI), firm size (FS), leverage (LEV), property, plant, and equipment (PPE), etc.

α_0, β_0 and γ_0 are constant term. α_1, α_2, α_3, β_1, β_2, β_3 and γ_1, γ_2, γ_3, γ_4 are estimated coefficients of the independent variable, the moderator variable, and the cross multiplication of the independent variable and the moderator variable, respectively. θ_{it}, μ_{it} and ε_{it} represent the random disturbance terms. In this paper, the panel data model is used to estimate the coefficients.

3.2. Variables

3.2.1. Sustainable Growth (SG)

Sustainable growth (SG) is the result of a set of activities in the business process, including two dimensions: sustainable business growth potential, and sustainable business profitability. The capital approach to sustainable growth is known to be useful in explaining sustainable development.

This paper uses the sustainable growth rate as proposed by Javeed et al. (2020b) to measure SG [14].

3.2.2. Environmental Regulations (*ERs*)

Referring to Liu et al. (2018), environmental regulations are government-initiated environmental protection measures that are universally applicable to all businesses; this paper breaks down environmental regulations from three perspectives: economic, legal, and supervisory [4].

(1) Economical environmental regulation (*ERA*): *ERA* refers to the use of economic tools to reduce or eliminate negative external consequences produced by pollution and is a voluntary regulation. The *ERA* is represented in this article by the overall investment share of industrial added value in pollution control.
(2) Legal environmental regulation (*ERB*): *ERB* refers to a set of severe restrictions that limit the production and administration of businesses in order to safeguard the environment; this is a market-based regulation. If a company violates environmental regulations, the government can apply administrative fines. As an alternative indication of *ERB*, we use the number of administrative penalty cases involving the environment.
(3) Supervised environmental regulation (*ERC*): *ERC* refers to government departments' oversight of environmental contamination, forcing businesses to enhance current equipment and technology in order to achieve cleaner output, and is command-and-control regulation. The number of environmental protection agencies at the end of each year is used as a proxy variable for *ERC* in this study.

3.2.3. Resource Endowment (*RE*)

In this paper, based on the studies of Xu et al. (2019) and Mtaturu (2020), the resource endowment of firms is divided into five areas based on the source of resources: government, and financing institutions such as banks, suppliers, customers, and other firms [55,56]. In addition, for the aspect quantitative analysis, this paper uses the data of government subsidy income, short-term loans, accounts payable, accounts receivable, and long-term equity investment of enterprises to represent the resource endowment of these five aspects respectively. Different basic index weights were assigned using the principal component analysis (PCA) approach, and the composite index *RE* was created.

3.2.4. Control Variables

Both the influence on *RE* and TIE should be considered while selecting control variables. *AT, HHI, FDI, FS, LEV,* and *PPE* are used as control variables in this study (Hille et al., 2020) [57]. The definitions of variables, statistical descriptions of the variables, and correlation analyses are shown in Tables 1–3, respectively.

Table 1. Variables and definitions.

Variable	Abbreviations	Definition Description
Dependent Variable		
Sustainable growth	*SG*	PM × (1 − D) × (1 + L)/(T − (PM × (1 − D) × (1 + L)))
Independent Variables		
Economical environmental regulation	*ERA*	The ratio of environmental pollution treatment investment to the industrial added value
Legal environmental regulation	*ERB*	The number of year-end administrative penalty cases on the environment
Supervised environmental regulation	*ERC*	The number of year-end environmental protection agencies
Resource endowment	*RE*	The principal component analysis
Control Variables		
Asset turnover	*AT*	Ratio of total sales to total asset
Herfindahl–Hirschman Index	*HHI*	The HHI of industry
Foreign direct investment	*FDI*	Ratio between foreign direct investment and GDP
Firm Size	*FS*	Natural logarithm of total assets
Leverage	*LEV*	Ratio of total liabilities to total assets
Property, Plant, and Equipment	*PPE*	Ratio of property, plant, and equipment to total sales

Table 2. Descriptive statistics of variables.

Variables	Obs	Mean	Std. Dev.	Min	Max
SG	882	0.056	0.142	−0.673	0.491
ERA	882	0.060	0.047	0.014	0.246
ERB	882	11.36	1.098	8.617	13.61
ERC	882	9.359	0.689	7.403	10.22
RE	882	10.62	1.559	7.199	14.24
AT	882	0.390	0.238	0.044	1.462
HHI	882	0.065	0.008	0.052	0.076
FDI	882	7.182	1.373	3.866	9.864
FS	882	22.78	1.425	20.19	26.34
LEV	882	0.563	0.200	0.051	0.941
PPE	882	21.77	1.796	15.86	25.94

Table 3. Correlation analysis of variables.

Variables	*SG*	*ERA*	*ERB*	*ERC*	*RE*	AT	HHI	FDI	FS	LEV	PPE
SG	1										
ERA	−0.015	1									
ERB	−0.003	−0.338 ***	1								
ERC	0.011	−0.229 ***	0.251 ***	1							
RE	0.050	0.095 ***	0.004	−0.300 ***	1						
AT	0.048	−0.062 *	−0.069 **	0.074 **	−0.125 ***	1					
HHI	0.022	−0.020	0.087 ***	−0.075 **	−0.119 ***	0.181 ***	1				
FDI	−0.047	−0.462 ***	0.635 ***	0.191 ***	−0.088 ***	−0.004	0.297 ***	1			
FS	0.140 ***	0.141 ***	0.010	−0.278 ***	0.900 ***	−0.257 ***	0.206 ***	0.150 ***	1		
LEV	0.175 ***	0.165 ***	0.203 ***	−0.217 ***	0.481 ***	−0.022	0.086 **	0.362 ***	0.413 ***	1	
PPE	0.098 ***	0.166 ***	0.095 ***	−0.232 ***	0.772 ***	−0.206 ***	0.163 ***	0.238 ***	0.882 ***	0.472 ***	1

Note: *** $p < 0.01$, ** $p < 0.05$, * $p < 0.1$.

4. Empirical Analysis

4.1. Data Source

The data for 91 renewable energy businesses were chosen for this study. The China Statistical Yearbook and the RESSET database provided data for *ER* (*ERA*, *ERB*, and *ERC*), while the CSMAR database provided data for *RE* and the WIND database provided data for *SG*. The CSMAR database was used to obtain data for the control variables. Interpolation was employed to supplement the data due to a lack of data for particular *REs*. All data were tested for stability to determine that they are stationary.

4.2. Fitting Results for Hypothesis 1

The results of the empirical calculations around Hypothesis 1 are shown in Table 4. The Hausman test values for models were 19.27, 18.98, and 21.32, respectively, at the 1% significance level, thus allowing the FE regression to be selected. Specifically, the FE regression for model 1 shows a coefficient value of 0.019 for the *ERA* term at the 1% significance level and a coefficient value of 0.273 for the *ERA* term at the 1% significance level in the GMM regression. In the regression results for model 2, the *ERB* term has a coefficient value of −0.3 at the 1% significance level in the FE regression and a coefficient value of 0.165 in the GMM regression. The coefficient value was 0.165. The coefficient of the *ERC* term in the FE regression of model 3 is 0.038 at the 1% level of significance, and in the GMM regression, the coefficient of the *ERC* term is 0.054 at the 1% level of significance. In summary, the results show that the economy, law, and supervision of environmental regulation all contribute to sustainable development, so Hypothesis 1 holds. Among, them *ERA* has the most significant contribution to *SG* compared to *ERB* and *ERC*.

Table 4. Measurement results of the relationship between *ER* and *SG*.

Variables	Model 1		Model 2		Model 3	
	FE	GMM	FE	GMM	FE	GMM
ERA	0.019 ***	0.273 ***				
ERB			0.300 ***	0.165 ***		
ERC					0.038 ***	0.054 ***
AT	0.120 **	0.092 *	0.120 **	−0.029 *	0.126 **	0.089 *
HHI	2.536	1.505	2.523	1.001	2.596	2.268
FDI	−0.011	−0.001	−0.010	−0.005	−0.013	−0.005
FS	0.057 ***	0.068 *	0.057 ***	−0.016	0.057 ***	0.040 *
LEV	−0.396 ***	−0.116	−0.396 ***	−0.379 ***	−0.388 ***	−0.222
PPE	−0.006	0.003 *	−0.005	0.004 *	−0.006	0.005 *
C	−1.048 *	−0.971 **	−1.018 *	−2.178	−1.381 **	1.211
R^2	0.321		0.452		0.476	
Hausman test	19.27 ***		18.98 ***		21.32 ***	
Arellano–Bond test						
AR (1)		0.051		0.036		0.027
AR (2)		0.427		0.432		0.733
Sargan test		0.913		0.907		0.946
Observations	790	790	790	790	790	790
Number of id	91	91	91	91	91	91

Note: *** $p < 0.01$, ** $p < 0.05$, * $p < 0.1$.

4.3. Fitting Results for Hypothesis 2

The results of the empirical calculations around Hypothesis 2 are shown in Table 5. Overall, the Hausman test values for Models 1, 2, and 3 were 21.22, 20.19, and 18.99 at the 1% significance level, respectively, indicating that the FE regression could be chosen. Specifically, in Model 1, the *RE* term has a coefficient of 0.414 in the FE regression and 0.284 in the GMM regression at the 10% significance level, indicating that *RE* has a significant contribution to *SG*. The paper then divides *RE* into SOEs and non-SOEs, which are analyzed in Model 2 and Model 3, respectively. According to Model 2 for SOEs, *RE* has a coefficient of 0.389 in the FE regression and 0.179 in the GMM regression at the 10% significance level. According to the results of Model 3 for non-SOEs, the *RE* term has a coefficient of 0.402 in the FE regression and 0.277 in the GMM regression at the 10% significance level. In summary, the results can be that Hypothesis 2 holds and the contribution of resource endowment to *RE* firms is more significant in non-state-owned firms.

Table 5. Measurement results of the relationship between *RE* and *SG*.

Variables	Model 1		Model 2		Model 3	
	FE	GMM	FE	GMM	FE	GMM
RE	0.414 *	0.284 **	0.389 *	0.179 *	0.402 *	0.277 *
AT	0.679 **	1.021 **	1.089 ***	2.315 *	0.857 ***	2.299 **
HHI	0.024 *	0.126	0.026	0.187	0.029	0.154 *
FDI	−0.093	−0.105 *	−0.087	−0.127	−0.089 **	−0.166
FS	0.209 **	0.106 *	0.233	0.176	0.231	0.119 *
LEV	−0.325	−0.421	−0.450	−0.298 *	−0.312	−0.206 **
PPE	0.546 *	0.219	0.444 *	0.398	0.590 *	0.265
C	1.508 *	0.882	3.001 *	0.832	2.098 *	0.891
R^2	0.545		0.547		0.550	
Hausman test	21.22 ***		20.19 ***		18.99 ***	
Arellano–Bond test						
AR (1)		0.039		0.027		0.025
AR (2)		0.593		0.642		0.589
Sargan test		0.899		0.915		0.907
Observations	790	790	588	588	202	202
Number of id	91	91	66	66	25	25

Note: *** $p < 0.01$, ** $p < 0.05$, * $p < 0.1$.

4.4. Fitting Results for Hypothesis 3

The results of the empirical calculations around Hypothesis 3 are shown in Table 6. This paper determines the effect of *RE* on the relationship between *ER* and *SG* by measuring the interaction term between *ER* and *RE*. Overall, the Hausman test values for Models 1, 2, and 3 were 28.08, 21.67, and 20.59, respectively, at the 1% significance level, allowing for the choice of FE regression. Specifically, in Model 1, the *ERA* × *RE* term has a coefficient of −0.016 for the FE regression and −0.31 for the GMM regression at the 5% level of significance. Model 2 shows that the *ERB* × *RE* term has a coefficient of −0.007 for the FE regression and −0.259 for the GMM regression at the 1% level of significance. Model 3 has the *ERC* × *RE* term at the 5% level of significance. In summary, Hypothesis 3 holds that *RE* is able to inhibit the facilitative effect between *ER* and *SG*, and *RE* has the most pronounced inhibitory effect on *ERA* and *SG* compared to *ERB* and *ERC*.

Table 6. Measurement results of the relationship between *ER* and *SG* with moderating effect of *RE*.

Variables	Model 1		Model 2		Model 3	
	FE	GMM	FE	GMM	FE	GMM
ERA	0.011 **	0.065 **				
ERB			0.005 **	0.072 **		
ERC					0.037 ***	0.180 **
RE	0.020 **	0.021 ***	0.021 ***	0.062 ***	0.020 **	0.133 ***
ERA × *RE*	−0.016 **	−0.310 **				
ERB × *RE*			−0.007 ***	−0.259 ***		
ERC × *RE*					−0.012 **	−0.312 **
AT	0.126 **	0.532 ***	0.126 **	0.561 ***	0.126 **	0.641 ***
HHI	−4.152	−7.343	−3.967	−7.483	−3.823	−7.591
FDI	2.915	5.024	2.918	5.109	2.958 *	5.336 *
FS	0.076 ***	0.678 ***	0.076 ***	0.725 ***	0.074 ***	0.702 ***
LEV	−0.366 *	−6.368 *	−0.359 *	−5.319 *	−0.365 *	−6.160 *
PPE	−0.006 *	−3.517	−0.007	−2.980	−0.007	−3.874
C	−1.307 *	−4.271 *	−1.632 *	−5.295 **	−2.165 *	−4.553 *
R^2	0.449		0.453		0.446	
Hausman test	28.08 ***		21.67 ***		20.59 ***	
Arellano–Bond test						
AR (1)		0.060		0.051		0.059
AR (2)		0.588		0.546		0.607
Sargan test		0.919		0.923		0.904
Observations	882	882	882	882	882	882
Number of id	91	91	91	91	91	91

Note: *** $p < 0.01$, ** $p < 0.05$, * $p < 0.1$.

4.5. Robustness Test

To ensure that the empirical findings are accurate, this paper instead measures *SG* using ROE (Cao and Wang, 2017; Song and Wang, 2018) [58,59]. The association between *ER*, *RE*, and *SG* is then measured again and the test results are shown in Table 7. Models 1, 2, and 3 measured the relationship between *ERA*, *ERB*, *ERC*, and *SG*, respectively, and the results showed that the coefficients of the *ERA*, *ERB*, and *ERC* terms were 0.216, 0.224, and 0.219, respectively, at the 5% level of significance. The relationship between the interaction terms of the *ER* series variables and *RE* variables on *SG* was measured in Models 4, 5, and 6, respectively, and the results showed that at the 5% level of significance the coefficients of the *ERA* × *RE*, *ER* × *RE*, and *ERC* × *RE* terms are −0.298, −0.344, and −0.316, respectively. In summary, it can be seen that the results of the robustness tests are basically consistent with the previous empirical results and the empirical findings are stable and valid.

Table 7. Robustness test.

Variables	Model 1	Model 2	Model 3	Model 4	Model 5	Model 6
	GMM	GMM	GMM	GMM	GMM	GMM
ERA	0.216 ***			0.358 ***		
ERB		0.224 ***			0.217 ***	
ERC			0.219 **			0.311 ***
RE	0.258 **	0.299 **	0.287 **	0.375 **	0.402 ***	0.395 ***
ERA × *RE*				−0.298 **		
ERB × *RE*					−0.344 **	
ERC × *RE*						−0.316 **
AT	0.172 *	0.176 *	0.178 *	0.098	0.099	0.090
HHI	5.565	5.284	5.223 *	9.517	8.346 *	8.885 *
FDI	0.121	0.199 **	0.124 **	0.548	0.827	0.718
FS	0.049	0.048	0.0499 *	0.272 *	0.286 **	0.262 *
LEV	−1.746 *	−1.934 *	−1.689 **	−2.778 **	−2.899 **	−2.957 **
PPE	0.065	0.067	0.068 *	0.421	0.398 *	0.413
C	7.169 **	7.206 **	7.308 **	5.477 **	6.103 **	6.890 *
Arellano–Bond test						
AR (1)	0.028	0.036	0.027	0.035	0.028	0.031
AR (2)	0.335	0.401	0.334	0.402	0.335	0.405
Sargan test	0.876	0.907	0.896	0.899	0.918	0.923
Observations	882	882	882	882	882	882
Number of id	91	91	91	91	91	91

Note: *** $p < 0.01$, ** $p < 0.05$, * $p < 0.1$.

5. Conclusions and Policy Recommendations

This paper studied the relationships between environmental regulation, resource endowment, and sustainable growth by using a fixed-effects model and system GMM method, and the research sample is the panel data of new energy enterprises from 2010 to 2019. This paper focuses on the impact of environmental regulation on the sustainable growth of Chinese renewable energy enterprises, and introduced resource endowment to examine its moderating effects. After empirical analysis, we obtained the following conclusions: economical environmental regulations, legal environmental regulations, and supervised environmental regulations are positively associated with sustainable growth. Compared with legal environmental regulations and supervised environmental regulations, economic environmental regulations have the greatest impact on sustainable growth. The resource endowment is positively associated with sustainable growth, especially for non-state-owned renewable energy enterprises, and resource endowment plays a moderating role between environmental regulations and sustainable growth. Furthermore, resource endowment has the greatest moderating effect on the relationship between economical environmental regulations and sustainable growth. According to the above conclusion, we propose the following suggestions.

The first one is based on the effectiveness of economic environment regulations. Economic environmental regulations intuitively discourage the consumption of traditional energy sources and promote the development of renewable energy from an economic perspective. Governments can implement flexible and effective economic policies that take into account local conditions, such as increasing taxes on traditional energy sources while providing policy subsidies and tax incentives for renewable energy. By reducing the financial pressure on renewable energy companies, new energy innovations can be promoted to achieve sustainable growth.

Secondly, based on the effectiveness of legal environmental regulations. The use of traditional energy sources inevitably leads to the consumption of environmental resources and environmental pollution. For government policies, on the one hand, through the establishment of a sound legal environmental regulation system, the use of traditional energy sources and the treatment and discharge of pollution can be regulated in order to curb the consumption of natural resources and mitigate environmental pollution. On the other hand, legal environmental regulations are also conducive to the management of renewable energies, as they regulate the research and development and production of renewable energies and promote the sustainable growth of new energy enterprises.

Thirdly, based on the effectiveness of supervisory environmental regulations. Strict and effective regulation, based on sound laws and regulations, can ensure that legal provisions are implemented. For example, strict monitoring of energy consumption and the treatment and discharge of pollutants by energy companies can effectively force traditional energy companies to transform and promote technological progress in the field of new energy. Thus, strengthening supervisory environmental regulation is beneficial to the sustainable growth of renewable energy companies.

Fourth, based on the effectiveness of resource endowment. Environmental endowments are inherently conducive to sustainable growth, so governments should actively guide energy companies to develop and build local resources with local characteristics, and sufficient environmental resources to ensure sustainable growth. However, given the 'resource curse' effect, local energy development should not be overly dependent on the benefits of environmental endowments and should focus on the long-term benefits of renewable energy. In addition, as resource endowments have a negative impact on the influence of environmental regulations, as resource endowments increase, the role of environmental regulation in sustainable growth decreases. Therefore, government departments should be fully aware of the contradictions between resource endowments and environmental regulation, and the link between resource endowment and environmental regulation should be better coordinated. For example, for traditional energy sources, environmental regulations should be strengthened to avoid the "resource curse" brought about by overly strong resource endowments, while for renewable energy enterprises, resource endowments can be moderately strengthened through the creation of a favorable financing environment and research environment to promote sustainable growth. Whether starting from an environmental regulatory perspective or a resource endowment perspective, the final goal is to curb traditional energy sources and promote renewable energy development so as to achieve sustainable growth. The government should be fully aware that technological development is the basic driver of sustainable growth, reasonably integrating the local environmental and social resources, enhancing resource-use efficiency, and encouraging scientific and technological research and development so as to achieve sustainable growth.

Author Contributions: Methodology, H.Z.; Writing—original draft, H.H.C.; Data curation, K.L.; Writing—review & editing, Z.R. All authors have read and agreed to the published version of the manuscript.

Funding: This research received no external funding.

Institutional Review Board Statement: Not applicable.

Informed Consent Statement: Not applicable.

Data Availability Statement: Not applicable.

Conflicts of Interest: The authors declare that they have no known competing financial interests or personal relationships that could have appeared to influence the work reported in this paper.

References

1. Iram, R.; Zhang, J.; Erdogan, S.; Abbas, Q.; Mohsin, M. Economics of energy and environmental efficiency: Evidence from OECD countries. *Environ. Sci. Pollut. Res.* **2020**, *27*, 3858–3870. [CrossRef]
2. Wang, J.; Wang, S.; Li, S.; Feng, K. Coupling analysis of urbanization and energy-environment efficiency: Evidence from Guangdong province. *Appl. Energy* **2019**, *254*, 113650. [CrossRef]
3. Ren, S.; Li, X.; Yuan, B.; Li, D.; Chen, X. The effects of three types of environmental regulation on eco-efficiency: A cross-region analysis in China. *J. Clean. Prod.* **2018**, *173*, 245–255. [CrossRef]

4. Liu, Y.; Li, Z.; Yin, X. Environmental regulation, technological innovation and energy consumption—A cross-region analysis in China. *J. Clean. Prod.* **2018**, *203*, 885–897. [CrossRef]
5. Telle, K.; Larsson, J. Do environmental regulations hamper productivity growth? How accounting for improvements of plants' environmental performance can change the conclusion. *Ecol. Econ.* **2007**, *61*, 438–445. [CrossRef]
6. Xie, R.-H.; Yuan, Y.-J.; Huang, J.-J. Different Types of Environmental Regulations and Heterogeneous Influence on "Green" Productivity: Evidence from China. *Ecol. Econ.* **2017**, *132*, 104–112. [CrossRef]
7. Wang, Y.; Deng, X.; Zhang, H.; Liu, Y.; Yue, T.; Liu, G. Energy endowment, environmental regulation, and energy efficiency: Evidence from China. *Technol. Forecast. Soc. Change* **2022**, *177*, 121528. [CrossRef]
8. Fu, T.; Cai, C.; Jian, Z. The illusion of "win–win" solution: Why environmental regulation in china promotes firm performance? *Struct. Change Econ. Dyn.* **2020**, *52*, 366–373. [CrossRef]
9. Shuai, S.; Fan, Z. Modeling the role of environmental regulations in regional green economy efficiency of China: Empirical evidence from super efficiency DEA-Tobit model. *J. Environ. Manag.* **2020**, *261*, 110227. [CrossRef]
10. Elkington, J. Towards the Sustainable Corporation: Win-Win-Win Business Strategies for Sustainable Development. *Calif. Manag. Rev.* **1994**, *36*, 90–100. [CrossRef]
11. Das, M.; Rangarajan, K.; Dutta, G. Corporate sustainability in SMEs: An Asian perspective. *J. Asia Bus. Stud.* **2020**, *14*, 109–138. [CrossRef]
12. Salzmann, O.; Ionescu-Somers, A.; Steger, U. The Business Case for Corporate Sustainability: Literature Review and Research Options. *Eur. Manag. J.* **2005**, *23*, 27–36. [CrossRef]
13. Engert, S.; Rauter, R.; Baumgartner, R.J. Exploring the integration of corporate sustainability into strategic management: A literature review. *J. Clean. Prod.* **2016**, *112*, 2833–2850. [CrossRef]
14. Javeed, S.A.; Latief, R.; Lefen, L. An analysis of relationship between environmental regulations and firm performance with moderating effects of product market competition: Empirical evidence from Pakistan. *J. Clean. Prod.* **2020**, *254*, 120197. [CrossRef]
15. Curtis, E.M.; Lee, J.M. When do environmental regulations backfire? Onsite industrial electricity generation, energy efficiency and policy instruments. *J. Environ. Econ. Manag.* **2019**, *96*, 174–194. [CrossRef]
16. Zhai, X.; An, Y. Analyzing influencing factors of green transformation in China's manufacturing industry under environmental regulation: A structural equation model. *J. Clean. Prod.* **2020**, *251*, 119760. [CrossRef]
17. Mickiewicz, T.; Nyakudya, F.W.; Theodorakopoulos, N.; Hart, M. Resource endowment and opportunity cost effects along the stages of entrepreneurship. *Small Bus. Econ.* **2017**, *48*, 953–976. [CrossRef]
18. Wang, S.; Wang, X.; Lu, B. Is resource abundance a curse for green economic growth? Evidence from developing countries. *Resour. Policy* **2022**, *75*, 102533. [CrossRef]
19. Yang, Q.; Song, D. How does environmental regulation break the resource curse: Theoretical and empirical study on China. *Resour. Policy* **2019**, *64*, 101480. [CrossRef]
20. Song, Y.; Li, G.; Wang, Q.; Meng, X.; Wang, H. Scenario analysis on subsidy policies for the uptake of electric vehicles industry in China. *Resour. Conserv. Recycl.* **2020**, *161*, 104927. [CrossRef]
21. Söderblom, A.; Samuelsson, M.; Wiklund, J.; Sandberg, R. Inside the black box of outcome additionality: Effects of early-stage government subsidies on resource accumulation and new venture performance. *Res. Policy* **2015**, *44*, 1501–1512. [CrossRef]
22. Lu, J.; Li, B.; Li, H.; Zhang, Y. Sustainability of enterprise export expansion from the perspective of environmental information disclosure. *J. Clean. Prod.* **2020**, *252*, 119839. [CrossRef]
23. Wu, L.; Sun, L.; Qi, P.; Ren, X.; Sun, X. Energy endowment, industrial structure upgrading, and CO_2 emissions in China: Revisiting resource curse in the context of carbon emissions. *Resour. Policy* **2021**, *74*, 102329. [CrossRef]
24. Wang, Q.; Wang, L.; Li, R. Renewable energy and economic growth revisited: The dual roles of resource dependence and anticorruption regulation. *J. Clean. Prod.* **2022**, *337*, 130514. [CrossRef]
25. Eiadat, Y.; Kelly, A.; Roche, F.; Eyadat, H. Green and competitive? An empirical test of the mediating role of environmental innovation strategy. *J. World Bus.* **2008**, *43*, 131–145. [CrossRef]
26. López-Gamero, M.D.; Molina-Azorín, J.F.; Claver-Cortés, E. The potential of environmental regulation to change managerial perception, environmental management, competitiveness and financial performance. *J. Clean. Prod.* **2010**, *18*, 963–974. [CrossRef]
27. Pargal, S.; Wheeler, D. Informal Regulation of Industrial Pollution in Developing Countries: Evidence from Indonesia. *J. Political Econ.* **1996**, *104*, 1314–1327. [CrossRef]
28. Li, R.; Ramanathan, R. Exploring the relationships between different types of environmental regulations and environmental performance: Evidence from China. *J. Clean. Prod.* **2018**, *196*, 1329–1340. [CrossRef]
29. Wang, X.; Shao, Q. Non-linear effects of heterogeneous environmental regulations on green growth in G20 countries: Evidence from panel threshold regression. *Sci. Total Environ.* **2019**, *660*, 1346–1354. [CrossRef]
30. Wang, Y.; Yan, W.; Ma, D.; Zhang, C. Carbon emissions and optimal scale of China's manufacturing agglomeration under heterogeneous environmental regulation. *J. Clean. Prod.* **2018**, *176*, 140–150. [CrossRef]
31. Song, M.; Wang, S.; Sun, J. Environmental regulations, staff quality, green technology, R&D efficiency, and profit in manufacturing. *Technol. Forecast. Soc. Change* **2018**, *133*, 1–14. [CrossRef]
32. Zheng, T.; Zhao, Y.; Li, J. Rising labour cost, environmental regulation and manufacturing restructuring of Chinese cities. *J. Clean. Prod.* **2019**, *214*, 583–592. [CrossRef]

33. Zhao, X.; Fan, Y.; Fang, M.; Hua, Z. Do environmental regulations undermine energy firm performance? An empirical analysis from China's stock market. *Energy Res. Soc. Sci.* **2018**, *40*, 220–231. [CrossRef]
34. Yuan, B.; Zhang, Y. Flexible environmental policy, technological innovation and sustainable development of China's industry: The moderating effect of environment regulatory enforcement. *J. Clean. Prod.* **2020**, *243*, 118543. [CrossRef]
35. Ramanathan, R.; He, Q.; Black, A.; Ghobadian, A.; Gallear, D. Environmental regulations, innovation and firm performance: A revisit of the Porter hypothesis. *J. Clean. Prod.* **2017**, *155*, 79–92. [CrossRef]
36. Dasgupta, S.; Laplante, B.; Mamingi, N.; Wang, H. Inspections, pollution prices, and environmental performance: Evidence from China. *Ecol. Econ.* **2001**, *36*, 487–498. [CrossRef]
37. Zhou, X.; Feng, C. The impact of environmental regulation on fossil energy consumption in China: Direct and indirect effects. *J. Clean. Prod.* **2017**, *142*, 3174–3183. [CrossRef]
38. Auty Richard, M. *Sustaining Development in Mineral: The Resources Curse Thesis*; Taylor & Francis Group: Abingdon, UK, 1993.
39. Badeeb, R.A.; Lean, H.H.; Clark, J. The evolution of the natural resource curse thesis: A critical literature survey. *Resour. Policy* **2017**, *51*, 123–134. [CrossRef]
40. Hilmawan, R.; Clark, J. An investigation of the resource curse in Indonesia. *Resour. Policy* **2019**, *64*, 101483. [CrossRef]
41. Barney, J.B. Resource-based theories of competitive advantage: A ten-year retrospective on the resource-based view. *J. Manag.* **2001**, *27*, 643–650. [CrossRef]
42. Barney, J. Firm resources and sustained competitive advantage. *J. Manag.* **1991**, *17*, 99–120. [CrossRef]
43. Chakravarty, S.; Scott, J.S. Relationships and Rationing in Consumer Loans. *J. Bus.* **1999**, *72*, 523–544. [CrossRef]
44. Peng, H.; Liu, Y. How government subsidies promote the growth of entrepreneurial companies in clean energy industry: An empirical study in China. *J. Clean. Prod.* **2018**, *188*, 508–520. [CrossRef]
45. Claro, S. Supporting inefficient firms with capital subsidies: China and Germany in the 1990s. *J. Comp. Econ.* **2006**, *34*, 377–401. [CrossRef]
46. Hu, A.G. Ownership, government R&D, private R&D, and productivity in Chinese industry. *J. Comp. Econ.* **2001**, *29*, 136–157. [CrossRef]
47. Yang, X.; He, L.; Xia, Y.; Chen, Y. Effect of government subsidies on renewable energy investments: The threshold effect. *Energy Policy* **2019**, *132*, 156–166. [CrossRef]
48. Cutillas Gomariz, M.F.; Sánchez Ballesta, J.P. Financial reporting quality, debt maturity and investment efficiency. *J. Bank. Financ.* **2014**, *40*, 494–506. [CrossRef]
49. Wang, Y. What are the biggest obstacles to growth of SMEs in developing countries?—An empirical evidence from an enterprise survey. *Borsa Istanb. Rev.* **2016**, *16*, 167–176. [CrossRef]
50. Beck, T.; Demirguc-Kunt, A. Small and medium-size enterprises: Access to finance as a growth constraint. *J. Bank. Financ.* **2006**, *30*, 2931–2943. [CrossRef]
51. Ferris, S.P.; Javakhadze, D.; Rajkovic, T. CEO social capital, risk-taking and corporate policies. *J. Corp. Financ.* **2017**, *47*, 46–71. [CrossRef]
52. Fan, Y.; Zhu, L. A real options based model and its application to China's overseas oil investment decisions. *Energy Econ.* **2010**, *32*, 627–637. [CrossRef]
53. Driouchi, T.; Bennett, D.J. Real Options in Management and Organizational Strategy: A Review of Decision-making and Performance Implications. *Int. J. Manag. Rev.* **2012**, *14*, 39–62. [CrossRef]
54. Xu, X.-L.; Chen, H.H. Exploring the relationships between environmental management and financial sustainability in the energy industry: Linear and nonlinear effects. *Energy Environ.* **2020**, *31*, 1281–1300. [CrossRef]
55. Xu, H.; Jilenga, M.T.; Deng, Y. Institutional Quality, Resource Endowment, and Economic Growth: Evidence from Cross-Country Data. *Emerg. Mark. Financ. Trade* **2019**, *55*, 1754–1775. [CrossRef]
56. Mtaturu, J. Foreign Direct Investment Inflows to Sub Saharan Africa: Do Resource Endowment and Market Size Matter? *Bus. Educ. J.* **2020**, *4*, 1–12.
57. Hille, E.; Althammer, W.; Diederich, H. Environmental regulation and innovation in renewable energy technologies: Does the policy instrument matter? *Technol. Forecast. Soc. Change* **2020**, *153*, 119921. [CrossRef]
58. Cao, B.; Wang, S. Opening up, international trade, and green technology progress. *J. Clean. Prod.* **2017**, *142*, 1002–1012. [CrossRef]
59. Song, M.; Wang, S. Market competition, green technology progress and comparative advantages in China. *Manag. Decis.* **2018**, *56*, 188–203. [CrossRef]

energies

MDPI

Article

A Hybrid Algorithm-Level Ensemble Model for Imbalanced Credit Default Prediction in the Energy Industry

Kui Wang [1], Jie Wan [2,3], Gang Li [4] and Hao Sun [2,3,*]

1 School of Economics and Management, University of Chinese Academy of Sciences, Beijing 100190, China; wangkui19@mails.ucas.ac.cn
2 Institutes of Science and Development, Chinese Academy of Sciences, Beijing 100190, China; wanjie16@mails.ucas.ac.cn
3 School of Public Policy and Management, University of Chinese Academy of Sciences, Beijing 100049, China
4 School of Business Administration, Northeastern University, Shenyang 110819, China; ligang@neuq.edu.cn
* Correspondence: sunhao201@mails.ucas.ac.cn

Abstract: Credit default prediction for the energy industry is essential to promoting the healthy development of the energy industry in China. While previous studies have constructed various credit default prediction models with brilliant performance, the class-imbalance problem in the credit default dataset cannot be ignored, where the numbers of credit default cases are usually much smaller than the number of non-default ones. To address the class-imbalance problem, we proposed a novel CT-XGBoost model, which adds to XGBoost with two algorithm-level methods for class imbalance, including the cost-sensitive strategy and threshold method. Based on the credit default dataset consisting of energy corporates in western China, which suffers from the class-imbalance problem, the CT-XGBoost model achieves better performance than the conventional models. The results indicate that the proposed model can efficiently alleviate the inherent class-imbalance problem in the credit default dataset. Moreover, we analyze how the prediction performance is influenced by different parameter settings in the cost-sensitive strategy and threshold method. This study can help market investors and regulators precisely assess the credit risk in the energy industry and provides theoretical guidance to solving the class-imbalance problem in credit default prediction.

Keywords: credit default prediction; energy industry; class imbalance; cost-sensitive; threshold method

Citation: Wang, K.; Wan, J.; Li, G.; Sun, H. A Hybrid Algorithm-Level Ensemble Model for Imbalanced Credit Default Prediction in the Energy Industry. *Energies* **2022**, *15*, 5206. https://doi.org/10.3390/en15145206

Academic Editor: Štefan Bojnec

Received: 21 June 2022
Accepted: 14 July 2022
Published: 18 July 2022

1. Introduction

In recent years, energy corporates experienced rapid development with continually increasing investment and became one of the most important markets of the global economy [1]. As the China Energy and Carbon Report 2050 [2] states, the demand for investment in Chinese new energy, energy conservation, etc., is about 7 trillion yuan. Under the massive stress of funding needs, the most common financing method for China's energy corporate is bank credit [3]. The essential risk for the creditors is corporate credit default, which means a firm fails to meet periodic repayments on a loan [4]. The financial damage caused by corporate credit default cannot be ignored, which may be a severe negative social cost or even a recession [5]. Hence, in order to promote the healthy development of China's energy industry, it is worthy of constructing an accurate corporate credit default prediction model.

A crucial issue in credit default prediction is the class-imbalance problem, which may impact the efficiency of the model negatively [6]. In the real world, the frequency of default cases is usually much smaller than that of non-default ones. It is challenging to develop an effective default forecasting model if the class distribution is imbalanced, as rare default instances are harder to be identified compared with common non-default instances [7,8]. For instance, assume the imbalance ratio of the two-class dat set is 99, with the majority non-default class accounting for 99% and the minority default class accounting for 1%. In order to minimize the error rate, the credit default prediction algorithms may simply classify all of the samples into the non-default class, where the error rate is only 1%. In such a case, all the samples of a minority default class can certainly be recognized as being of an incorrect class. Nonetheless, such a credit default predicting model is of little value because the main aim is to correctly identify as many default instances as possible without misclassifying too many non-default instances. Thus, the purpose of this study was to construct a credit default prediction model which can efficiently assess the credit risk by solving the inherent class-imbalance problem in default prediction work.

To avoid the negative effect of the class imbalance problem on credit default prediction, previous studies have proposed various imbalance processing approaches, which can be generally grouped into data-level methods, algorithm-level methods, and hybrid methods [6]. Data-level methods focus on rebalancing the class distribution of the training dataset before constructing the models [7,9,10]. Algorithm-level methods involve modifying existing algorithms or proposing novel algorithms to directly tackle datasets with class imbalances, and such learning algorithms can outperform previously existing algorithms [11–13]. Recently, the hybrid methods have gained popularity for their superior performance in learning from class-imbalanced datasets. Given the strong classification ability of pure ensemble models, the hybrid methods usually incorporate the pure ensemble models with data-level methods to construct novel models to deal with the class imbalance problem [14,15]. However, data-level methods that are combined with ensemble models have some inherent limitations, which might impact the efficiency of the model. For instance, oversampling methods may increase the probability of overfitting when training the learning algorithms, whereas undersampling methods may eliminate too much helpful information from the majority class [16].

In this study, we propose a novel hybrid model to solve the class-imbalance problem in credit default prediction. The novel model is a combination of an ensemble model and algorithm-level methods for the class imbalance problem, which can avoid the limitations of data-level methods in handling the class imbalance problem. Due to the superior performance of XGBoost among common credit default prediction models [17–19], we selected it as the ensemble model to be embedded. Then, the novel model CT-XGBoost is proposed by combining the base XGBoost model with a cost-sensitive strategy that assigns more misclassification costs for minority classes and a threshold method that sets a more rational threshold for default classification. To assess the performance of our proposed CT-XGBoost model on credit default prediction for class imbalance problems, we constructed a database of credit defaults sourced from a commercial bank in western China. As in most previous studies [20], we used the financial variables from the financial statements as the predictors to assess whether the corporates (debtors) would default. As for the benchmark models, we select previously commonly used models: logistic regression, support vector machine (SVM), neural network, random forest, and XGBoost.

Our paper has the following contributions. First, this paper proposes a novel model CT-XGBoost, which is a modified version of XGBoost that attempts to solve the class-imbalance problem in credit default data. Over the years, the class imbalance in the credit default dataset has been a crucial problem, where the number of default classes is much smaller than that of non-default classes. Without considering the class-imbalance problem, the classification model may be overwhelmed by the majority class and neglect the minority class. Nevertheless, previous studies on class imbalance problems seldom combine the ensemble model with multiple algorithm-level methods. We modified the XGBoost model with both cost-sensitive strategy and threshold method and propose the new model CT-XGBoost. Compared with the conventional intelligent model, our proposed CT-XGBoost model has better performance in default prediction. Second, we also contribute to the interpretability of the credit default prediction by identifying the top 20 most important financial variables by measuring the variables' ability to discriminate between the default and non-default samples. In practice, a good default prediction model requires not only strong classification ability, but also acceptable interpretability. Considering that research has mainly focused on the accuracy of the model but ignores the interpretability, we calculated the importance values of financial features by measuring the contributions of these features to classification. The more critical a financial feature is, the more attention it should be paid when evaluating credit default probabilities.

2. Literature Review

In this paper, the primary purpose is credit default prediction with data suffering from the class imbalance problem, and two main fields of literature are involved: credit default prediction models and techniques for solving the class imbalance problem. Representative studies are presented in the following.

2.1. Credit Default Prediction Models

In the field of corporate credit default prediction, statistical methods are first employed. Date back to the work of Beaver [21], the univariate discriminant model was used for default prediction, and the results demonstrate that the univariate linear model can utilize financial information to forecast default effectively. The multivariate discriminant model was firstly used by Altman [22] to construct the famous Z-score model, and the result shows that its default predictive power is significantly better than that of single variable analysis. The logit regression model, which can transform the dependent variable of corporate default into a continuous one by logistic function, was more rational than the multivariate discriminant model for default prediction [23]. Nonetheless, it requires that there is no linear functional relationship among the predictor variables, which may cause a multi-collinearity problem [24]. To alleviate this problem, Serrano-Cinca and Gutiérrez-Nieto [25] proposed partial least square discriminant analysis (PLS-DA) for default prediction, which is not affected by multi-collinearity. Using classical statistical methods, researchers can identify the determinants most relevant to default prediction, which can help test default theories and guide regulations of credit markets.

A significant strand of literature has found that intelligent models in credit default prediction models are efficient in predicting corporate defaulting [20,26–28]. Without the strict assumptions of the traditional statistical models (e.g., independence and normality among predictor variables), intelligent techniques can automatically derive knowledge from training data [28–30]. In addition, the intelligent methods permit non-linear decision boundaries (e.g., neural networks and SVM with non-linear kernels), which provide better model flexibility and predictive performance. In general, relative to statistic models, the corporate default prediction performance of intelligent techniques is better. For instance, Kim et al. [20] found that the neural network model outperforms logit regression. Similarly, Lahmiri [31] documented that SVM is significantly more accurate than a linear discriminant analysis classifier.

A trend in recent literature is adopting ensemble learning, which has achieved notable success in real-world applications. Differently from the mechanisms of conventional machine learning methods (such as SVM), which consist of a single estimator, ensemble learning methods combine a number of base estimators to get better generalization ability and robustness. In the work of Moscatelli et al. [27], ensemble models, including random forest and gradient boosted trees, were applied to predict corporate defaults, and the results showed that ensemble models perform better than models with a single estimator. Compared with neural networks, the ensemble model named AdaBoost had better default prediction performance in both cross-validation and test set estimation of the prediction error [32].

Among the commonly used ensemble models, the decision-tree-based XGBoost recently spread rapidly and is widely utilized in the field of credit default risk assessment [10,33,34], achieving satisfactory prediction results with its strong learning ability. For instance, in the study of Wang et al. [35], the XGBoost model was used to predict the default risk of the Chinese credit bond market, and the results show that the XGBoost model can accurately predict the default cases. For the personal credit risk evaluation, Li et al. [36] compared XGBoost to logistic regression, decision tree, and random forest. Based on the dataset from the Lending Club Platform, the XGBoost model has better performance in both feature selection and classification.

2.2. Techniques for Solving the Class-Imbalance Problem

While previous studies could effectively predict corporate default by intelligent methods, an important problem that cannot be ignored is the class-imbalance in the default database. In the real world, the default class includes a small number of data points, and the non-default class includes a large number of data points. After ignoring the class-imbalance problem, the learning algorithms or constructed models for default prediction can be overwhelmed by the majority non-default class and ignore the minority default class [7]. As the primary purpose of the default predicting model is to identify default corporates among all the corporates, the class-imbalance problem cannot be ignored.

To overcome the limitation of the class-imbalance problem, various imbalance processing approaches have been proposed. Such approaches can be generally divided into three categories: data-level methods, algorithmic-level methods, and hybrid methods [14].

Data-level methods focus on processing the imbalanced dataset before the model's construction. As the stage of data preprocessing and the stage of model training can be independent, the data preprocessing methods resample the imbalanced training dataset before training the model. To create a balanced dataset, the original imbalanced dataset can be resampled by (1) oversampling the minority class, (2) under-sampling the majority class, or (3) a hybrid of the two methods [6]. A widely used data-level method is the synthetic minority over-sampling technique (SMOTE) [9]. SMOTE generates new artificial minority cases by inserting them between existing minority cases and their neighbors. In credit default prediction tasks, after preprocessing the imbalanced dataset with SMOTE, the model based on the processed balanced training dataset can perform better [10,37]. The simplest but most effective under-sampling method is random under-sampling (RUS) [38], which involves the random elimination of majority class samples and helps improve the performance of assessing credit risk [39]. Moreover, hybrid data preprocessing methods, which combine the oversampling and undersample methods, were suggested to be helpful by recent studies [14].

Algorithmic-level methods involve modifying existing learning algorithms or proposing novel ones to directly solve the class-imbalance problem of the dataset; such algorithms usually outperform previously existing algorithms [6]. Commonly used approaches in the literature include (1) the cost-sensitive method, (2) the threshold method, and (3) one-class learning. The most commonly used is the cost-sensitive method, which deals to the nature of class imbalance by defining different misclassification costs for different classes [14]. The threshold method focuses on setting different threshold values for different classes in the model learning stage [13]. The main idea of the one-class method is to train the classifier from a training set that contains only the minority class [12].

Recently, hybrid methods have gained more popularity in learning from imbalanced datasets because of their superior performance [6]. The main idea of hybrid methods is that ensemble methods, or individual classifiers, are coupled with data or algorithm-level approaches [16], such as balanced random forests, which apply a random under-sampling strategy to the majority class to create a balanced class dataset before training an ensemble classifier with decision trees as base models [11]. SMOTEBoost combines the SMOTE oversampling approach and a rule-based learner, which is a boosting procedure [40]. Similarly, RUSBoost, which combines the random under-sampling approach with a boosting procedure, performs simpler, faster, and less complexly than SMOTEBoost during the model training [15]. Moreover, several studies combined the cost-sensitive method with boosting models where different classes are assigned different misclassification costs [41].

In summary, previous literature on class imbalance learning has proposed various methods, and hybrid methods have better performance. However, previous studies on hybrid methods mainly focus on combining ensemble learners with data-level methods, and hybridization of ensemble models with algorithm-level approaches has rarely been considered. Compared with the data-level methods, the algorithmic-level methods may be more suitable to be combined with ensemble models for the class imbalance in credit default data. The main two reasons are: (1) First, the data-level methods can alter the shape of the original data, which may impact the efficiency of the model. The oversampling strategy may increase the possibility of overfitting during the model learning process, and the undersampling strategy might eliminate some valuable data present in the majority class [16]. (2) Second, relative to data-level methods, algorithm-level methods are more straightforward and efficient in computation, making them more appropriate for big-data streams [14].

Thus, in this paper, we propose a novel algorithm that combines the algorithm-level methods and the popular ensemble model XGBoost. The main reason to select XGBoost is the superior performance of XGBoost in the credit default prediction task [17]. As for the selection of algorithm-level methods, we selected the commonly used cost-sensitive methods to combine with XGBoost. This is because the cost-sensitive method is widespread in financial management, where businesses are usually driven by profit but not accuracy [6]. Moreover, we added the threshold method into the new model, where a more rational threshold is set to classify the samples into two groups. Details of modification will be explained in the next section.

3. Methods

In this section, we present the novel CT-XGBoost prediction model with cost-sensitive and threshold methods. Figure 1 shows how XGBoost is modified into CT-XGBoost in this paper. In XGBoost, the misclassification costs for both classes are the same, and the threshold is simply set as 0.5. Thus, we improved XGBoost, turning it into CT-XGBoost, by solving two challenges: How to assign misclassification costs for the two classes properly. How do you set the threshold rationally? In this paper, two corresponding strategies (cost-sensitive strategy and threshold method) are adopted to overcome the challenges, misclassification cost is determined based on the imbalance ratio of the dataset, and a threshold is set considering the number of different classes samples. Then, XGBoost is modified into CT-XGBoost systematically.

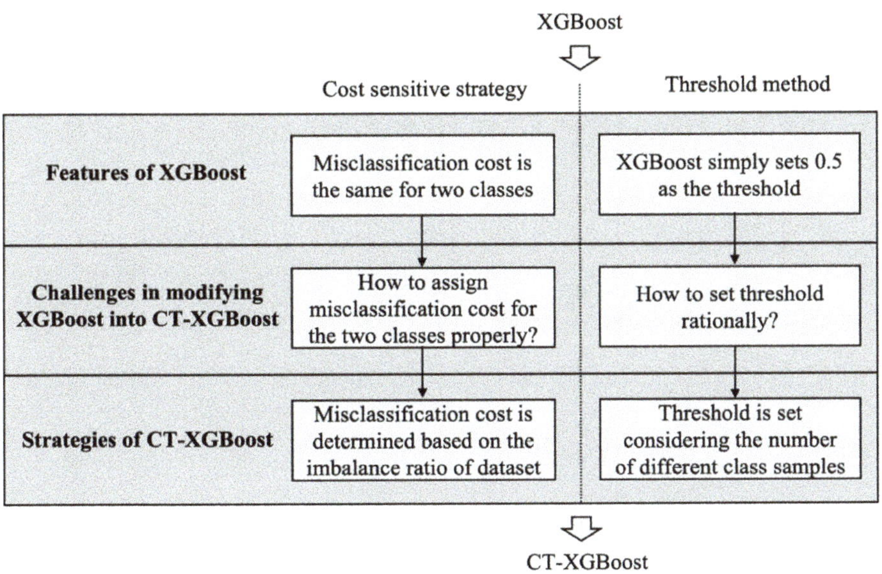

Figure 1. The process of modifying XGBoost into CT-XGBoost.

In order to introduce CT-XGBoost logically, we first explain the theory of XGBoost and then illustrate how we modify XGBoost into CT-XGBoost. After that, the commonly used default prediction models are introduced, which are used to compare with our proposed model. Lastly, performance evaluation methods of the credit default prediction are explained.

3.1. XGBoost

XGBoost [19], the full name of which is extreme gradient boost, is a distributed and efficient implementation of gradient boost tree. It is an improved model based on the gradient boosting decision tree (GBDT), which belongs to the family of boosting methods. The chief idea of XGBoost is to incorporate a series of weak learners into a strong learning algorithm [2]. By adding new weak learners, the probability of making mistakes is reduced continuously, and the final output value is the sum of the results of many weak learners. To better understand the mechanism of XGBoost, the prediction function, objective function, and optimization process are introduced as follows.

Considering a dataset with n substances and m features, where $D = \{(x_i, y_i)|x_i \in R^m, y_i \in R\}$ and $x_i = \{x_{i1}, x_{i2}, \ldots, x_{im}|i = 1, 2, \ldots, n\}$. The basic idea of XGboost is to iteratively construct t weak estimators to predict the output y_i by the predictor x_i.

$$\begin{gathered} \hat{y}_i^0 = 0 \\ \hat{y}_i^1 = f_1(x_i) = \hat{y}_i^0 + f_1(x_i) \\ \hat{y}_i^2 = f_1(x_i) + f_2(x_i) = \hat{y}_i^1 + f_2(x_i) \\ \cdots \\ \hat{y}_i^t = \sum_{k=1}^{t} f_k(x_i) = \hat{y}_i^{t-1} + f_t(x_i) \end{gathered} \tag{1}$$

Each weak estimator $f_k(x_i), k = 1, 2, \ldots t$ is generated from the iteration of the gradient boosting algorithm, and the output value $\hat{y}_i^t$ is the summation of the output value of previous iteration $\hat{y}_i^{t-1}$ and the present result $f_t(x_i)$. To learn the set of estimators, the objective function that needs to be minimized can be expressed as:

$$L^t(y, \hat{y}^t) = \sum_{i=1}^{n} l(y_i, \hat{y}_i^t) + \sum_{k=1}^{t} \Omega(f_k), \tag{2}$$

where $l(y_i, \hat{y}_i^t)$ is the loss function that measures the difference between the target value and the prediction value $\hat{y}_i^t$. The second term is the regularization of the model, which is used to penalize the complexity of the entire model, and it can be calculated as follows:

$$\Omega(f_k) = \gamma T_k + \frac{1}{2}\lambda \sum_{j=1}^{T_k} w_{kj}^2, \tag{3}$$

Here, T_k represents the number of leaf nodes in the $k-$th base tree estimator, and γ is the penalty parameter for the number of leaf nodes. Meanwhile, w_{kj} represents the weight of the $j-$th leaf node in the base tree estimator and λ is the penalty parameter for the leaf node weight.

Up to now, we have a basic idea about the chief goal of XGBoost [19]. Next, we will introduce the process of how to optimize the objective. First, considering the training process is an additive consideration, as in Equation (1), f_t is greedily added to minimize the objective, when predicting the output value $\hat{y}^t$ at the t-th iteration.

$$L^t = \sum_{i=1}^{n} l\left(y_i, \hat{y}_i^{t-1} + f_t(x_i)\right) + \Omega(f_t) \tag{4}$$

Using the second gradient approximation of the Taylor explosion, Equation (4) can be expanded as follows.

$$L^t \cong \sum_{i=1}^{n}\left[l(y_i,\ \hat{y}_i^{t-1}) + g_i f_t(x_i) + \frac{1}{2}h_i f_t^2(x_i)\right] + \Omega(f_k) \tag{5}$$

where g_i and h_i indicate first and second gradient statistics. By removing the constant term $l(y_i,\ \hat{y}_i^{t-1})$, we can obtain the simplified objective as follows.

$$\widetilde{L}^t \cong \sum_{i=1}^{n}\left[g_i f_t(x_i) + \frac{1}{2}h_i f_t^2(x_i)\right] + \Omega(f_k) \tag{6}$$

Define the set of samples of the j leaf node as $I_i = \{i|q(x_i) = j\}$ and then expand the regularization term.

$$\begin{aligned}\widetilde{L}^t &\cong \sum_{i=1}^{n}\left[g_i f_t(x_i) + \tfrac{1}{2}h_i f_t^2(x_i)\right] + \gamma T_t + \tfrac{1}{2}\lambda \sum_{j=1}^{T_t} w_{tj}^2 \\ &= \sum_{j=1}^{T_t}\left[G_j w_{tj} + \tfrac{1}{2}(H_j + \lambda) w_{tj}^2\right] + \gamma T_t\end{aligned} \tag{7}$$

where $G_j = \sum_{i\in I_j} g_i$ and $H_j = \sum_{i\in I_j} h_i$. Then, the optimal weight w_j^* of leaf j can be computed by

$$w_j^* = -G_j H_j + \lambda \tag{8}$$

and we get the corresponding optimal objective value by substituting $-\frac{G_j}{H_j+\lambda}$ for w_j^* in Equation (8).

$$\widetilde{L}^t(q) = -\frac{1}{2}\sum_{j=1}^{T_t}\frac{G_j^2}{H_j + \lambda} + \gamma T_t \tag{9}$$

where $\widetilde{L}^t(q)$ is used as the assessment function to evaluate the quality of the tree structure $q(x)$. Specifically, the smaller the value of $\widetilde{L}^t(q)$, the higher quality of the tree structure.

So far, the model with the T base estimator has been basically constructed and the prediction value of XGBoost is $\hat{y}_i^t$, which can represent the default probability of the i-th corporate in this paper.

3.2. CT-XGBoost

XGBoost is a strong approach for various tasks. Nonetheless, the efficiency of the model can be limited due to the class-imbalance problem in the credit default data. Thus, it may be a good idea to modify XGBoost to adapt to the class imbalance of the credit default dataset. Assume the credit default dataset for the training model contains N samples in total, where the number of credit default samples is N_d and the number of non-default samples is N_n. In the real world, the number of default samples is hugely greater than the number of non-default samples, which causes the class-imbalance problem with the imbalance ratio defined as $\frac{N_n}{N_d}$. To solve the problem, we proposed a novel CT-XGBosot, which is modified from the XGBoost model.

Specifically, we modified the XGBoost in two aspects: (1) The cost-sensitive strategy is employed to assign more misclassification costs for default class samples relative to non-default class samples. During the calculation of the loss function, a novel parameter, called the penalty ratio in this paper, is added to control the ratio of misclassification costs for different classes. (2) We set a more reasonable threshold considering the class imbalance, which is used to classify the samples into two groups based on the predicting default probabilities. The corporates with default probabilities above the threshold are classified as the default group, and those with default probabilities below the threshold are classified as the non-default group. The modification will be explained in detail as follows.

3.2.1. Cost-Sensitive Strategy

In the process of default prediction model training, an important step is to calculate the objective function. Equation (2) is the objective function of XGBoost. In Equation (2), the first term $\sum_{i=1}^{n} l(y_i,\ \hat{y}_i^t)$ is the loss function, which measures the disparity between the prediction results and the true results [23].

However, the importance of each sample to the loss function is the same. The misclassification costs of default class and non-default class samples are equal. Due to the class imbalance problem where the non-default samples are the majority, the contribution of non-default samples to the loss may be larger than that of default samples. The model may wrongly take the chief aim of correctly classifying the non-default samples. Thus, it is important to modify XGBoost by assigning more misclassification costs to default class samples in the training process.

In CT-XGBoost, to increase the misclassification cost of default class samples, we modify the loss function $\sum_{i=1}^{n} l(y_i,\ \hat{y}_i^t)$ as follows.

$$\sum_{i=1}^{n}[y_i * C_d * l(y_i,\ \hat{y}_i^t) + (1 - y_i) * C_n * l(y_i,\ \hat{y}_i^t)] \tag{10}$$

where C_d, C_n are the weights of misclassification costs for default and non-default class samples. Since the magnitudes of C_d, C_n do not influence the training process, we define a new parameter p, called penalty ratio, which equals to $\frac{C_d}{C_n}$. In this paper, we set penalty ratio p as the dataset imbalance ratio $\frac{N_n}{N_d}$. Then, the loss contributed by default samples will be larger than before.

3.2.2. Threshold Method

Considering the default prediction is essentially a binary classification, a threshold is crucial to be set to determine the predicted default probability should be divided into which category. Corporates with default probabilities higher than the threshold are regarded as default class, and those with default probabilities lower than the threshold are regarded as the non-default class.

However, most of the previous prediction methods simply set 0.5 as the threshold, which is not suitable for imbalanced data [42]. For instance, if the default probability generated by the prediction model is a uniform distribution of $[0, 1]$ and the threshold is set as 0.5, half of the samples will be classified as a default class, which results in many non-default samples being misclassified. Thus, how to set a rational threshold is an important problem for default prediction.

In the CT-XGBoost model, we set a rational threshold which equals the N_d-th highest default probability in the training dataset. After the default probability of the testing dataset is predicted, corporates with default probabilities higher than the threshold are classified as default corporates, and those with default probabilities lower than the threshold are classified as non-default corporates.

3.3. *Benchmark Prediction Models*

For the performance evaluation of our proposed model, we compared its default predictive ability to those of other models widely used in the literature. Thus, we constructed a statistical method with logit regression, and intelligent techniques, including support vector machine and neural network. Moreover, ensemble models, random forest and XGBoost [42], were also constructed as benchmark models. The following content will simply introduce these benchmark models, except XGBoost, which has been explained in Section 3.1.

3.3.1. Logistic Regression

Logistic regression is one of the most popular models in credit default prediction due to its simplicity and interpretability [3]. Logistic regression overcomes the limitation of the linear regression model, which requires that the explained variables obey a normal distribution and be continuous. To design a failure prediction model, this method aims to estimate the probability of corporate failure based on the explanatory variables. The model can be expressed as follows:

$$P(Y=1|X) = \frac{e^{\beta_0+\beta_1 X}}{1+e^{\beta_0+\beta_1 X}} \tag{11}$$

where X is the vector of explanatory variables, Y is the indicator of corporate failure, β_1 is a vector of coefficients, and β_0 is a scale parameter. The parameters β_0, β_1 are estimated by the maximum likelihood method. With this method, we can forecast corporate failure by comparing the possibility to a threshold and further interpret the variables by the coefficients of each variable. To prevent overfitting, we apply l_1 and l_2 regularization.

3.3.2. Support Vector Machine

As a distribution-free and robust machine learning method, SVM has been commonly applied in the domain of credit default risk assessment [31]. In brief, SVM is a generalized linear model which constructs an optimal hyperplane as a decision boundary. The decision boundary ensures the accuracy of correct classification while maximizing the separation between the boundary and the closest samples. The samples nearest to the optimal hyperplane are called support vectors [3]. All other training samples are irrelevant for determining the optimal hyperplane. The optimized strategy of SVM is to address a convex quadratic programming problem. To separate samples with non-linear relationships, non-linear kernel functions are adopted to project input vectors into a high-dimensional vector space in which the samples become more separable [42]. To avoid overfitting, we adjust the penalty for misclassification.

3.3.3. Neural Network

Neural network, also called deep learning, is one of the most popular artificial intelligence techniques and has also been commonly used in the field of corporate failure prediction. This model operates analogously to human neural processing and consists of numerous neurons. When tacking the binary classification tasks, the neural network typically includes three layers of network: (1) the input layer consists of as many neurons as the dimensionality of input variables, (2) hidden layers consist of a given number of neurons that is set by user, and (3) the output layer consists of one neuron which is used to divide the input sample [5]. The neurons in a particular layer are linked to both the preceding and the following layer. For every single neuron, the corresponding value is calculated by the sum of its inputs with weights and a given non-linear function. During the training procedure, the weight parameters in a neural network are adjusted step by step by back-propagation to narrow the differences between outputs and true values [5]. When the epoch set beforehand arrives, the training process stops and the output value is divided into a specific category according to a threshold. For the overfitting problem, we employ the dropout method, which randomly switches off portions of the connection during the training process.

3.3.4. Random Forest

Random forest is a supervised machine learning technique that consists of multiple decision trees. It is a modification of the bagging ensemble learning approach, and the classification process is determined by the integration of the categories output by a series of individual trees. In this research, random forest built a number of decision tree classifiers that were trained step by step on bootstrap replicates of the credit default dataset through randomly selecting explanatory variables. According to the majority voting result from the decision trees, the model provides the classification of observations. Moreover, the model can identify the importance of each variable based on its information gain. Importantly, to obtain the generalization performance, we adopted the commonly used method in decision trees by controlling the number of trees in the random forest.

3.4. Performance Evaluation for Credit Default Prediction

To assess the out-of-sample performances of credit default prediction models, we adopted the split methods used by previous research [5]. We randomly divided the dataset into a training dataset and test dataset in an 80% to 20% ratio. Due to the class imbalance of the dataset in which non-default samples represent the majority group, we used a stratified sampling method for splitting to ensure the same population structure of training data and testing data.

Considering that credit default prediction is a binary classification problem, we can evaluate the out-of-sample performance via the metrics for classification models. One common metric is *overall accuracy*, defined as follows:

$$Overall\ accuracy = \frac{TP + TN}{TP + FN + FP + TN} \tag{12}$$

where *TP* (true positive) is the number of default companies which are correctly classified as default; *FN* (false negative) is the number of default companies which are wrongly classified as non-default; *TN* (true negative) is the number of non-default companies which are correctly classified as non-default; and *FP* (false positive) is the number of non-default companies that are wrongly classified as default.

Given the class-imbalance problem for the credit default prediction, the prediction performances of the two classes needed to be evaluated separately. For this purpose, *type I accuracy* and *type II accuracy* were taken into account. *Type I accuracy* (or sensitivity) is defined as the proportion of default samples predicted by the model correctly, and *type II accuracy* (or specificity) is defined as the proportion of non-default samples correctly predicted by the model.

$$Type\ I\ accuracy = \frac{TP}{TP + FN} \tag{13}$$

$$Type\ II\ accuracy = \frac{TN}{TN + FP} \tag{14}$$

Moreover, the area under the receiver operating characteristic curve (AUC) is a popular estimation of a classification model's overall performance [5]. The ROC curve is a graph consisting of two-dimensionality, on which one axis is the true positive rate (sensitivity) and the other axis is the false positive rate (1-specificity). While changing the default probability threshold, the curve would plot each point representing the true positive rate and false positive rate. For the reason that AUC is a part of the unit square area, its value shall always range from 0 to 1.0 [37]. In addition, AUC should be more than 0.5 for the model to be realistic, and the closer it is to 1, the better the prediction performance of the default prediction model.

4. Empirical Results

4.1. Data

We used a database of bank-loan defaults of firms in the west region of China for 2017–2021. The database was sourced from a bank in Xinjiang province of China. The database consists of the loan information and the financial statements of firms that are the debtors of the bank. According to the Industrial classification for national economic activities in China (GB/T 4754), we selected the companies in the energy sector. A firm is defined as defaulting if it fails to pay the loan periodically. The remaining companies are defined as non-default. The number of default firms is 205, and the number of non-default firms is 33, making the imbalance ratio about 6.21.

In determining the variables used to assess credit default risk, the majority of academic studies use financial variables as predictors of the default prediction models [43,44].

For instance, representative work by Beaver [21] constructed 30 financial variables from the financial statement, and the results demonstrated that the financial variables could provide a superior ability to predict corporate default. Thus, in this paper, we construct a comprehensive list of financial variables, including all accounting items in the financial statements. The reason for the selection of all accounting items but not a portion of accounting items was to avoid eliminating potentially useful information after discarding the unselected variables.

4.2. Credit Default Prediction Performance

In this section, we present a comparative analysis between our proposed CT-XGBoost model and other commonly used models, including logistic regression, SVM, neural network, random forest, and XGBoost. Table 1 summarizes the average prediction results of different models with ten times 5-fold cross-validation. Among conventional models, XGBoost showed superior performance with an AUC value of 95.44%, and its other evaluation results are also superior. Similarly, Zhang and Chen [10] found that, compared with logistic regression, SVM, random forest, and et al., XGBoost achieves better credit default prediction performance: 91.4% AUC. Moreover, Wang et al. [35] constructed the XGBoost model for default prediction and found that the prediction performance in terms of AUC was 88.07%. By comparison, the credit default prediction performance in our study is better than in previous studies. The main reason for the difference between the results of this study and previous studies is the different default datasets used. We focused on the credit default companies in the energy industry. Overall, these results demonstrate that the XGBoost model is an efficient algorithm for credit default prediction, and it is rational to select XGBoost as the model to be modified for its superior performance.

Table 1. Credit default prediction performance comparison analysis of different prediction models.

Models	Logistic Regression	SVM	Neural Network	Random Forest	XGBoost	CT-XGBoost
Overall accuracy (%)	94.34	93.05	94.17	94.17	**94.54**	89.58
Type I accuracy (%)	65.25	68.37	54.29	65.71	71.07	**91.43**
Type II accuracy (%)	96.36	98.24	98.03	98.53	**99.02**	89.27
AUC (%)	92.86	94.36	90.35	94.70	95.44	**96.38**

Note: The number in bold-face indicates the best performance for each metric.

It is notable that the chief aim of credit default prediction is to accurately identify as many default samples as possible without misclassifying too many non-default samples. However, the prediction performance of XGBoost has not met our expectations, with the type I accuracy value of only 71.07%, whereas the type II accuracy value is 99.02%. The reason for this phenomenon is the class imbalance problem, which causes the prediction model to be overwhelmed by the majority of non-default samples. Thus, this study proposes a novel CT-XGBoost to solve the class imbalance problem.

First, comparing the type I, type II, and overall accuracy, we can see that the CT-XGBoost model is more rational than other conventional models. We can see that the type I accuracy value of CT-XGBoost is 91.43%, which is 20.36% higher than that of the representative XGBoost model. The result implies that our proposed model has a superior ability to identify default class samples. Second, while the type II and overall accuracy values of our proposed model are lower than those of other models, the accuracy values of sacrifice are 9.75% and 4.96%, respectively, which are lower than the benefit in type I accuracy. In addition, the main aim of credit default prediction is to accurately identify the default class samples. Finally, as for the AUC value, which can evaluate the comprehensive performance of the prediction model, we can notice that our proposed model is better than benchmark models. The average AUC value of CT-XGBoost was 96.38%, which is better than the AUC values of other default prediction models, which ranged from 90.35% to 95.44%. These results suggest that our proposed model, which modifies the XGBoost model with cost-sensitive and threshold methods, outperforms other benchmark models when dealing with the class-imbalance problem.

4.3. The Importance of Predictor

A practical default prediction model should have not only good accuracy, but also a clear, interpretable result. To make the model acceptable for users, transparency in the decision process is indispensable. For instance, according to the Equal Credit Opportunity Act of the U.S., the creditors are mandated to provide applicants, upon request, with specific reasons underlying credit denial decisions. In previous studies [35,36], some methods are proposed to identify the significant performance drivers of the XGBoost model in default prediction. In this study, we applied the "Feature Importance" function to estimate the importance of the financial features used in our proposed CT-XGBoost model.

Before introducing the "Feature Importance" function, the splitting mode of leaf nodes in CT-XGBoost needs to be explained. First, a portion of features is selected as a candidate set. Then, determine a split point of the leaf node in the tree by using a greedy algorithm and calculating the Gini score to determine the best splitting point. Define I_L and I_R as the sample sets of leaf nodes and right nodes after splitting. Assume $I = I_L + I_R$; then, the objective function value $\widetilde{L}^t_{no-split}$ before splitting and the objective function value $\widetilde{L}^t_{split}$ can be obtained as follows:

$$\widetilde{L}^t_{no-split} = -\frac{1}{2}\frac{(G_L + G_R)^2}{H_L + H_R + \lambda} + \gamma T_{no-split} \tag{15}$$

$$\widetilde{L}^t_{split} = -\frac{1}{2}\left[\frac{G_L^2}{H_L + \lambda} + \frac{G_R^2}{H_R + \lambda}\right] + \gamma T_{split} \tag{16}$$

where G, H are the first derivatives and the second derivatives after splitting, and subscripts L, R indicate the left and the right node. Then, loss Gain value for leaf nodes in the t-th tree can be calculated, and the node with the highest Gain value is determined as the splitting point.

$$Gain = \frac{1}{2}\left[\frac{G_L^2}{H_L + \lambda} + \frac{G_R^2}{H_R + \lambda} - \frac{(G_L + G_R)^2}{H_L + H_R + \lambda}\right] - \gamma \tag{17}$$

The Gain value can be used to estimate the importance of features, which measures the ability to classify the default and non-default samples. Considering that CT-XGBoost is a model where a number of trees should be simultaneously considered, we calculated the "Feature Importance" function for the r-th feature as follows:

$$Importance_r = \frac{\sum_{k=1}^{t} Gain_r^k}{\sum_{r=1}^{m}\left(\sum_{k=1}^{t} Gain_r^k\right)} \tag{18}$$

$Gain_r^k$ is the Gain value for the r-th feature in the k-th tree, t is the number of trees, and m is the number of features. So far, the "Feature Importance" function has been explained, and the importance of financial variables can be calculated with Equation (15).

Table 2 represents the feature importance results of the top 20 most important financial variables, ranked based on the feature importance values from highest to lowest. Starting with the most important, the ten features that contribute to the CT-XGBoost model's credit default prediction ability are: (1) other receivables, (2) sales expense, (3) long-term deferred, (4) non-operating income, (5) accounts receivable, (6) taxes, (7) prepaid accounts, (8) liabilities and owner's equity, (9) capital reserves, and (10) cash flow generated from operating activities net amount. The higher the feature's importance, the stronger ability of the financial variable to classify the default and non-default samples. The results may be of great worth for practitioners, as they can help explain why an applicant is classified as a credit default class.

Table 2. Feature importance of the top 20 important financial variables in the CT-XGBoost model.

Rank	Financial Features	Feature Importance
1	Other receivables	0.237242
2	Sales expense	0.080695
3	Long-term deferred	0.051185
4	Non-operating income	0.048665
5	Long-term equity investment	0.044733
6	Accounts receivable	0.038403
7	Taxes	0.034495
8	Prepaid accounts	0.034344
9	Liabilities and owners' equity	0.034228
10	Capital reserves	0.032203
11	Cash flow generated from operating activities net amount	0.030894
12	Intangible assets	0.026951
13	Operating costs	0.025119
14	Inventories	0.024920
15	Construction work in process	0.023612
16	Net increase in cash and cash equivalents	0.020937
17	Cash flow generated from investing activities net amount	0.019130
18	Advance from customers	0.017562
19	Operating revenue	0.016754
20	Bill receivable	0.015882

4.4. The Influence of the Parameter Setting in CT-XGBoost

As mentioned in Section 3.2, our proposed CT-XGBoost model has modifications in the form of two algorithm-level methods: the cost-sensitive strategy and the threshold method. The parameter in the cost-sensitive strategy is the penalty ratio, which can assign different misclassification costs to different class samples, and the parameter in the threshold method is the threshold value, which can be used to classify the default probabilities into two classes. Considering that these two parameters can influence the performance of CT-XGBoost, we further analyzed how the credit default performance changes with different parameters and found the best parameter settings.

4.4.1. Parameter Setting for Cost-Sensitive Strategy

The chief aim of the cost-sensitive strategy in the CT-XGBoost model is to assign different misclassification costs to different class samples. The parameter for the cost-sensitive strategy is the penalty ratio p, which is the misclassification cost ratio between the default class and the non-default class. In Section 3.2.1, we set parameter p as $\frac{N_n}{N_d}$, where N_n, N_d are the numbers of non-default and default samples in the training dataset, respectively. The results in Section 4.2 demonstrate that a cost-sensitive strategy in CT-XGBoost is helpful for class imbalance credit default prediction. In this section, we investigate the influence of penalty ratio p in the cost-sensitive strategy on the prediction performance of the CT-XGBoost model. We set the penalty ratio p to range from 1 to 10 with increments of 1, and also 6.21 (the imbalance ratio of the dataset). The higher p is, the more misclassification costs are assigned to the default class samples. For fixing the parameters of the threshold method to those in Section 3.3.2, the figure shows the results.

First, we can notice that there are fluctuations in the prediction performance at different values of penalty ratio p. As the value of p increases from 1 to 10, the curves of the four performance metrics changes with similar trends, which are roughly upward and then downward. The results suggest that the default prediction performance can be better when more misclassification costs are assigned to default class samples; at the same time, high misclassification costs may not benefit the prediction model. This means that an appreciable penalty ratio p is important for default prediction. As shown by the dotted line in Figure 2, the prediction performance of the CT-XGBoost model was best when the penalty ratio p was set as 6.21, which is the imbalance ratio in the training dataset. Thus, it is crucial to consider the class distribution in the dataset when setting the penalty ratio p for the cost-sensitive method.

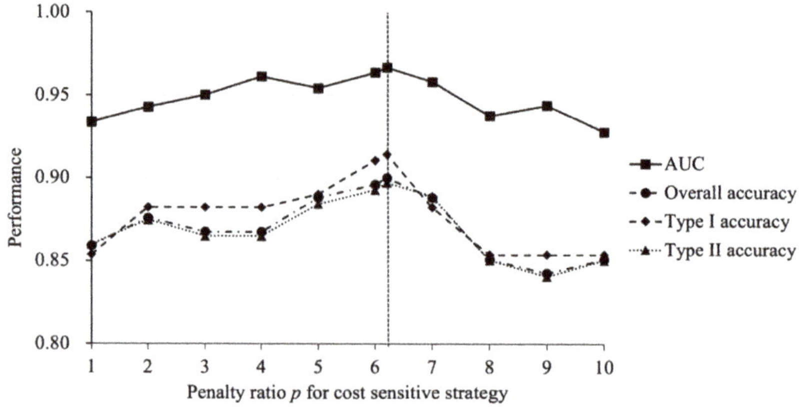

Figure 2. Performance of CT-XGBoost with different parameters for the cost-sensitive strategy.

4.4.2. Parameter Setting for Threshold Method

The role of threshold setting in CT-XGBoost is to classify samples with the predicting default probabilities into two groups. The sample is considered as default when the default probability is higher than the threshold, and non-default in reverse. As mentioned in Section 3.2.2, we set the threshold value as the default probability value of the N_d-th sample in the training dataset. In practice, the threshold determination is useful for controlling credit risk, and the creditors, such as banks, can control the number of debtors by adjusting the threshold for deciding whether to approve a loan. A higher threshold value means more applicants will be considered as non-default and approved for a loan. Meanwhile, the creditors will face higher credit risk. Therefore, investigating the influence of threshold setting on the default prediction performance is very important. In this studies, we varied the threshold value according to the predicting probabilities of samples in the training dataset. For fixing the penalty ratio p to the optimal value of 6.21, the results are presented in Figure 3.

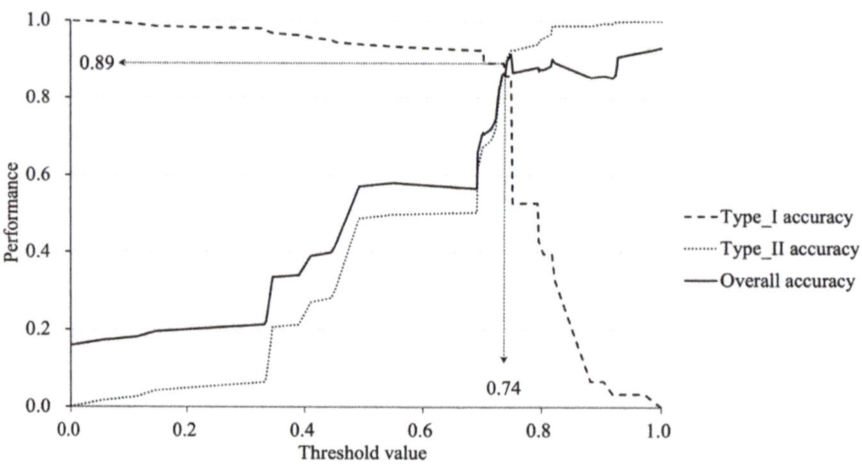

Figure 3. Performance of CT-XGBoost with different parameters for threshold method. (The AUC performance is not shown because its value was unchanged with different thresholds).

We can see that as the threshold increases from 0 to 1, the curve of *type I accuracy* shows a downward trend, and the curves of *type II* and *overall accuracy* show similar upward trends. These results demonstrate that the prediction performance can be significantly influenced by threshold setting. When setting a lower threshold value, more potential credit defaults can be identified, but more true non-default cases can be mis-considered as default. In addition, the three curves in Figure 3 intersect when the threshold value is 0.74; default and non-default samples can be identified equally accurately. When the threshold value increases based on 0.74, the *type I accuracy* decreases rapidly, but the *type II accuracy* increases slightly. Thus, it is proper to set the threshold to around 0.74 in this case.

Moreover, the creditor can find the optimal threshold value based on its credit risk tolerance ability. When the creditor has weak credit risk tolerance ability, the threshold can be set low to obtain a high *type I accuracy*, which means the majority of potential default applicants are identified. However, we should notice that the *type II accuracy* can be low caused by a low threshold, which means a large number of risk-free clients would be turned away. To avoid losing huge benefits, assuming that the creditor can tolerate about 10% of default cases, the proper *type II accuracy* threshold is about 0.74, which means that about 89% of potential credit default applicants can be accurately identified. At the same time, the *type II accuracy* can be limited to 89%, which means that the creditor would only lose about 11% of free-risk clients.

5. Conclusions

In order to accurately predict the credit defaulting of energy corporates, the class-imbalance problem in the default dataset cannot be ignored. To tackle the problem, this paper proposed a novel and efficient default prediction model, CT-XGBoost, which was modified from the strong classification model XGBoost with the cost-sensitive strategy and threshold method. In the empirical analysis, we constructed a corporate credit default dataset from a commercial bank in China, which suffers from the class-imbalance problem. In order to evaluate the performance of our proposed CT-XGBoost, we selected five commonly used credit default prediction models as benchmark models, including logistic regression, SVM, neural network, random forest, and XGBoost. The empirical results demonstrate that our proposed CT-XGBoost outperforms the benchmark models. Therefore, the novel model CT-XGBoost can be helpful to solve the class-imbalance problem and assess the credit risk of energy companies efficiently.

We further analyzed the feature importance of the input financial variables, in order to identify the significant drivers which contribute to identifying the corporate defaults in the energy industry. The results show the top 10 most important features are: (1) other receivables, (2) sales expense, (3) long-term deferred, (4) non-operating income, (5) accounts receivable, (6) taxes, (7) prepaid accounts, (8) liabilities and owner's equity, (9) capital reserves, and (10) cash flow generated from operating activities' net amount. In practice, these financial variables in the company's financial statements might be the key information for creditors to estimate the credit risk in the energy industry.

Moreover, we conducted sensitivity analysis to investigate how the different parameter settings in CT-XGBoost influence the prediction performance. The results show that the parameter in the cost-sensitive strategy, which represents the attention focused on the minority default companies, should be determined according to the actual ratio between the number of credit default and non-default companies. In addition, as the threshold value in the threshold method is set lower, *type I accuracy* decreases and *type II accuracy* increases. In practice, the threshold value represents the percentage of loan applications approved by creditors. According to their risk tolerance, the creditors can find the optimal threshold, which not only can control real losses caused by credit default but also the opportunity cost of rejecting too many loan applications.

In general, the novel model proposed in this study can efficiently estimate the credit risk of bank loans for energy companies, which is helpful for creditors who are making decisions. According to the results, this study proposes some recommendations: (1) As the crucial industry for economic development, energy companies should make the most of loan funds and avoid credit risk arising from cash flow problems. Meanwhile, energy companies should disclose more transparent information in timely manner, to help investors comprehensively understand the company's operation and accurately assess the company's credit risk. (2) In the credit loan market, the credit rating institutions should improve the credit rating system, which not only can efficiently assess the credit default probability but also can provide explainable reasons to ensure the reliability of the system. (3) The government and regulators should establish sound laws and regulations to promote a healthy development environment for the energy industry, including policy-based financial support, financial subsidies, fair credit law, etc.

Nonetheless, our research has several limitations which could promote future research. First, the class-imbalanced problem not only exists for credit default but also for financial fraud, as the fraudsters make up a minority of whole samples. Thus, it would be interesting to investigate whether our proposed model can help solve the class-imbalance problem in the default identification task. Second, in this paper, the information used to predict credit default is the financial variables of corporates, which is in a structured form. Future research can extend the horizons to investigate the default predicting ability of other unstructured information, such as news reports and meeting audio.

Author Contributions: K.W.: conceptualization, investigation, resources, data curation. J.W.: Writing—review and editing. G.L.: formal analysis, funding acquisition. H.S.: methodology, software, validation, writing—original draft preparation. All authors have read and agreed to the published version of the manuscript.

Funding: This research was funded by the grants from MOE Social Science Laboratory of Digital Economic Forecasts and Policy Simulation at UCAS; the General Project of Hebei Natural Science Foundation (G2019501105); the "Three Three Three Talent Project" funded project in Hebei Province (A202001067); the 2019 Hebei Provincial Colleges and Universities Youth Top Talent Program Project (BJ2019213).

Institutional Review Board Statement: Not applicable.

Informed Consent Statement: Not applicable.

Data Availability Statement: Not applicable.

Conflicts of Interest: The authors declare no conflict of interest.

References

1. Wei, L.; Li, G.W.; Zhu, X.Q.; Sun, X.L.; Li, J.P. Developing a Hierarchical System for Energy Corporate Risk Factors Based on Textual Risk Disclosures. *Energy Econ.* **2019**, *80*, 452–460. [CrossRef]
2. Development Research Center of State Council, NDRC, Tsinghua University. *China Energy and Carbon Report 2050*; Science Press: Beijing, China, 2009; pp. 306–319.
3. Liu, Y.; Li, F.; Yu, X.; Yuan, J.; Zhou, D. Assessing the Credit Risk of Corporate Bonds Based on Factor Analysis and Logistic Regress Analysis Techniques: Evidence from New Energy Enterprises in China. *Sustainability* **2018**, *10*, 1457. [CrossRef]
4. Habib, A.; D'Costa, M.; Huang, H.J.; Bhuiyan, M.B.U.; Sun, L. Determinants and Consequences of Financial Distress: Review of the Empirical Literature. *Financ. Account.* **2020**, *60*, 1023–1075. [CrossRef]
5. Mai, F.; Tian, S.; Lee, C.; Ma, L. Deep Learning Models for Bankruptcy Prediction Using Textual Disclosures. *Eur. J. Oper. Res.* **2019**, *274*, 743–758. [CrossRef]
6. Guo, H.; Li, Y.; Shang, J.; Gu, M.; Huang, Y.; Bing, G. Learning from Class-Imbalanced Data: Review of Methods and Applications. *Expert Syst. Appl.* **2017**, *73*, 220–239. [CrossRef]
7. Lin, W.; Tsai, C.; Hu, Y.; Jhang, J. Clustering-Based Undersampling in Class-Imbalanced Data. *Inf. Sci.* **2017**, *409*, 17–26. [CrossRef]
8. Zhu, X.; Ao, X.; Qin, Z.; Chang, Y.; Liu, Y.; He, Q.; Li, J. Intelligent Financial Fraud Detection Practices in Post-Pandemic Era: A Survey. *Innovation* **2021**, *2*, 100176. [CrossRef]
9. Chawla, N.V.; Bowyer, K.W.; Hall, L.O.; Kegelmeyer, W.P. SMOTE: Synthetic Minority Over-Sampling Technique. *J. Artif. Intell.* **2002**, *16*, 321–357. [CrossRef]
10. Zhang, Y.; Chen, L. A Study on Forecasting the Default Risk of Bond Based on XGboost Algorithm and Over-Sampling Method. *Theor. Econ. Lett.* **2021**, *11*, 258–267. [CrossRef]
11. O'Brien, R.; Ishwaran, H. A Random Forests Quantile Classifier for Class Imbalanced Data. *Pattern Recogn.* **2019**, *90*, 232–249. [CrossRef]

12. Raskutti, B.; Kowalczyk, A. Extreme Re-Balancing for SVMs: A Case Study. *ACM SIGKDD Explor. Newsl.* **2004**, *6*, 60–69. [CrossRef]
13. Weiss, G.M. Mining with Rarity: A Unifying Framework. *ACM SIGKDD Explor. Newsl.* **2004**, *6*, 7–19. [CrossRef]
14. Chen, Z.; Duan, J.; Kang, L.; Qiu, G. A Hybrid Data-Level Ensemble to Enable Learning from Highly Imbalanced Dataset. *Inf. Sci.* **2021**, *554*, 157–176. [CrossRef]
15. Seiffert, C.; Khoshgoftaar, T.M.; Hulse, J.V.; Napolitano, A. RUSBoost: A Hybrid Approach to Alleviating Class Imbalance. *IEEE Trans. Syst. Man Cybern. Part A Syst. Hum.* **2010**, *40*, 185–197. [CrossRef]
16. Sun, Z.; Song, Q.; Zhu, X.; Sun, H.; Xu, B.; Zhou, Y. A Novel Ensemble Method for Classifying Imbalanced Data. *Pattern Recogn.* **2015**, *48*, 1623–1637. [CrossRef]
17. Li, H.; Cao, Y.; Li, S.; Zhao, J.; Sun, Y. XGBoost Model and Its Application to Personal Credit Evaluation. *IEEE Intell. Syst.* **2019**, *35*, 52–61. [CrossRef]
18. Mushava, J.; Murray, M. A Novel XGBoost Extension for Credit Scoring Class-Imbalanced Data Combining a Generalized Extreme Value Link and a Modified Focal Loss Function. *Expert Syst. Appl.* **2022**, *202*, 117233. [CrossRef]
19. Chen, T.; Guestrin, C. XGBoost: A Scalable Tree Boosting System. In Proceedings of the 22nd ACM SIGKDD International Conference on Knowledge Discovery and Data Mining, San Francisco, CA, USA, 13–17 August 2016; ACM: San Francisco, CA, USA, 2016; pp. 785–794.
20. Kohv, K.; Lukason, O. What Best Predicts Corporate Bank Loan Defaults? An Analysis of Three Different Variable Domains. *Risks* **2021**, *9*, 29. [CrossRef]
21. Beaver, W. Financial Ratios as Predictors of Failure. *J. Acc. Res.* **1966**, *4*, 71–111. [CrossRef]
22. Altman, E. Financial Ratios, Discriminant Analysis and the Prediction of Corporate Bankruptcy. *J. Financ.* **1968**, *23*, 589–609. [CrossRef]
23. Ohlson, J. Financial Ratios and the Probabilistic Prediction of Bankruptcy. *J. Account. Res.* **1980**, *18*, 109–131. [CrossRef]
24. Sun, J.; Li, H.; Huang, Q.H.; He, K.Y. Predicting Financial Distress and Corporate Failure: A Review from the State-of-the-Art Definitions, Modeling, Sampling, and Featuring Approaches. *Knowl.-Based Syst.* **2014**, *57*, 41–56. [CrossRef]
25. Serrano-Cinca, C.; Gutiérrez-Nieto, B. Partial Least Square Discriminant Analysis for Bankruptcy Prediction. *Decis. Support Syst.* **2013**, *54*, 1245–1255. [CrossRef]
26. Kim, H.; Cho, H.; Ryu, D. Corporate Default Predictions Using Machine Learning: Literature Review. *Sustainability* **2020**, *12*, 6325. [CrossRef]
27. Moscatelli, M.; Parlapiano, F.; Narizzano, S.; Viggiano, G. Corporate Default Forecasting with Machine Learning. *Expert Syst. Appl.* **2020**, *161*, 113567. [CrossRef]
28. Lahmiri, S.; Bekiros, S. Can Machine Learning Approaches Predict Corporate Bankruptcy? Evidence from a Qualitative Experimental Design. *Quant. Financ.* **2019**, *19*, 1569–1577. [CrossRef]
29. Zhao, Y.; Li, J.; Yu, L. A Deep Learning Ensemble Approach for Crude Oil Price Forecasting. *Energ. Econ.* **2017**, *66*, 9–16. [CrossRef]
30. Liu, M.; Li, G.; Li, J.; Zhu, X.; Yao, Y. Forecasting the Price of Bitcoin Using Deep Learning. *Financ. Res. Lett.* **2021**, *40*, 101755. [CrossRef]
31. Lahmiri, S. Features Selection, Data Mining and Finacial Risk Classification: A Comparative Study. *Intell. Syst. Acc. Financ. Manag.* **2016**, *23*, 265–275. [CrossRef]
32. Alfaro, E.; García, N.; Gámez, M.; Elizondo, D. Bankruptcy Forecasting: An Empirical Comparison of AdaBoost and Neural Networks. *Decis. Support Syst.* **2008**, *45*, 110–122. [CrossRef]
33. Liu, J.; Zhang, S.; Fan, H. A Two-Stage Hybrid Credit Risk Prediction Model Based on XGBoost and Graph-Based Deep Neural Network. *Expert Syst. Appl.* **2022**, *195*, 116624. [CrossRef]
34. Lei, X.; Lin, L.; Xiao, B.; Yu, H. Re-Exploration of Small and Micro Enterprises' Default Characteristicsbased on Machine Learning Models with SHAP. *Chin. Manag. Stud.* 2022, *in press*. (In Chinese) [CrossRef]
35. Wang, J.; Rong, W.; Zhang, Z.; Mei, D. Credit Debt Default Risk Assessment Based on the XGBoost Algorithm: An Empirical Study from China. *Wirel. Commun. Mob. Comput.* **2022**, *2022*, 1–14. [CrossRef]
36. Li, Y.; Stasinakis, C.; Yeo, W.M. A Hybrid XGBoost-MLP Model for Credit Risk Assessment on Digital Supply Chain Finance. *Forecast* **2022**, *4*, 184–207. [CrossRef]
37. Lu, W. Imbalanced Credit Risk Prediction Based on SMOTE and Multi-Kernel FCM Improved by Particle Swarm Optimization. *Appl. Soft. Comput.* **2022**, *114*, 108153. [CrossRef]
38. Tahir, M.A.; Kittler, J.; Yan, F. Inverse Random under Sampling for Class Imbalance Problem and Its Application to Multi-Label Classification. *Pattern Recogn.* **2012**, *45*, 3738–3750. [CrossRef]
39. Chi, J.; Zeng, G.; Zhong, Q.; Liang, T.; Feng, J.; Xiang, A.; Tang, J. Learning to Undersampling for Class Imbalanced Credit Risk Forecasting. In Proceeding of the 2020 IEEE International Conference on Data Mining (ICDM), Sorrento, Italy, 17–20 November 2020; pp. 72–81. [CrossRef]
40. Galar, M.; Fernández, A.; Barrenechea, E.; Bustince, H.; Herrera, F. A Review on Ensembles for the Class Imbalance Problem: Bagging-, Boosting-, and Hybrid-Based Approaches. *IEEE Trans. Syst. Man Cybern. Part C Appl. Rev.* **2012**, *42*, 463–484. [CrossRef]
41. Fernández-Baldera, A.; Buenaposada, J.M.; Baumela, L. BAdaCost: Multi-Class Boosting with Costs. *Pattern Recogn.* **2018**, *79*, 467–479. [CrossRef]

42. Zhu, X.; Li, J.; Wu, D.; Wang, H.; Liang, C. Balancing Accuracy, Complexity and Interpretability in Consumer Credit Decision Making: A C-TOPSIS Classification Approach. *Knowl.-Based Syst.* **2013**, *52*, 258–267. [CrossRef]
43. Yu, S.; Chi, G.; Jiang, X. Smal Enterprise Facility Rating Based on the Maximum Discrimination of Indicator System. *Chin. Manag. Stud.* **2020**, *28*, 38–50. (In Chinese) [CrossRef]
44. Wang, Y.; Yang, S. Corporate Financial Distress Prediction Based on Multi-Dimensional Efficiency Indicators. *Chin. Manag. Stud.* **2021**, *29*, 32–41. (In Chinese) [CrossRef]

energies

MDPI

Article

Measuring Pollution Control and Environmental Sustainable Development in China Based on Parallel DEA Method

Ying Feng [1], Chih-Yu Yang [2], Ching-Cheng Lu [3,*] and Pao-Yu Tang [2]

[1] Business College, Northwest University of Political Science and Law, No. 558 West Chang'an Road, Chang'an District, Xi'an 710122, China
[2] Department of Economics, Soochow University, 56, Kueiyang St., Sec. 1, Taipei 100, Taiwan
[3] Department of Business, National Open University, No. 172, Zhongzheng Road, Luzhou District, New Taipei City 247, Taiwan
* Correspondence: join1965@gmail.com

Abstract: The purpose of this study is to explore the impact of pollution control on industrial production efficiency in 31 provinces and cities in the Yellow River and Non-Yellow River basins in China from 2013 to 2017, using the methods of the directional distance function (hereinafter referred to as DDF) and the technology gap ratio (hereinafter referred to as TGR) in parallel, while taking the industrial production sector (labor force, total capital formation, energy consumption and industrial water consumption) and the pollution control sector (wastewater treatment funds and waste gas treatment funds) as input variables. Undesirable outputs (total wastewater discharge, lead, SO_2 and smoke and dust in wastewater) and an ideal output variable (industrial output value) are taken as output variables. It is found that the total efficiency of DDF in the Non-Yellow River Basin is 0.9793, which is slightly better than 0.9688 in the Yellow River Basin. Among the 17 provinces and cities with a total efficiency of 1, only Shandong and Sichuan are located in the Yellow River Basin. The TGR values of 31 provinces, cities and administrative regions are less than 1, and the average TGR value of the Yellow River Basin is 0.3825, which is lower than the average TGR value of the Non-Yellow River Basin of 0.5234. We can start by improving the allocation of manpower and capital, implementing the use of pollution prevention and control funds, improving the technical level of industrial production, improving pollutant emission, and increasing output value to improve overall efficiency performance. This study uses the parallel method, taking the industrial production department and the pollution control department as inputs, to objectively evaluate the changes in industrial production efficiency and technology gap in the Yellow River and Non-Yellow River basins, which is conducive to mastering the situation of pollution control and industrial production efficiency, and provides the reference for SDG-6- and SDG-9-related policy making.

Keywords: DDF; TGR; wastewater; waste gas; treatment funds; Yellow River

Citation: Feng, Y.; Yang, C.-Y.; Lu, C.-C.; Tang, P.-Y. Measuring Pollution Control and Environmental Sustainable Development in China Based on Parallel DEA Method. *Energies* **2022**, *15*, 5697. https://doi.org/10.3390/en15155697

Academic Editors: Junpeng Zhu and Xinlong Xu

Received: 3 July 2022
Accepted: 3 August 2022
Published: 5 August 2022

1. Introduction

While pursuing industrial and economic development, wastewater and air pollution have short-term and long-term impacts on the environment and human beings [1,2]. Countries around the world have invested a lot of money and resources to try to solve the problems of wastewater and air pollution caused by production and manufacturing. Human beings need to take sustainable actions within the existing environmental resources [3]. Therefore, the agenda for sustainable development sets out 17 sustainable development goals to be achieved by 2030 and will mobilize countries around the world to incorporate sustainable development goals into their national development strategies. Sustainable development goals SDG-6 (sustainable development of water resources) and SDG-9 (development of inclusive sustainable industry) play a vital role in environmental protection, economic development and the promotion of human well-being in achieving these sustainable development goals. In China, since the implementation of the reform and opening-up policy after 1980, prosperity and affluence have gone deep into the mainland from the early coastal areas. According to data released by the World Bank (as shown in Figure 1), we can see the impact of the 2008 financial tsunami and the 2019 COVID-19 pandemic on the global GDP growth rate. In addition, China's GDP growth rate is better than that of the world.

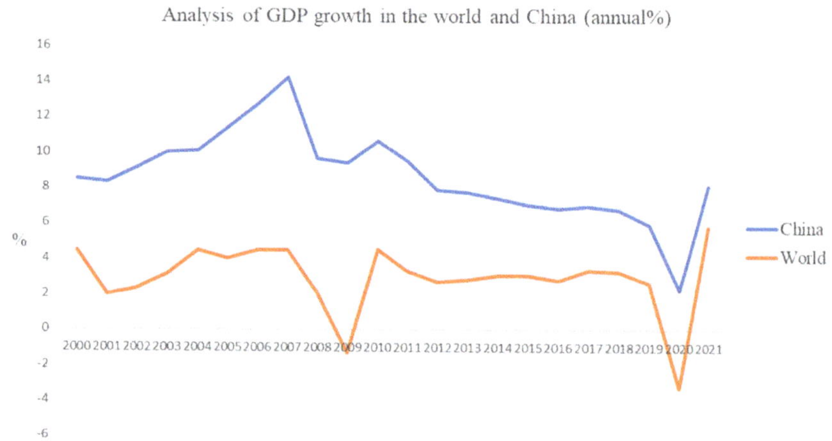

Figure 1. Analysis of GDP growth in the world and China (annual%).

The destruction of environmental resources caused by development has directly affected national health and the environment for human survival. According to China's "industrial classification of national economy", industries are divided into three categories: the primary industry is mainly agriculture, the secondary industry is mainly industry and the tertiary industry is mainly the service industry. According to the data of the National Bureau of Statistics of China (National Bureau of Statistics of China: http://www.stats.gov.cn/tjsj/ndsj/, accessed on 1 February 2022) (as shown in Figure 2), from 1978 to 2020, the fastest growth of China's GDP is in the secondary industry, followed by the service industry. The rapid development of China's economy largely depends on energy consumption, which has caused serious pollution [4,5]. In order to achieve energy conservation and emission reduction and strengthen pollution control [6], we must pay attention to the relevant issues of sustainable development goals SDG-6 and SDG-9. In order to improve environmental quality, improve national health and well-being, maintain environmental resources and pursue sustainable development, which has become a universal common value, the State Council of China put forward the outline of the Yellow River Basin Ecological Protection and High-Quality Development Plan in 2021 (Outline of Ecological Protection and High-Quality Development Plan for the Yellow River Basin (2021): http://www.gov.cn/zhengce/2021-10/08/content_5641438.htm, accessed on 1 February 2022). In addition to investing in pollution prevention and control funds, it also standardized the high energy consumption and high-pollution enterprises in the region. It includes various pollutant discharge standards and monitoring systems to ensure that significant progress will be made in the ecology and development of the Yellow River Basin by 2025.

As mentioned above, the government's policy and financial expenditure on environmental protection have a certain input–output relationship and impact mechanism between regional energy use and pollutant emission (Figure 3). When the industrial production department promotes economic development due to the investment of labor, capital, energy and water resources, the pollution control department is due to the investment of government prevention and control funds. It is beneficial to improve the unintended substances discharged from the production process, such as SO_x and smoke dust in waste gas, heavy metal lead in wastewater, etc., to maintain the natural environment and the health of people. Therefore, this study uses DDF and TGR methods to objectively evaluate the impact of pollution control on the production efficiency of the Yellow River Basin and Non-Yellow River Basin in China from 2013 to 2017. The structure of this study is as follows: Section 2 analyzes the literature on industrial production efficiency, energy efficiency, water efficiency, air pollution emission and treatment; Section 3 introduces the methods; and Section 4 introduces the data, narrative statistics and empirical result analysis. The last part puts forward conclusions and suggestions for future research.

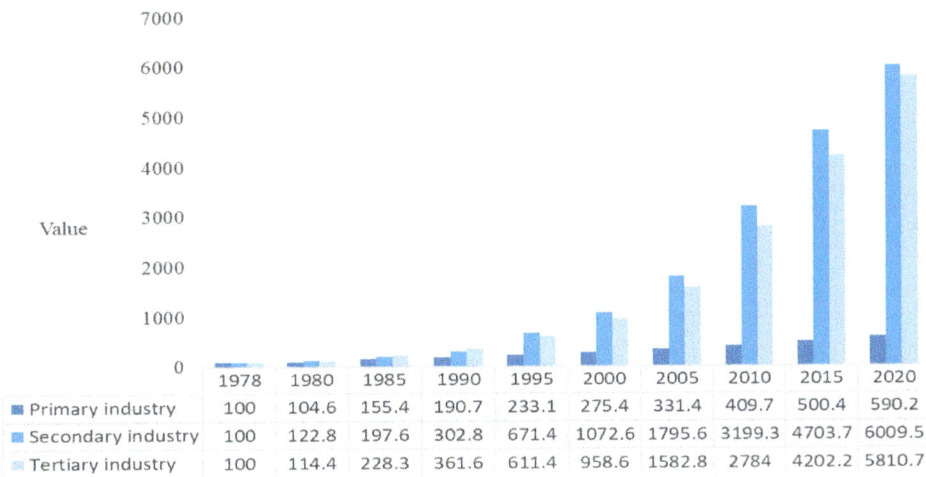

Figure 2. China's GDP index from 1978 to 2020.

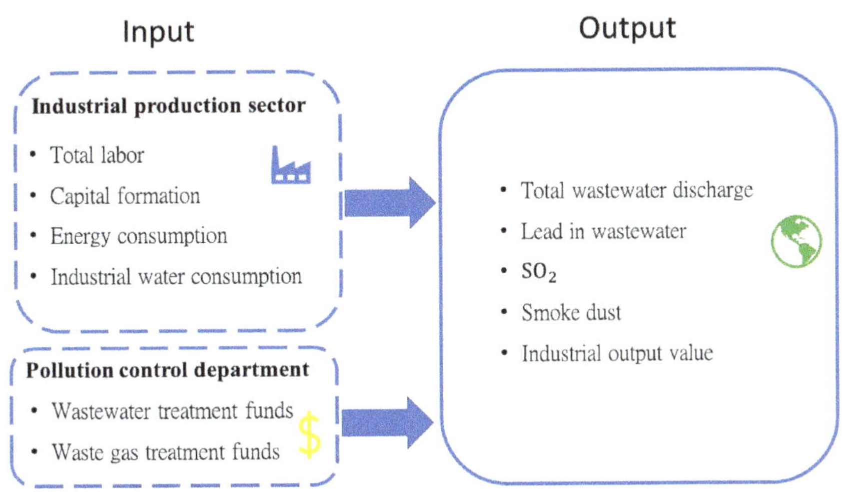

Figure 3. Input and output process of variables in this study.

2. Literature

Previous studies on industrial production efficiency, such as the one conducted by [7], used the SDG-9 index to assess the degree of industrialization of countries, as well as social inclusiveness, less use of natural resources and environmental impact. Ref. [8] using the DEA method, discusses the relationship between the American manufacturing industry and environmental performance. The unintended output is reported as SO_x, NO_x, Co, etc. It is found that air pollution is mainly a by-product of manufacturing activities. The share of the manufacturing industry in the total amount of state-owned products and the share of the polluting industry in the total amount of manufacturing activities are two important factors determining the intensity of pollution. Using DEA, Ref. [9] discuss the energy conservation and carbon reduction efficiency of China's industrial production from 2006 to 2010. The input variables are labor, capital and energy consumption, and the output variables are SO_2, wastewater and GDP. It is found that the energy conservation and emission reduction efficiency in East China is the best. Ref. [10] using the DEA method, evaluated the environmental efficiency of 46 countries in 2002, 2007 and 2011. The input variables are labor, capital and energy use, and the output variables are GDP, CO_2 and NO_x. The study found that the energy efficiency of countries rich in oil and natural gas resources is relatively poor. Ref. [11] using the DEA method, discuss the analysis of the energy and environmental efficiency of two petrochemical plants in China from 2012 to 2013, and divide the output into expected output and unexpected output. It was found that by analyzing the energy efficiency and environmental efficiency of the ethylene production process in complex chemical processes, the energy saving and emission reduction potential of ethylene plants can be obtained, and the efficiency performance of DMU can be improved by improving energy efficiency and reducing carbon emission. Research on energy efficiency by [11] evaluated the efficiency of the water, food and energy (WEF) relationship in 30 provinces and municipalities in China from 2005 to 2017. Inputs were labor force, water resource use, energy use, food consumption and other variables, and outputs were social benefits, wastewater discharge and solid discharge. The researchers analyzed the weight of the WEF relationship, and put forward the strategy of sustainable resource management. Ref. [12] discussing the research results of DEA application in the field of energy and environment from the 1980s to 2010, found that the development process will produce various pollutants to air, water and other types of pollutants which are related to health and climate change. Therefore, it is necessary to strike a balance between economic growth and pollution mitigation. Ref. [13] using the DEA method to explore the impact of U.S. economic growth on the environmental efficiency of the power sector, found that there is a stable n-shape relationship between environmental efficiency and regional economic growth, while in the case of local pollutants, there is an inverted n-shape relationship between environmental efficiency and regional economic growth. For policymakers, climate change needs to consider the relationship between economy, environment and society at the same time. On the research related to water use efficiency, Ref. [14] evaluated the efficiency of SDG-6 and a serious water shortage in the Medjerda Basin in Tunisia. Ref. [15] used the DEA method to explore the water use efficiency of 10 cities in the Minjiang River Basin in China in 2018. The research found that the input of social water and economic water are different, and the output of GDP and unintended wastewater are the factors affecting water use efficiency. Ref. [16] using TFP and Tobit models, discuss the water use efficiency of 30 provinces and municipalities in China from 2006 to 2015. The study found that the efficiency of water use in the administrative regions of provinces and cities is low, so we should establish the awareness of water conservation from the investment of education, so as to balance economic development and water use efficiency. Ref. [17] used the DEA method to explore China's regional ecological efficiency from 2003 to 2014. The input variables were labor force, water consumption, energy consumption, etc., and the output variables were GDP, SO_2, smoke and dust, industrial wastewater, household waste, etc. The study found that the efficiency and progress rate of the eastern region are better than other regions, and there is still room for improvement in China's overall environmental efficiency. Ref. [18] used DDF to evaluate the water resources and wastewater discharge efficiency of China's industrial sector. The input variables were labor, capital and industrial water consumption, and the output variables were industrial output value, chemical oxygen demand, etc. The study found that the eastern region has made progress in science and technology, and the pollutants discharged by industrial production in the western region are more serious. Ref. [19] using the DDF model, evaluated the efficiency of administrative water removal in 31 provinces and cities in China from 2011 to 2015. The study found that there were significant differences between the efficiency performance and technology gap in Eastern, Central and Western China. Ref. [20] evaluating the relationship between China's industrial water efficiency and regional differences from 2005 to 2015, found that the industrial water efficiency values of administrative regions in 31 provinces and cities are less than 1, among which the per capita water resources, R&D investment, regulation formulation, GDP and industrial structure will affect the industrial water efficiency. Ref. [5] using the SBM model, studied the economic production and sewage treatment efficiency of 30 provinces and cities in China from 2011 to 2017. The input variables were labor force, domestic and industrial water, investment in sewage treatment projects and the number of sewage treatment plants. The output variables were GDP, chemical oxygen demand of wastewater discharge and heavy metal pollution. The study found that there are great differences in inefficiency in different regions of China. The efficiency in the economic production stage is significantly higher than that in the sewage treatment stage. The sewage treatment efficiency is the main drag factor of the overall efficiency. Ref. [21] assessed the regional differences of China's provincial air pollution efficiency from 2006 to 2015. The study found that there were significant differences in air pollution emission efficiency in various regions. Air pollution emission efficiency was significantly positively correlated with economic development level, industrial structure optimization, technological innovation and foreign direct investment (FDI), and negatively correlated with energy consumption structure. Ref. [22] used DEA and regression analysis to explore China's energy efficiency performance from 2001 to 2013. Input variables included labor, capital and energy use, and output variables were GDP, industrial wastewater, solid waste and air pollutants. The study found that technological innovation has a positive impact on TFEE. The government should pay attention to technological innovation, which will be conducive to the effectiveness of energy conservation and emission reduction and environmental pollution prevention and control. Research on pollution control costs, such as [23], discusses the efficiency of China's iron and steel industry and pollution control. It is found that the production efficiency of China's iron and steel industry is low and causes serious pollution to the environment. Enterprises must improve the overall efficiency by increasing environmental protection investment, introducing foreign advanced technology and strengthening the R&D of pollutant management.

As for the discussion on energy consumption and pollution control technology, for example, in a paper by [24], it is estimated that Beijing, China, will improve its air quality by adjusting its industrial structure, controlling pollutant emissions, controlling vehicle pollution emissions and other measures and regulations due to rapid industrialization, urbanization and motorization, the continuous growth of energy consumption and the resulting emissions of a variety of pollutants. Ref. [25] assessing the impact of foreign investment on greenhouse gas emissions in developing countries, found that foreign investment enabled technology transfer, improved labor, reduced greenhouse gas emissions, improved energy efficiency and achieved sustainable development goals.

As mentioned above, most previous studies focused on industrial production efficiency, energy efficiency, pollutant emission and control. Therefore, this study uses the DDF method to explore the impact of pollution control on the production efficiency of 31 provinces and municipalities in the Yellow River Basin and Non-Yellow River Basin in China from 2013 to 2017, and uses TGR to measure the change in the technology gap. We objectively evaluate the efficiency difference of pollution control in different provinces and cities to provide an effective reference basis for policy formulation and budget control.

The main contributions of this study are as follows:

(1) Different from the previous literature results, this study uses the parallel method and takes the industrial production department and the pollution control department as

input variables to objectively evaluate the impact of pollution prevention and control funds on industrial production efficiency in 31 provinces and municipalities in China;

(2) This study compares the changes in industrial production efficiency and the technology gap between the Yellow River Basin and Non-Yellow River Basin, which is conducive to mastering the situation of pollution control and production efficiency in 31 provinces and municipalities in China, and provides objective suggestions as a reference for SDG-6- and SDG-9-related policy making.

3. Research Method

Ref. [26] first put forward the concept of a deterministic nonparametric front in 1957. It is used to measure the production level of a decision-making unit. Then, Ref. [27] proposed the CCR model. Ref. [28] proposed the BCC model. Over time, Ref. [29] (1996) proposed the directional distance function (DDF). In addition, Ref. [30] introduced the VRS super-efficiency Nerlove–Luenberger (N–L) model to solve the unreasonable problem. This method can adjust the input and output levels in the same proportion, and the efficiency value obtained under the VRS super-efficiency of DDF can be used for ranking all DMUs. The directional distance function model under variable return to scale (VRS) and the calculation method of efficiency values used in this study are as follows:

3.1. Directional Distance Function, DDF

This study uses [31] to extend the non-oriented method in the DDF model based on the SBM described by [32]. All models can evaluate the general efficiency value (≤1) at the same time, and its calculation method is as follows:

Non-oriented DD model

In this case, we have

$$\begin{aligned} &\max \beta \\ &\text{s.t. } X\lambda + \beta g_x \leq x_k \\ &Y\lambda - \beta g_y \geq y_k \\ &\textstyle\sum \lambda = 1 \\ &\lambda \geq 0 \end{aligned} \tag{1}$$

$$(d^{(I)}, d^{(IN)}, d^{(O)}, d^{(ON)}, d^{(OBad)}) = (x_o^{(I)}, 0, y_o^{(O)}, 0, y_o^{(OBad)}) \tag{2}$$

[DD-C]

$$\begin{aligned} &\xi^* = MAX\xi \\ &\text{st.}X^{(I)}\lambda + \xi x_o^{(I)} + s^{(I)} = x_o^{(I)} \\ &X^{(IN)}\lambda + s^{(IN)} = x_o^{(IN)} \\ &Y^{(O)}\lambda - \xi y_o^{(O)} - s^{(O)} = y_o^{(O)} \\ &Y^{(ON)}\lambda - s^{(ON)} = y_o^{(ON)} \\ &Y^{(OBad)}\lambda + \xi y_o^{(OBad)} + x^{(OBad)} = y_o^{(OBad)} \\ &\xi \geq 0, \lambda \geq 0, s^{(I)} \geq 0, s^{(IN)} \geq 0, s^{(O)} \geq 0, s^{(ON)} \geq 0, s^{(OBad)} \geq 0. \end{aligned} \tag{3}$$

We define the efficiency value of DMU(x_o, y_o) as

$$\theta^* = 1 - \xi^*.$$

3.2. Technology Gap Ratio, TGR

Since the production boundary of g groups is included in the common production boundary, the technical efficiency under the common boundary must be less than that under the group boundary. The ratio of the two is called the technical efficiency gap ratio (TGR), as follows:

$$\text{TGR} = \frac{\text{Technical efficiency under common boundary}}{\text{Technical efficiency under group boundary}} \quad (4)$$

4. Data Analysis and Empirical Results

4.1. Selection of Data Sources and Variables

This study evaluates the impact of pollution control in 31 provinces and municipalities of China on China's industrial production efficiency from 2013 to 2017. The publicly quantifiable data are obtained from the statistical yearbook of China's National Bureau of Statistics (National Bureau of Statistics of China: http://www.stats.gov.cn/tjsj/ndsj/, accessed on 1 February 2022) from 2013 to 2017, and the efficiency is analyzed through open and objective data. The relevant contents of the selected variables are as follows:

Labor force: including manufacturing, power, heat, gas and water production and supply, and the number of employed persons in urban units. Employed persons refer to persons aged 16 and above who engage in certain social work and obtain labor remuneration or business income. Unit: 10,000 persons.

Total capital formation: refers to the total value of fixed assets acquired by permanent residents less fixed assets disposed of in a certain period of time. Fixed assets are assets produced through production activities with a service life of more than one year and a unit value of more than the specified standard, excluding natural assets. It can be divided into total tangible fixed capital formation and total intangible fixed capital formation. Unit: 100 million yuan.

Energy consumption: electricity consumption by region. Unit: 100 million kWh.

Industrial water consumption: industrial water consumption by region. Unit: 10,000 tons.

Wastewater treatment fund: the completion of wastewater treatment investment generated by industrial pollution. Unit: 10,000 yuan.

Waste gas treatment funds: the completion of waste gas treatment investment generated by industrial pollution. Unit: 10,000 yuan.

Total wastewater discharge: total wastewater discharge by region. Unit: 10,000 tons.

Lead in wastewater: the discharge of main pollutants in wastewater. Unit: kg.

SO_2: emission of sulfur dioxide in waste gas by region. Unit: 10,000 tons.

Smoke and dust: emission of smoke (powder) dust in waste gas by region. Unit: 10,000 tons.

Industrial output value: regional industrial output value. Unit: 100 million yuan.

4.2. Input and Output Variables Statistical Analysis

As shown in the narrative analysis of various variables from 2013 to 2017 (Table 1), the average part shows a growth trend in labor force, total capital formation, energy consumption, waste treatment funds and industrial output value. The amount of industrial wastewater, wastewater treatment funds, lead, SO_2, smoke and dust in wastewater show a downward trend. The total amount of wastewater discharge has little change. In the largest part, labor force, total capital formation, energy consumption, waste gas treatment funds and industrial output value show a growth trend. Lead, SO_2, smoke and dust in wastewater show a downward trend, and other variables change little. In the minimum part, total capital formation, energy consumption, total wastewater discharge and industrial output value show a growth trend, the wastewater treatment funds and waste gas treatment funds show a downward trend, and the other variables have little change.

Table 1. Input–output variables from 2013 to 2017 statistical analysis.

		Labor Force	Total Capital Formation	Energy Consumption	Total Industrial Water Consumption	Wastewater Treatment Funds	Waste Gas Treatment Funds
Average	2013	148.6032	11,812.7645	1723.3352	459,258.0645	45,272.5484	83,133.5484
	2014	182.6548	12,682.8452	1794.7323	453,709.6774	40,284.6129	206,745.4194
	2015	182.1581	13,043.6935	1836.5487	437,548.3871	37,176.5806	254,643.0645
	2016	176.2806	13,773.8355	1927.3248	430,580.6452	38,198.0645	168,325
	2017	170.4032	14,564.3548	2034.7742	421,935.4839	34,916	181,119.3871
Max	2013	561	30,952.9	4956.62	1,931,000	263,797	303,865
	2014	1052.4	33,780.8	5235.23	2,201,000	150,634	701,240
	2015	1046	35,587.4	5310.69	2,380,000	175,141	1,281,351
	2016	1011.7	34,647.1	5610.13	2,390,000	164,863	781,673
	2017	991	39,657.5	5959	2,486,000	158,518	966,722
Min	2013	1.5	899.1	30.65	17,000	922	174
	2014	2.1	1052.1	33.98	17,000	572	466
	2015	2.1	1032	40.53	17,000	90	1453
	2016	2.3	1162.8	49.22	14,000	893	273
	2017	1.6	1376.1	58	16,000	15	47
St. Dev	2013	134.3751	7620.3602	1242.815	440,588.3164	51,282.9423	73,775.0457
	2014	209.0452	8142.8745	1289.1049	460,820.1524	38,242.0501	162,597.069
	2015	211.8303	8407.9079	1365.0845	475,238.4551	37,554.2495	260,514.5884
	2016	205.8528	8924.2812	1451.4981	479,438.3363	41,606.7403	159,393.1626
	2017	200.9428	9999.0218	1520.3991	490,473.7122	38,104.0369	197,602.603

		Total Wastewater Discharge	Lead in WASTEWATER	SO_2	Smoke and Dust	Industrial Output Value
Average	2013	224,336.5161	2455.2355	85.181	41.231	8629.4897
	2014	231,024.1613	2360.7935	63.6906	56.1532	8946.4632
	2015	237,200.8387	2562.2484	59.9719	49.6135	8874.8148
	2016	229,385.5806	1707.4355	35.5765	32.6023	9199.2548
	2017	225,697.129	1237.0387	28.2394	25.6861	9731.27
Max	2013	862,471	24,318.6	663	131.33	27,426.26
	2014	905,082	21,609.3	159.02	179.77	29,144.15
	2015	911,523	18,172.8	152.57	157.54	30,259.49
	2016	938,261	14,564.8	113.45	125.68	32,650.89
	2017	882,020	7656.9	73.91	80.37	35,291.83
Min	2013	5005	2.6	0.42	0.68	61.16
	2014	5450	2.5	0.42	1.39	66.16
	2015	5883	3.6	0.54	1.71	69.88
	2016	6143	5.1	0.54	1.65	86.44
	2017	7176	3.7	0.35	0.66	102.16
St. Dev	2013	184,430.1127	4632.3228	114.938	30.1237	7120.6401
	2014	190,473.1071	4170.2846	39.6557	42.5921	7468.1265
	2015	195,601.8016	4068.4679	37.3853	38.3135	7741.992
	2016	194,225.7237	3159.2792	25.0893	26.6502	8400.5834
	2017	185,112.1309	1898.2696	19.5832	18.5803	9102.946

4.3. Empirical Results

In this study, 31 provinces and municipalities in China were divided into two groups: the Yellow River Basin and the Non-Yellow River Basin. The DDF method was used to evaluate the difference in industrial production efficiency between the two groups. The common boundary efficiency and group boundary efficiency of the two groups are evaluated by the TGR method to find the technology gap ratio. The results and analysis are as follows.

(1) Industrial production efficiency of DDF in the Yellow River and Non-Yellow River Basins

The empirical results show that (as shown in Figure 4 and Appendix A) the best average value of the total efficiency of the Yellow River and Non-Yellow River basins is 1 in 17 provinces and cities, including Beijing, Tianjin and Hebei, of which only Shandong and Sichuan are located in the Yellow River Basin, and a total of 15 provinces and cities are located in the Non-Yellow River Basin. The average total efficiency of the Non-Yellow River Basin is 0.9793, and the three worst-performing regions are Yunnan (0.7804), Xinjiang (0.9188) and Guizhou (0.9257). The average value of the total efficiency of the Yellow River Basin is 0.9688, which is slightly lower than that of the Non-Yellow River Basin. The three regions with the worst performance of the total efficiency are Gansu (0.8604), Shanxi (0.9417) and Ningxia (0.9592). We further explore the period efficiency of each year in the Non-Yellow River Basin, with the best performance in 2015 and 2016, the efficiency value is 0.982, the worst performance is 0.9741 in 2013, of which Yunnan (0.7197) has the worst efficiency performance. In the part of efficiency in each year of the Yellow River Basin, only 0.9863 performed best in 2013, slightly higher than 0.9741 in the Non-Yellow River Basin. In the next four years, the overall efficiency performance lagged behind the Non-Yellow River Basin. The overall efficiency performance was the worst in 2015 (0.9561), of which Gansu (0.7593) performed the worst in 2015.

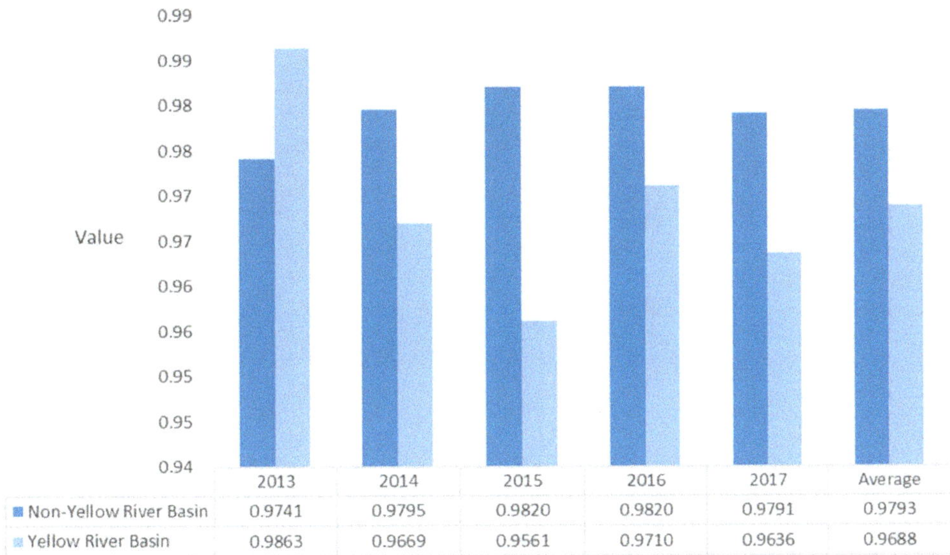

Figure 4. Efficiency of DDF in the Yellow River and Non-Yellow River Basins from 2013 to 2017.

This study further uses the Wilcoxon rank sum test to make $\alpha = 0.05$; the confidence interval is 95%, and the result shows that $z = -3.517$, which indicates that there are regional differences in DDF efficiency between the Yellow River Basin and the Non-Yellow River Basin, and the efficiency value of the Non-Yellow River Basin is better than that of the Yellow River Basin.

(2) Analysis of TGR technology gap ratio between the Yellow River and Non-Yellow River Basins

We use TGR to objectively measure the level of industrial production efficiency. When the TGR value is closer to 1, it means that the industrial production efficiency is relatively high and the efficiency is better. On the contrary, the lower or closer the TGR value is to

0, the more it indicates that there is still room for significant improvement. According to the TGR of 31 provinces and cities in China from 2013 to 2017 (Figure 5 and Appendix B), the TGR values of 22 provinces and cities in the Non-Yellow River Basin are less than 1, indicating that the technical level has not reached the technical level on the common boundary, which can improve the efficiency of industrial production and pollution control. The average value of TGR is 0.5234, and a total of 12 regions are higher than the average value. The better-performing regions are Tibet (0.9876), Hainan (0.8965) and Liaoning (0.8675), the three worst-performing regions are Hubei (0.1413), Guangxi (0.1321) and Hunan (0.1156). The TGR values of nine provinces, cities and administrative regions in the Yellow River Basin are also less than 1. The average TGR value is 0.3825, which is lower than the average TGR value of the Non-Yellow River Basin by 0.5234. In total, the four regions are higher than the average value. The better-performing regions are Ningxia (0.7545), Qinghai (0.5708) and Shandong (0.5411), and the three worst-performing regions are Sichuan (0.1965), Shaanxi (0.1822) and Inner Mongolia (0.1152).

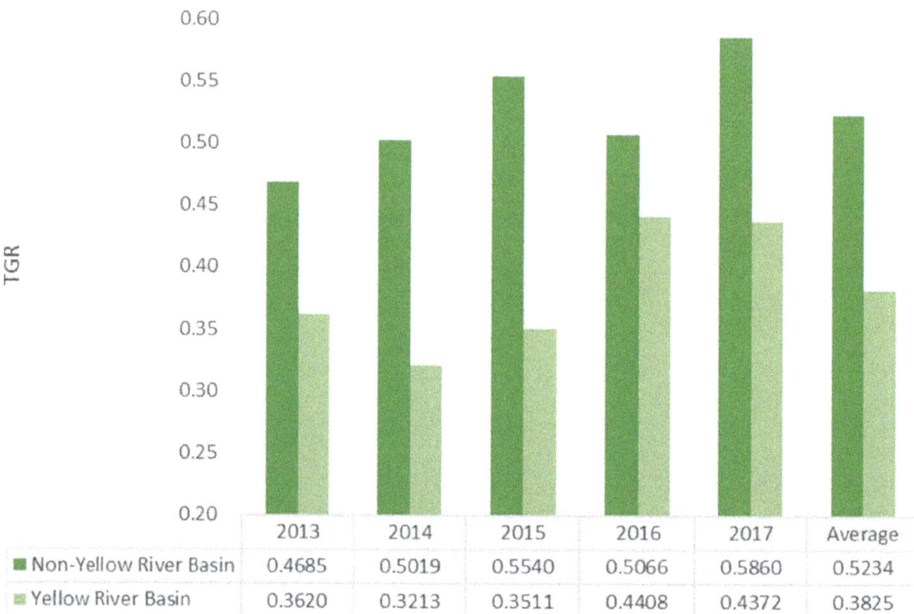

Figure 5. TGR analysis of the Yellow River and Non-Yellow River basins from 2013 to 2017.

Based on the combined analysis of DDF efficiency and TGR results of the Yellow River and Non-Yellow River basins from 2003 to 2007 (Figure 6), the overall efficiency performance of the two regions has little change during the study period. Among them, the Yellow River Basin was only slightly better than the Non-Yellow River basin (0.9863) in 2013 (0.9741), but in the TGR part, the performance of the two regions still has room for significant improvement. Among them, the TGR of the Yellow River Basin is significantly behind the Non-Yellow River Basin. Through reasonable human and capital allocation, we can implement the use of pollution prevention and control funds, and improve the technical level of industrial production to improve pollutant emission and increase output value, to improve the overall efficiency performance.

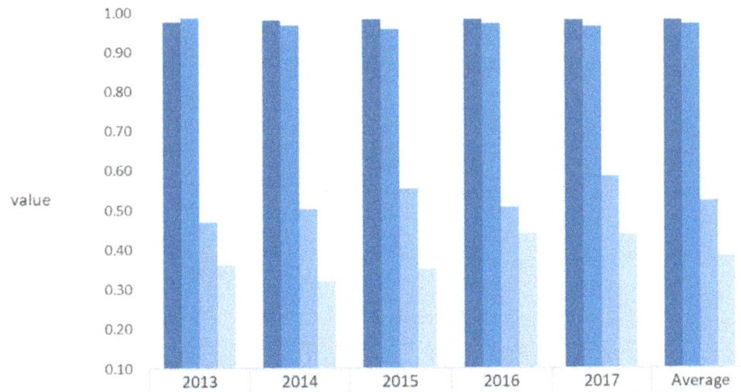

	2013	2014	2015	2016	2017	Average
DDF efficiency in Non-Yellow River Basin	0.9741	0.9795	0.9820	0.9820	0.9791	0.9793
DDF efficiency in the Yellow River Basin	0.9863	0.9669	0.9561	0.9710	0.9636	0.9688
TGR in Non-Yellow River Basin	0.4685	0.5019	0.5540	0.5066	0.5860	0.5234
TGR in the Yellow River Basin	0.3620	0.3213	0.3511	0.4408	0.4372	0.3825

Figure 6. Analysis of DDF efficiency and TGR in the Yellow River and Non-Yellow River basins from 2013 to 2017.

5. Conclusions and Suggestions

Using DDF and TGR methods, this study invested industrial production departments and pollution control departments in parallel to explore the impact of pollution control on industrial production efficiency in 31 provinces and municipalities in the Yellow River and Non-Yellow River basins of China.

5.1. Conclusions

(1) During the study period, the total efficiency of the Non-Yellow River Basin was 0.9793, slightly better than that of the Yellow River Basin of 0.9688. Among the 17 provinces and cities with a total efficiency of 1, only Shandong and Sichuan were located in the Yellow River Basin, and the other 15 provinces and cities were located in the Non-Yellow River Basin, indicating that the industrial production efficiency still had significant regional differences due to the input of production factors and pollution control funds.

(2) During the study period, the TGR values of 31 provinces and municipalities in the Yellow River Basin and Non-Yellow River Basin were less than 1, while the average TGR value of the Yellow River Basin was 0.3825, which was lower than the average TGR value of Non-Yellow River Basin by 0.5234, indicating that the technical level did not reach the technical level on the common boundary, and there is still room for substantial improvement. In order to achieve the sustainable development goals of SDG-6 and SDG-9, in addition to the cost of pollution prevention and control, clean energy should be developed to reduce pollution, and rational allocation of resources should be used to improve industrial production technology and overall efficiency.

(3) The main contribution of this study is in introducing the method of parallel DEA; in addition to many input variables in the industrial production sector, we also discuss the impact of the financial input of the pollution control department on wastewater, exhaust emissions and total efficiency. In addition, this study covers the research scope of the Yellow River Basin and the Non-Yellow River Basin, which helps to provide broader policy recommendations.

5.2. Research Recommendations

The open and quantifiable data of the Yellow River and Non-Yellow River basins in this study are taken from the database of the National Bureau of Statistics of China. The pollution control is carried out through open and objective data. The analysis of China's industrial production efficiency has restrictions on the selection of input- and output-related variables due to the difficulty in obtaining and omission of some data. It is suggested that, in the future, scholars more widely consider relevant data and extend the observation period to make the research results more objective. In addition, this study mainly focuses on the Yellow River and Non-Yellow River basins. It is suggested that different basins such as the Yangtze River and the Pearl River be added as object of discussions in the future to make a longer-cycle cross-basin comparison with each other, to understand China's efforts in industrial production and pollution control, and to provide analysis and basis for SDG-6 and SDG-9 sustainable development goals and policies.

Author Contributions: Conceptualization, Y.F. and C.-C.L.; methodology, C.-C.L.; software, C.-Y.Y.; validation, P.-Y.T. and Y.F.; formal analysis, Y.F. and C.-Y.Y.; investigation, Y.F.; resources, Y.F.; data curation, C.-Y.Y. and P.-Y.T.; writing—original draft preparation, C.-Y.Y. and C.-C.L.; writing—review and editing, C.-Y.Y. and C.-C.L. All authors have read and agreed to the published version of the manuscript.

Funding: This research was funded by the projects as below: Shaanxi Province Innovation Capability Support Program Soft Science Project (2022KRM107), Shaanxi Province Innovation Capability Support Program (2020KJXX-038), National Social Science Foundation of China (21BJY138), Shaanxi Province Innovation Capability Support Program Soft Science Project (2022KRM045), National Natural Science Foundation of China(42171281), Shaanxi Province Science and Technology Innovation Team Project (2021TD-35), And The APC was funded by [Ying Feng].

Institutional Review Board Statement: Not applicable.

Informed Consent Statement: Not applicable.

Data Availability Statement: The data presented in this study are available on request from the corresponding author.

Conflicts of Interest: The authors declare no conflict of interest.

Appendix A

Table A1. Efficiency of DDF in the Yellow River and Non-Yellow River basins from 2013 to 2017.

	DMU	2013	2014	2015	2016	2017	Average
Non-Yellow River Basin	Beijing	1	1	1	1	1	1
	Tianjin	1	1	1	1	1	1
	Hebei	1	1	1	1	1	1
	Liaoning	1	1	1	1	1	1
	Jilin	1	1	1	1	1	1
	Black Dragon River	1	1	0.9822	1	0.9153	0.9795
	Shanghai	1	1	1	1	1	1
	Jiangsu	1	1	1	1	1	1
	Zhejiang	1	1	1	1	1	1
	Anhui	1	1	1	1	1	1
	Fujian	0.9452	0.9623	0.9628	0.9739	1	0.9688
	Jiangxi	1	1	1	1	1	1
	Hubei	0.9752	1	1	1	1	0.9950
	Hunan	1	1	1	1	1	1
	Guangdong	1	1	1	1	1	1
	Guangxi	0.9851	0.9599	0.9883	1	0.9535	0.9774
	Hainan	1	1	1	1	1	1
	Chongqing	1	1	1	1	1	1
	Guizhou	0.8202	0.8524	0.956	1	1	0.9257
	Yunnan	0.7197	0.7832	0.8194	0.8017	0.7779	0.7804
	Tibet	1	1	1	1	1	1
	Xinjiang	0.9856	0.9906	0.8959	0.829	0.8928	0.9188
	Average	0.9741	0.9795	0.9820	0.9820	0.9791	0.9793

Table A1. *Cont.*

	DMU	2013	2014	2015	2016	2017	Average
Yellow River Basin	Shanxi	0.9803	0.9184	0.8889	0.9211	1	0.9417
	Inner Mongolia	1	1	1	1	0.9912	0.9982
	Shandong	1	1	1	1	1	1
	Henan	1	0.9553	0.9696	1	1	0.9850
	Sichuan	1	1	1	1	1	1
	Shaanxi	1	1	0.9871	1	1	0.9974
	Gansu	0.9196	0.9101	0.7593	0.8367	0.8763	0.8604
	Qinghai	1	1	1	0.9816	0.9041	0.9771
	Ningxia	0.9771	0.9183	1	1	0.9005	0.9592
	Average	0.9863	0.9669	0.9561	0.9710	0.9636	0.9688

Appendix B

Table A2. TGR analysis of the Yellow River and Non-Yellow River basins from 2013 to 2017.

	DMU	2013	2014	2015	2016	2017	Average
Non-Yellow River Basin	Tibet	0.9999	0.9905	0.9475	0.9999	1.0000	0.9876
	Hainan	0.8629	1.0000	1.0000	0.6278	0.9917	0.8965
	Liaoning	0.7303	0.9488	0.9319	0.8129	0.9134	0.8675
	Black Dragon River	1.0000	0.8539	0.7762	0.8059	0.9007	0.8673
	Beijing	0.3992	0.8811	1.0000	0.9602	1.0000	0.8481
	Tianjin	0.8239	0.8445	0.8448	0.6499	0.9749	0.8276
	Chongqing	0.7837	0.6317	0.8055	0.7210	1.0000	0.7884
	Shanghai	0.6246	0.8022	0.6441	0.5595	1.0000	0.7261
	Hebei	0.5811	0.6070	0.5647	0.5945	0.6271	0.5949
	Jilin	0.6069	0.6110	0.5334	0.4927	0.7060	0.5900
	Jiangsu	0.4275	0.4250	0.4565	0.6617	0.8183	0.5578
	Zhejiang	0.5043	0.5831	0.4352	0.5733	0.5748	0.5341
	Guizhou	0.2885	0.2245	0.6365	0.8179	0.3949	0.4725
	Xinjiang	0.4264	0.4511	0.4149	0.2540	0.2649	0.3623
	Yunnan	0.1555	0.1228	1.2204	0.1098	0.1180	0.3453
	Guangdong	0.2396	0.2759	0.2457	0.2527	0.3327	0.2693
	Fujian	0.1190	0.1098	0.1074	0.5034	0.2717	0.2223
	Anhui	0.1420	0.1688	0.1492	0.2067	0.3441	0.2022
	Jiangxi	0.2279	0.1759	0.1545	0.1403	0.1336	0.1664
	Hubei	0.1083	0.1171	0.1063	0.1685	0.2065	0.1413
	Guangxi	0.1500	0.1202	0.1116	0.1289	0.1498	0.1321
	Hunan	0.1053	0.0971	0.1026	0.1043	0.1685	0.1156
	Average	0.4685	0.5019	0.5540	0.5066	0.5860	0.5234
	DMU	**2013**	**2014**	**2015**	**2016**	**2017**	**Average**
Yellow River Basin	Ningxia	0.7257	0.5888	0.9243	1.0000	0.5335	0.7545
	Qinghai	0.5908	0.5105	0.6123	0.5184	0.6219	0.5708
	Shandong	0.4784	0.5117	0.5014	0.7260	0.4879	0.5411
	Shanxi	0.5274	0.3898	0.2413	0.6932	0.8063	0.5316
	Gansu	0.2921	0.2695	0.2941	0.2438	0.2970	0.2793
	Henan	0.1332	0.1277	0.2076	0.2589	0.6282	0.2711
	Sichuan	0.2343	0.2222	0.1036	0.2253	0.1971	0.1965
	Shaanxi	0.1672	0.1651	0.1600	0.1811	0.2375	0.1822
	Inner Mongolia	0.1090	0.1059	0.1154	0.1205	0.1254	0.1152
	Average	0.3620	0.3213	0.3511	0.4408	0.4372	0.3825

Note: the total average value of TGR in the Yellow River and non yellow river basins is 0.4825.

References

1. Mahmoud AE, D.; Hosny, M.; El-Maghrabi, N.; Fawzy, M. Facile synthesis of reduced graphene oxide by Tecoma stans extracts for efficient removal of Ni (II) from water: Batch experiments and response surface methodology. *Sustain. Environ. Res.* **2022**, *32*, 22. [CrossRef]
2. Chen, S.; Oliva, P.; Zhang, P. The effect of air pollution on migration: Evidence from China. *J. Dev. Econ.* **2022**, *156*, 102833. [CrossRef]
3. Mahmoud AE, D.; Umachandran, K.; Sawicka, B.; Mtewa, T.K. Water resources security and management for sustainable communities. In *Phytochemistry, the Military and Health*; Elsevier: Amsterdam, The Netherlands, 2021; pp. 509–522.
4. Hu, H.; Jin, Q.; Kavan, P. A study of heavy metal pollution in China: Current status, pollution-control policies and countermeasures. *Sustainability* **2014**, *6*, 5820–5838. [CrossRef]
5. Lin, H.; Chen, H.; Zhang, L.; Luo, Y.; Shi, Y.; Zou, W. Energy consumption, air pollution, and public health in China: Based on the Two-Stage Dynamic Undesirable DEA model. *AirQual. Atmos. Health* **2021**, *14*, 1349–1364. [CrossRef]
6. Miao, C.-L.; Meng, X.-N.; Duan, M.-M.; Wu, X.-Y. Energy consumption, environmental pollution, and technological innovation efficiency: Taking industrial enterprises in China as empirical analysis object. *Environ. Sci. Pollut. Res.* **2020**, *27*, 34147–34157. [CrossRef]
7. Kynčlová, P.; Upadhyaya, S.; Nice, T. Composite index as a measure on achieving Sustainable Development Goal 9 (SDG-9) industry-related targets: The SDG-9 index. *Appl. Energy* **2020**, *265*, 114755. [CrossRef]
8. Zaim, O. Measuring environmental performance of state manufacturing through changes in pollution intensities: A DEA framework. *Ecol. Econ.* **2004**, *48*, 37–47. [CrossRef]
9. Wu, J.; Lv, L.; Sun, J.; Ji, X. A comprehensive analysis of China's regional energy saving and emission reduction efficiency: From production and treatment perspectives. *Energy Policy* **2015**, *84*, 166–176. [CrossRef]
10. Valadkhani, A.; Roshdi, I.; Smyth, R. A multiplicative environmental DEA approach to measure efficiency changes in the world's major polluters. *Energy Econ.* **2016**, *54*, 363–375. [CrossRef]
11. Geng, Z.; Dong, J.; Han, Y.; Zhu, Q. Energy and environment efficiency analysis based on an improved environment DEA cross-model: Case study of complex chemical processes. *Appl. Energy* **2017**, *205*, 465–476. [CrossRef]
12. Sueyoshi, T.; Yuan, Y.; Goto, M. A literature study for DEA applied to energy and environment. *Energy Econ.* **2017**, *62*, 104–124. [CrossRef]
13. Halkos, G.E.; Polemis, M.L. The impact of economic growth on environmental efficiency of the electricity sector: A hybrid window DEA methodology for the USA. *J. Environ. Manag.* **2018**, *211*, 334–346. [CrossRef] [PubMed]
14. Fehri, R.; Khlifi, S.; Vanclooster, M. Disaggregating SDG-6 water stress indicator at different spatial and temporal scales in Tunisia. *Sci. Total Environ.* **2019**, *694*, 133766. [CrossRef] [PubMed]
15. Hu, Z.; Yan, S.; Yao, L.; Moudi, M. Efficiency evaluation with feedback for regional water use and wastewater treatment. *J. Hydrol.* **2018**, *562*, 703–711. [CrossRef]
16. Song, M.; Wang, R.; Zeng, X. Water resources utilization efficiency and influence factors under environmental restrictions. *J. Clean. Prod.* **2018**, *184*, 611–621. [CrossRef]
17. Yang, L.; Zhang, X. Assessing regional eco-efficiency from the perspective of resource, environmental and economic performance in China: A bootstrapping approach in global data envelopment analysis. *J. Clean. Prod.* **2018**, *173*, 100–111. [CrossRef]
18. Li, J.; See, K.F.; Jin, C. Water resources and water pollution emissions in China's industrial sector: A green-biased technological progress analysis. *J. Clean. Prod.* **2019**, *229*, 1412–1426. [CrossRef]
19. Zhou, Z.; Wu, H.; Song, P. Measuring the resource and environmental efficiency of industrial water consumption in China: A non-radial directional distance function. *J. Clean. Prod.* **2019**, *240*, 118169. [CrossRef]
20. Chen, Y.; Yin, G.; Liu, K. Regional differences in the industrial water use efficiency of China: The spatial spillover effect and relevant factors. *Resour. Conserv. Recycl.* **2021**, *167*, 105239. [CrossRef]
21. Wang, K.-L.; Miao, Z.; Zhao, M.-S.; Miao, C.-L.; Wang, Q.-W. China's provincial total-factor air pollution emission efficiency evaluation, dynamic evolution and influencing factors. *Ecol. Indic.* **2019**, *107*, 105578. [CrossRef]
22. Wang, H.; Wang, M. Effects of technological innovation on energy efficiency in China: Evidence from dynamic panel of 284 cities. *Sci. Total Environ.* **2020**, *709*, 136172. [CrossRef] [PubMed]
23. Chen, L.; He, F.; Zhang, Q.; Jiang, W.; Wang, J. Two-stage efficiency evaluation of production and pollution control in Chinese iron and steel enterprises. *J. Clean. Prod.* **2017**, *165*, 611–620. [CrossRef]
24. Zhang, H.; Wang, S.; Hao, J.; Wang, X.; Wang, S.; Chai, F.; Li, M. Air pollution and control action in Beijing. *J. Clean. Prod.* **2016**, *112*, 1519–1527. [CrossRef]
25. Sarkodie, S.A.; Strezov, V. Effect of foreign direct investments, economic development and energy consumption on greenhouse gas emissions in developing countries. *Sci. Total Environ.* **2019**, *646*, 862–871. [CrossRef] [PubMed]
26. Farrell, M.J. The Measurement of Productive Efficiency. *J. R. Stat. Soc.* **1957**, *120*, 253–290. [CrossRef]
27. Charnes, A.; Cooper, W.W.; Rhodes, E. Measuring the efficiency of decision making units. *Eur. J. Oper. Res.* **1978**, *2*, 429–444. [CrossRef]
28. Banker, R.D.; Charnes, A.; Cooper, W.W. Some Models for Estimating Technical and Scale Inefficiencies in Data. Envelopment Analysis. *Manag. Sci.* **1984**, *30*, 1078–1092. [CrossRef]
29. Chambers, R.G.; Chung, Y.; Fare, R. Benefit and distance functions. *J. Econ. Theory* **1996**, *70*, 407–419. [CrossRef]

30. Ray, S.C. The directional distance function and measurement of super-efficiency: An application to airlines data. *J. Oper. Res. Soc.* **2008**, *59*, 788–797. [CrossRef]
31. Färe, R.; Grosskopf, S. Directional distance functions and slacks-based measures of efficiency. *Eur. J. Oper. Res.* **2010**, *200*, 320–322. [CrossRef]
32. Tone, K. A slacks-based measure of efficiency in data envelopment analysis. *Eur. J. Oper. Res.* **2001**, *130*, 498–509. [CrossRef]

energies

Article

The Carbon Emission Reduction Effect of City Cluster—Evidence from the Yangtze River Economic Belt in China

Xin Li [1], Chunlei Huang [2,3,*], Shaoguo Zhan [4] and Yunxi Wu [5]

1 Graduate Institute for Taiwan Studies, Xiamen University, Xiamen 361005, China
2 School of International Trade and Economics, University of International Business and Economics, Beijing 100029, China
3 School of Business, Huanggang Normal University, Huanggang 438000, China
4 School of Public Policy and Management, Tsinghua University, Beijing 100084, China
5 School of Economics and Management, Fuzhou University, Fuzhou 350108, China
* Correspondence: huchle1234@163.com

Abstract: Climate anomalies are affecting the world. How to reduce carbon emissions has become an important issue for governments and academics. Although previous researchers have discussed the factors of carbon emission reduction from environmental regulation, economic development, and industrial structure, limited studies have explored the carbon emission reduction effect of a city's spatial structure. Based on 108 Chinese cities from the Yangtze River Economic Belt between 2003 and 2017, this paper examines the impact of the city cluster policy on city carbon emissions using the difference-in-differences (DID) method. We find that: (1) The city cluster policy has significantly reduced the cities' carbon emissions by 7.4%. Furthermore, after a series of robust and endogenous tests, such as parallel trend and PSM-DID, the core conclusion still remains. (2) We further identify possible economic channels through this effect, and find that city cluster policy would increase city productivity, city technological innovation, and industrial structure optimization. The conclusions of this paper have important practical significance for China to achieve carbon neutrality and facilitate future deep decarbonization.

Keywords: carbon emission reduction; city cluster policy; difference-in-difference method

Citation: Li, X.; Huang, C.; Zhan, S.; Wu, Y. The Carbon Emission Reduction Effect of City Cluster—Evidence from the Yangtze River Economic Belt in China. *Energies* **2022**, *15*, 6210. https://doi.org/10.3390/en15176210

Academic Editor: Donato Morea

Received: 22 July 2022
Accepted: 19 August 2022
Published: 26 August 2022

1. Introduction

Air pollution, from sources such as climate anomalies, melting glaciers, and haze, has become an important issue around the world. China, as the world's largest emitter of CO_2 and SO_2 [1,2], is facing serious domestic environmental pollution problems and tremendous international community pressure [3]. In 2021, China committed to peaking its carbon emissions by 2030 and becoming carbon neutral by 2060. Therefore, the question of how to balance economic development and ecological protection has become an important issue. The city cluster policy can provide a potential solution. the city cluster is not only the most promising and dynamic area in China's economic development, but also the area with the highest concentration of ecological and environmental problems. The city cluster will have a profound impact on regional resource utilization and environmental protection, but it is not clear through which specific path. These are issues of interest to many cities and economists.

How to reduce carbon emissions is a crucial issue for sustainable development. The Chinese government has established a series of environmental regulation policies, such as the Air Pollution Prevention and Control Action Plan in 2013, which is also noted as China's Clean Air Action [4–6], carbon emission trading system in 2014 [7–10], and so on. Although previous scholars have discussed the air governance effect of environmental regulation [11,12], they have seldom exploited the environmental effect of a city's spatial layout [13,14], especially the impact of the city cluster on carbon emission reduction. Therefore, to fill academic gaps, this paper will examine the impact of the city cluster policy on carbon emissions, based on the sample of China's Yangtze River Economic Belt.

Based on city-level panel data from 108 Chinese cities in the Yangtze River Economic Belt between 2003 and 2017, this paper examines the impact of city cluster policy on carbon emissions using the difference-in-differences (DID) method. We found that: (1) The city cluster policy has significantly reduced the cities' carbon emissions, with an average reduction of 7.4% in city carbon emissions; furthermore, after a series of robust and endogenous tests, such as parallel trend and PSM-DID, the core conclusion still remains. (2) We further identify possible economic channels through this effect, and find that city cluster policy would increase city productivity, city technological innovation, and industrial structure optimization. (3) The emission reduction effect of the city cluster policy only exists in the nation's city clusters.

There are three reasons why we choose China as the background to study the impact of city clusters on carbon emission reduction. First, China is the world's most populous country and the world's second-largest economy. It is of practical significance to evaluate the impact of carbon trading pilot policies. Secondly, China is the largest carbon emission emitter and an emerging country [15]. Conclusions from such research on China may provide a useful reference for other developing countries to implement carbon trading pilot programs. Finally, China is a centralized country that adopts a vertical management and organizational structure. The environmental policies are initially formulated by the central government of China, and then implemented by local governments, which ensures the exogenous nature of the policy. In addition, there are 30 provinces in China, allowing us to use this cross-sectional variation to determine the policy effect of the city cluster.

There are two main contributions to this paper. For one thing, this paper has a potential academic contribution. Previous scholars have discussed the air governance effect from the perspective of environmental regulation [4–6], but they have seldom discussed the city's spatial layout. In this paper, we use China's city cluster policy in 2011 to examine the effect of city integration on carbon emissions. For another thing, this paper has strong policy implications. Based on China's city cluster policy in 2011, this paper examines the effects of the city cluster on carbon emissions. We found that the city cluster will reduce carbon emissions. The conclusions provide useful policy implications for policy-makers to reduce city carbon emissions. Rich results from a heterogeneity analysis provide policy-makers with an understanding of economic facts, and point out the direction for improving the carbon trading system.

The rest of the paper is organized as follows: the second part is the theoretical analysis, the third part presents the data and empirical design, the fourth part presents the empirical results, the fifth part is the further discussion, and the sixth part consists of the conclusions and policy implications.

2. Theoretical Analysis

A large number of theoretical and empirical studies show that the city cluster can have a positive effect on economic productivity, technological innovation capacity, and industrial structure optimization through the agglomeration of factors. Details are as follows.

First, the city cluster can reduce carbon emissions by improving city productivity. Generally speaking, the public infrastructure construction of member cities in the city cluster can be further improved, by, for example, improving the railway station [16]. Infrastructure improvements can help foster city networks that enable the flow of people and capital among cities [17]. The formation of a city cluster can improve the capacity of resource allocation in a larger region. It helps the flow of production elements from large to small and medium-sized cities, and improves the aggregation economic and ecological efficiency of small and medium-sized cities [18]. In addition, the scale effect generated by the city cluster has contributed to an increase in city productivity.

With the implementation of the city cluster, the production costs and price index of products would decrease, leading to the expansion of local demand and market size. The increased returns to scale resulting from this expansion will further promote agglomeration and thus increase regional productivity. The increasing returns to scale generated by expansion further promote agglomeration and improve regional production productivity.

Secondly, the city cluster can reduce carbon emissions through technology. On the one hand, economic growth in cities is accompanied by the build-up of human capital and the overflow of knowledge. The city cluster can increase opportunities for the inter-regional exchange of people and learning, and promote collaborative research and development [19], thus accelerating the diffusion and application of knowledge and new technologies within the region and promoting technological progress. Therefore, the clustering spillover effects of sharing, matching, and learning mechanisms within large cities are more obvious than those in small cities [20], which will accelerate the dissemination and application of knowledge and new technologies within the region and thus promote technological progress.

On the other hand, the city cluster significantly improves market openness and facilitates the aggregation of high-quality factors in the city cluster. Non-local enterprises bring the cross-regional flow of enterprise innovation factors, creating favorable conditions for inter-regional knowledge spillover and improving innovation efficiency [21]. In conclusion, the knowledge spillover brought by the city cluster can not only promote local technological progress, but also have a significant impact on the technological progress of neighboring regions as well through the cross-regional flow of innovation factors.

Finally, the city cluster can reduce carbon emissions through industrial structure upgrading. The optimization of industrial structure and energy efficiency are key factors in carbon reduction [22,23]. The city cluster reduces barriers for non-local enterprises and foreign investment to enter the local market by lowering trade barriers within cities, which accelerates competition among enterprises within the market. To survive, enterprises will eventually choose industrial structure upgrading through the market competition mechanism. At the same time, the hierarchy produced in the development of the city cluster is inevitably accompanied by various degrees of specialization. Diversified metropolises will take on the role of incubators for innovative industries, while small and medium-sized cities will use their comparative advantage in production factors to reduce production costs and become agglomerations for some industries [24]. Referring to Ó Huallacháin and Lee (2011) [25], specialized production facilitates eco-efficiency through channels such as the promotion of economic factor aggregation, technological progress, resource intensification, and Marshallian externalities. Therefore, the optimization of a city system stemming from the development of the city cluster will promote the industrial structure upgrading.

Therefore, based on the above analysis, this paper puts forward three assumptions, as shown in Figure 1.

Hypothesis 1. *City cluster pilot policy can reduce carbon emissions by increasing city productivity.*

Hypothesis 2. *City cluster pilot policy can reduce carbon emissions by improving the level of technological innovation.*

Hypothesis 3. *City cluster pilot policy can reduce carbon emissions through optimizing industrial structures.*

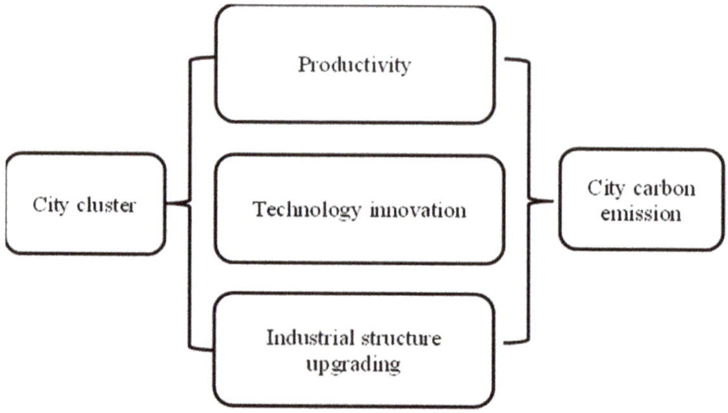

Figure 1. Theoretical hypothesis framework.

3. Data and Empirical Design

3.1. Data

We investigate the impact of city cluster policy on carbon emissions according to the panel data of 108 Chinese cities in the Yangtze River economic belt from 2003–2017. Our carbon emissions data were obtained from the Carbon Emission Accounts and Database https://www.ceads.net/ (accessed on 5 May 2022); other city data are from the China City Statistics Yearbook. The carbon emissions data are only updated to 2017. Our final sample consists of 1620 city-year observations covering the 2003–2017 period.

3.2. Empirical Design

The purpose of this study is to examine the effect of city cluster policy on CO_2 emissions. As a classical method for policy evaluation, the difference-in-differences (DID) model has been widely adopted by most scholars, and we also use this method. This method can examine the difference in CO_2 emissions before and after the city cluster policy implementation, and assess the average effect of city cluster policy on carbon emissions. The benchmark model is as Formula (1).

$$LnCO_{2c,t} = \alpha + \beta \times Treat_c \times Post_t + \varnothing \times Control_{c,t} + \delta_c + \mu_t + \varepsilon_{c,t} \tag{1}$$

where c is the city and t is the year. Independent variable $LnCO_{2c,t}$ indicates the carbon emissions of city c in year t. Our dependent variable is the city cluster policy (Treat*Post). The coefficient on Treat*Post, β, is the one with the main interest. Thus, β reflects the impact of the city cluster policy on carbon emissions. A negative β implies that a city cluster policy will reduce carbon emissions in cities. Control is our control variable. δ_c and μ_t are city-fixed effect and year-fixed effect, respectively, and $\varepsilon_{c,t}$ is a random error term.

3.3. Variables

3.3.1. Independent Variables

Our independent variable consists of city-level CO_2 emissions ($LnCO_2$). Unfortunately, only provincial-level CO_2 data and county-level CO_2 data are available in the Carbon Emission Accounts & Database. Meanwhile, we have also noticed that county-level CO_2 is measured by light intensity, not real CO_2 emissions. Given such two dimensions of CO_2 data structure, this paper uses two methods to construct city-level CO_2 emissions. One method uses county-level CO_2 emission data directly added to the city level [26]. Another method uses county-level CO_2 data to calculate the proportion of each city in its province. Based on this proportion, provincial-level CO_2 is allocated to each city by this proportion weight, and the weighted city CO_2 is constructed. We would use weighted city

CO_2 emissions from the second method in the benchmark regression. We also use the CO_2 emissions from the first method in the robustness test.

3.3.2. Dependent Variables

Our dependent variable is the city cluster policy (Treat*Post). It is an interaction item between Treat and Post. The Treat variable equals one in the city cluster list. The city cluster list in the Yangtze River economic belt contains the Chengyu city cluster (Nation), the Dianzhong city cluster (Region), the Yangtze River city cluster (World), the Yangtze middle river city cluster (Nation), the Qianzhong city cluster(Region), and zero otherwise. Post equals one in a year that is equal to or larger than 2011, and zero otherwise.

3.3.3. Control Variables

Following prior research, we add several control variables to the model, which include: city economic development (LnGDP), city openness (Open), city financial development (Finance), city government scale (Gov), the ratio of city secondary industry (Sec_Ind), and the ratio of city tertiary industry (Ter_Ind). Table 1 provides detailed definitions of all variables. The definitions of the main variables are shown in Table 1.

Table 1. Variables Definition.

Variable	Definition
$LnCO_2$	The logarithm of city CO_2
Treat*Post	An indicator variable that equals one if the city is eventually included in the low-carbon city pilot list by the end of our sample period, and zeroes otherwise
LnGDP	the logarithm of city GDP
Open	The ratio of the city actual utilization of foreign direct investment to city GDP
Finance	The ratio of the city balance of bank deposits and loans to city GDP
Gov	The ratio of the city government public finance expenditure to city GDP
Sec_Ind	The ratio of the city added value of the secondary industry to city GDP
Ter_Ind	The ratio of the city added value of the tertiary industry to city GDP

3.4. Descriptive Statistics

The dependent variable consists of city-level CO_2 emissions ($LnCO_2$). The average $LnCO_2$ is 2.704, the standard deviation is approximately 0.830, the average Treat*Post value is 0.321, and the standard deviation is 0.467. This indicates that there are significant differences in various cities. In terms of the standard deviation of the control variables, there is a degree of variation among the cities; city-level CO_2 emissions may be affected by these differences. The descriptive statistics of the main variables are shown in Table 2.

Table 2. Descriptive statistics of the variables.

Variables	Observations	Mean	Sd	Min	Max
		Dependent Variables			
$LnCO_2$	1620	2.704	0.830	0.250	5.128
		Independent Variables			
Treat*Post	1620	0.321	0.467	0	1
		Control Variables			
LnGDP	1620	16.02	1.081	12.93	19.46
Open	1620	0.023	0.022	0	0.201
Finance	1620	2.084	0.873	0.764	6.255
Gov	1620	0.165	0.090	0.049	1.485
Sec_Ind	1620	0.482	0.093	0.187	0.759
Thi_Ind	1620	0.371	0.075	0.207	0.698

4. Empirical Results

4.1. The Effect of City Cluster Policy on CO_2 Emissions

The estimated results of Equation (1) are shown in Table 3. It can be seen that the coefficient estimates of Treat*Post are significantly negative, suggesting that city carbon emissions have decreased after the city cluster policy. This negative impact has economic implications. For example, during our sample period, carbon emissions from cities declined by an average of 7.4% after cities were classified as a city cluster. This result supports our previous hypothesis. The path of the effect may come from the positive impact of the city cluster on production efficiency, technological innovation ability and the rationalization of industrial structure, which will be further examined in this paper. To sum up, the city cluster policy helps cities reduce carbon emissions.

Table 3. The effect of city cluster policy on CO2 emissions.

	(1)	(2)
	$LnCO_2$	**$LnCO_2$**
Treat*Post	−0.099 ***	−0.074 ***
	(−6.235)	(−5.032)
LnGDP		0.405 ***
		(5.486)
Open		0.580 *
		(1.746)
Finance		0.056 *
		(1.879)
Gov		0.196
		(1.249)
Sec_Ind		0.070
		(0.409)
Thi_Ind		0.043
		(0.202)
Constant	2.736 ***	−3.975 ***
	(418.071)	(−3.311)
City FE	YES	YES
Year FE	YES	YES
Observations	1620	1620
Adj_R2	0.957	0.959

Note: *t* statistics are shown in parentheses; *** and * represent significance at the 1%, and 10% levels, respectively. All variable definitions are in Table 1. The sample period covers 2003 through 2017. Regressions in all columns control for year-fixed effects and city-fixed effects.

4.2. Parallel Trend Analysis

A parallel trend is a prerequisite for the DID model. It means that there is no systematic difference in carbon emission trends between the two groups before the policy, or, even if there are differences, the differences are fixed. Therefore, we followed Li et al. (2016) [27] and Zhu and Xu (2022) [5], and constructed our model as:

$$LnCO_{2c,t} = \alpha + \beta_t \times \text{Treat} \times D^{\text{Year}(t)} + \varnothing \times \text{Control}_{c,t} + \delta_c + \mu_t + \varepsilon_{c,t} \quad (2)$$

$D^{\text{Year}(t)}$ is year dummy variables, and it is equal to 1 when year is t. For example, D^{2006} is equal to 1 when year is 2006, and 0 otherwise. Therefore, the parameters of β_t identify t year policy effects. To avoid Treat $\times$ $D^{\text{Year}(t)}$ collinearity, we use the policy year (i.e., 2011) as the base year. The estimation results are presented in Figure 2. We can see that there is no pre-policy effect (before 2011), indicating that our identification satisfies the parallel trend assumption. Furthermore, the policy has a strong continuity effect.

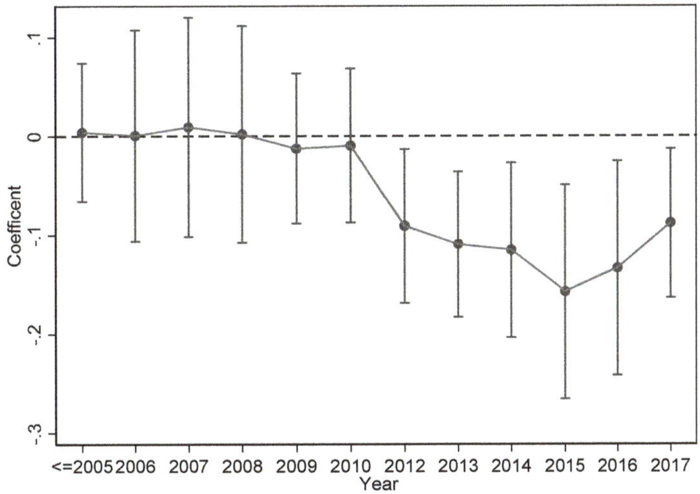

Figure 2. Dynamic effects of the City Cluster.

4.3. Robustness Test

4.3.1. Propensity Score Matching DID (PSM-DID)

Of course, the city cluster policy is not a perfect quasi-natural experiment. There is a certain degree of randomness in the selection of the city cluster. Strictly speaking, whether a city can be selected as one of the city clusters is not a completely random selection process. It will be disturbed by economic factors, political factors, and human factors.

These differences will affect the validity of the DID model. In order to reduce the interference caused by these differences in model estimation, we will use the propensity score matching (PSM) method proposed by Heckman et al. (1998) [28] to select comparable treatment and control groups, and then use a DID model to estimate the policy effects [26]. We adopt the 1:1 nearest neighbor matching method. The estimation result is shown in column (1) of Table 4. It can be seen that the Treat*Post is still significantly negative with the city's carbon emissions (lnCO_2), indicating that our core findings remain valid after alleviating the problem of sample selection bias.

Table 4. Robustness check of the effect of city cluster policy on CO_2 emissions(PSM-DID).

	(1)	(2)	(3)	(4)	(5)	(6)
	LnCO_2		LnCO_2_2		LnCO_2	
Treat*Post	−0.026 *	−0.064 ***	−0.025 **	−0.069 ***	−0.060 ***	−0.056 ***
	(−1.698)	(−2.653)	(−2.159)	(−4.787)	(−4.460)	(−4.216)
PSM	YES					
Two-Stage		YES				
Replace Y			YES			
CAA Policy				YES		YES
CET Policy					YES	YES
Control Vars	YES	YES	YES	YES	YES	YES
City FE	YES	YES	YES	YES	YES	YES
Year FE	YES	YES	YES	YES	YES	YES
Observations	727	216	1620	1620	1620	1620
R-squared	0.987	0.996	0.969	0.962	0.963	0.963
Adj_R2	0.985	0.992	0.967	0.959	0.960	0.960

Note: *t* statistics are shown in parentheses; ***, **, and * represent significance at the 1%, 5%, and 10% levels, respectively.

4.3.2. Two-Period DID

The regression coefficients in the baseline regression may be overestimated due to sequential correlation issues. To solve this problem, we will adopt a two-period estimation strategy according to Bertrand et al. (2004) [29]. The data will be divided into two periods based on the point in time of the policy. That is, the variables in the two periods are averaged to construct a two-period DID sample.

The estimation result is shown in column (2) of Table 4. We find that the coefficient on Treat*Post is significantly negative at the 1% level, suggesting that the city cluster policy can reduce city carbon emissions even after considering potential serial correlation issues.

4.3.3. Alternating the Explained Variable

The estimated results may be sensitive to different definitions of critical variables. To ensure whether the measurement of carbon emissions is robust, we use unweighted CO_2 emissions (LnCO_2_2) as alternative measurements. The estimation result is shown in column (3) of Table 4. It can be seen that the coefficient on Treat*Post is significantly negative with LnCO_2_2, suggesting that the basic conclusion remains unchanged even with the replacement of the core explanatory variables.

4.3.4. The Impact of Related Environmental Policies

During our sample period, some environmental regulatory policies occurred in China, which may affect carbon emissions. To eliminate the impact of these environmental policies on city carbon emissions, in this section we will further control the effect of these policies.

Two major environmental regulatory policies were instituted during the sample period. The first is the Clean Air Action policy (CAA) in 2013. Following Zhu and Xu (2022), we manually collect cities' air pollution targets and generate the variable CAA [5]. Following Zhu and Xu (2022), we measure CAA: CAA = Ln(air pollution targets) * Post(year $\geq$ 2013) [5]. Adding CAA variables to the baseline model (1), the result is shown in column (4) of Table 5. The Treat*Post is still significantly negative with the city carbon emissions (LnCO_2), suggesting that the basic conclusion is robust even if we control the effect of the Clean Air Action policy on city carbon emissions.

Table 5. Other robustness check of the effect of city cluster policy on CO_2 emissions.

	(1)	(2)	(3)
		LnCO_2	
Treat*Post	−0.057 ***	−0.074 ***	−0.054 ***
	(−3.615)	(−4.661)	(−3.936)
Control Pro_Trend	YES		
Control 2008 Finance Crisis		YES	
Control outlier			YES
Control Vars	YES	YES	YES
City FE	YES	YES	YES
Year FE	YES	YES	YES
Observations	1620	1512	1620
Adj_R2	0.960	0.958	0.964

Note: *t* statistics are shown in parentheses; *** represent significance at the 1% levels, respectively.

The second is the provincial carbon emissions trading policy implemented in 2013 http://www.tanpaifang.com/tanjiaoyi/2012/0219/41.html (accessed on 5 May 2022). We set a policy dummy variable CET that equals one if the province is eventually included in the carbon emission system list by the end of our sample period, and zero otherwise. Adding CET variables to the baseline model (1), the result is shown in column (5) of Table 5. The Treat*Post is still significantly negative with the city's carbon emissions

($LnCO_2$), suggesting that the basic conclusion is robust even if we control the effect of carbon emissions trading policy on city carbon emissions.

Finally, we control the impact of both the Clean Air Action (CAA) and the carbon emissions trading policy (CET). The result in column (6) of Table 5 shows that the core explanatory variable Treat*Post is still significantly negative with the city carbon emissions. Although considering the interference of these two environmental regulatory policies, the city cluster policy can still significantly reduce the city carbon emissions.

4.3.5. Other Robustness Check

In addition to the above four robustness tests, other robustness tests are discussed to ensure the robustness of the results in this paper.

Control provincial trend. To exclude the impact of the variation of some characteristics of provinces over time trend on the city carbon emission, we add to control the provincial trend. The result is shown in column (1) of Table 5. The Treat*Post is still significantly negative with the $LnCO_2$.

Eliminate the impact of the financial crisis. A financial crisis affects economic development, which affects the city's carbon emissions. Therefore, we should exclude the 2008 sample, which would eliminate the impact of the 2008 financial crisis on city carbon emissions. The result is shown in column (2) of Table 5. The Treat*Post is still significantly negative with $LnCO_2$.

Winsorize the data. In baseline regression, some variables may lead to extreme values in the data. Therefore, to alleviate the impact of extreme values on the estimated results in this paper, we process the data with 1% winsorizing. The result is shown in column (3) of Table 5. The Treat*Post is still significantly negative with $LnCO_2$.

Placebo test of the experimental group. Following Li et al. (2016) [27], we randomly select a city cluster for placebo testing. Figure 3 shows the distribution of the regression coefficients of the "artificial" processing variable Treat*Post in the simulation. It can be observed that the randomly assigned estimated values are concentrated around zero, while the truly estimated coefficients are on the left side of Figure 3. It verifies that the city cluster policy has significantly reduced the city's carbon emissions.

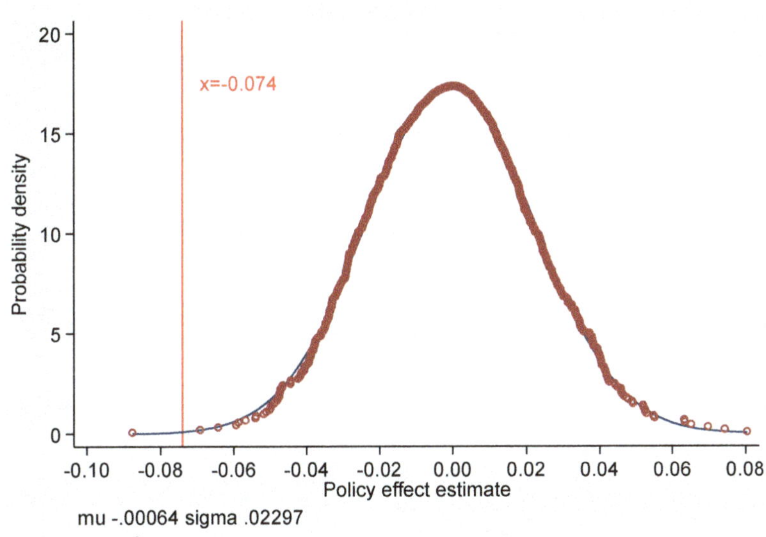

Figure 3. Placebo test of the experimental group.

In a word, the robustness above indicates that the core conclusion still remains when a series of possible and potential interference factors are excluded.

5. Further Discussion

5.1. Economic Channels

In this section, we will explore the three plausible underlying economic channels by which city cluster policy affects city carbon emissions. The economic channels build on existing theories, and factors such as productivity, technological innovation, and industrial structure optimization are important in reducing city carbon emissions.

5.1.1. Productivity Effect

In this section, we will examine whether the city cluster improves city productivity through the city scale effect, which reduces carbon emissions. Based on the article of Chen et al. (2022) [30], we measure an index of the city's total factor productivity (TFP) and examine whether the city cluster has an impact on city productivity based on model (3).

$$TFP_{c,t} = \alpha + \beta \times Treat_c \times Post_t + \varnothing \times Control_{c,t} + \delta_c + \mu_t + \varepsilon_{c,t} \quad (3)$$

The empirical results are shown in Column (1) of Table 6. The core explanatory variable Treat*Post is significantly positively correlated with the explained variable TFP at the confidence level of 5%. The TFP level of the city increased by 22% after cities were classified as the city cluster. Compared to other cities, China's Yangtze River Delta, Pearl River Delta, and Beijing–Tianjin–Hebei region have more advanced infrastructure development, providing a more favorable environment for the flow of production factors. This urban network further creates a scale effect and promotes urban productivity.

Table 6. Mechanism analysis of the effect of city cluster policy on CO_2 emissions.

	(1)	(2)	(3)
	TFP	LnRD	ISO
Treat*Post	0.22 **	0.611 ***	−0.006 *
	(2.515)	(11.235)	(−1.796)
Control Vars	YES	YES	YES
City FE	YES	YES	YES
Year FE	YES	YES	YES
Observations	1547	1584	1618
R-squared	0.845	0.951	0.971
Adj_R2	0.842	0.946	0.969

Note: *t* statistics are shown in parentheses; ***, **, and * represent significance at the 1%, 5%, and 10% levels, respectively.

5.1.2. Technological Innovation Effect

In this section, we will examine whether the city cluster enhances city innovation, which reduces carbon emissions, through city knowledge spillovers. Therefore, based on the article of Du et al. (2021) and Lyu et al. (2019) [31,32], we take city R&D input as a proxy variable of city innovation and will examine whether the city cluster has an impact on city innovation based on model (4).

$$LnRD_{c,t} = \alpha + \beta_t \times Treat_c \times Post_t + \varnothing \times Control_{c,t} + \delta_c + \mu_t + \varepsilon_{c,t} \quad (4)$$

The empirical results are shown in Column (2) of Table 6. It can be seen that the core explanatory variable Treat*Post is significantly positively correlated with the explained variable LnRD at the confidence level of 5%. The city R&D input increased by 61.1% after cities were classified as the city cluster. It can be seen that the flow of production factors brought by the city cluster does significantly enhance the knowledge spillover effect, and cross-regional knowledge spillover creates favorable conditions for improving innovation efficiency.

5.1.3. Industrial Structure Optimization

In this section, we will examine whether the city cluster improves city industrial structure upgrading by reducing the proportion of secondary industries, which reduces carbon emissions. Therefore, based on the article of Liu et al. (2021) [33], we measured an index of the rationalization of industrial structure, which can reflect the coupling degree of the element inputs and outputs.

Formula (5) will be used to measure the rationalization degree of industrial structure.

$$ISO_{i,t} = \sum_{j=1}^{3} \frac{y_{ijt}}{Y_{it}} \ln\left(\frac{y_{ijt}}{Y_{it}} / \frac{l_{ijt}}{L_{it}}\right) \quad (5)$$

where i is the city, t is the year, and j is the industry. Variable y_{ijt} indicates the carbon emissions of the industry j in city i and year t. Variable Y_{it} indicates the gross of the industry of city c in year t. Variable l_{ijt} indicates the number of employees of the industry j in city i and year t. Variable L_{it} indicates the total number of employees of city c in year t. Obviously, the closer ISO is to 0, the higher the coupling degree between the allocation and output ratio of employees in the three industries is and the more reasonable the industrial structure is. On the contrary, the industrial structure is unreasonable.

We will examine whether the city cluster has an impact on city industrial structure upgrading based on model (6).

$$ISO_{c,t} = \alpha + \beta_t \times Treat_c \times Post_t + \varnothing \times Control_{c,t} + \delta_c + \mu_t + \varepsilon_{c,t} \quad (6)$$

The empirical results are shown in Column (3) of Table 6. It can be seen that the core explanatory variable Treat*Post is significantly negatively correlated with the explained variable ISO at the confidence level of 10%. The ISO index decreased by 0.6% after cities were classified as the city cluster. Therefore, the optimization of a city system stemming from the development of the city cluster will promote the industrial structure upgrading.

5.2. The Effect of City Cluster Policy on CO_2 Emissions across City Cluster Positioning Level

The level of the city cluster positioning determines the resource allocation capacity of the city cluster. The city cluster with a high positioning level can provide favorable organizational leadership and abundant human, financial, and other resources, which support the implementation of the city cluster to reduce city carbon emissions. However, the regional-level city cluster governments have a limited ability to allocate resources. In this case, they cannot provide appropriate policies and funds to attract talents and promote the transformation and upgrading of enterprises. The world-level city cluster also does not affect the reduction of carbon emissions. The reason is that the goal of the world-class city cluster is to create a more open Chinese market, and attracting foreign investment is the top priority. As a result, the policy benefits of regional and world-level city cluster cannot be realized. The national-level city cluster has sufficient allocation capacity to reduce carbon emissions.

The result is shown in column (1) of Table 7. Only variable Nation*Treat*Post is significantly negative with $LnCO_2$, the regional and world city cluster positioning show weak policy effects, neither variables World*Treat*Post nor Region*Treat*Post are significantly affected.

Table 7. The effect of city cluster policy on CO_2 emissions across city cluster positioning level.

	(1)
	$LnCO_2$
World*Treat*Post	−0.020
	(−0.528)
Nation*Treat*Post	−0.103 ***
	(−6.854)
Region*Treat*Post	0.030
	(1.031)
Control Vars	YES
City FE	YES
Year FE	YES
Observations	1,620
R-squared	0.963
Adj_R2	0.959

Note: *t* statistics are shown in parentheses; *** represent significance at the 1% levels, respectively.

6. Conclusions and Policy Implications

6.1. Conclusions

Based on 108 Chinese cities from Yangtze River Economic Belt between 2003 and 2017, this study examines the impact of the city cluster policy on cities' carbon emissions using the difference-in-differences method. The main conclusions are as follows.

(1) The city cluster policy has significantly reduced the level of cities' carbon emissions. During our sample period, carbon emissions from cities declined by an average of 7.4% after cities were classified as the city cluster. After a series of robustness tests, the conclusion remains robust.

(2) Productivity, technological innovation, and industrial structure optimization are three essential mechanisms for the city cluster policy to reduce carbon emissions. We find that the TFP level of the city increased by 22%, the city R&D input increased by 61.1%, and the ISO index decreased by 0.6% after cities were classified as the city cluster. It means that cities are more productive, innovative, and have a more reasonable industrial structure.

(3) There is a difference in the effect of the positioning level of the city cluster on the reduction of carbon emissions. The effect of city cluster policies on carbon emission reduction is significant only in the national-level city cluster. The carbon emissions from the national-level city cluster declined by an average of 10.3%.

6.2. Policy Implications

This paper has the following three policy implications:

First, this paper finds that the city cluster will significantly reduce city carbon emissions. Therefore, the government should adopt a more diversified approach to air control. It can not only reduce air pollution through environmental regulation but also reduce carbon emissions by setting up the city cluster through city spatial layout. Policymakers should actively adhere to the city cluster model. They should not only continue to vigorously promote the development of mature city clusters in the Yangtze River Delta, Pearl River Delta, and Beijing–Tianjin–Hebei region, but also strengthen the concentration of city clusters in the Yangtze River, Chengdu–Chongqing, and Central Plains.

Second, based on the mechanism analysis, the city cluster can reduce city carbon emissions by improving productivity, improving innovation, and optimizing industrial structure. Therefore, the government can take "industrial transfer and innovation drive" as an opportunity to actively promote the transfer of traditional industries from core cities (or big cities) to non-core cities (or small and medium-sized cities). It can improve the efficiency of the utilization of production factors in the city cluster through specialization and industrial upgrading. At the same time, guide the core cities to build innovation systems. The government can achieve the goal of carbon emission reduction by optimizing the industrial structure of the city.

Third, based on the heterogeneity analysis, only the national-level city cluster can achieve the purpose of city emission reduction. Therefore, the government should set up more nation-level city clusters, rather than regional-level or world-level city clusters. It needs to further improve the resource support capacity of the national-level city cluster and promote the transformation and upgrading of enterprises through the introduction of talents and financial subsidies.

6.3. Limitations and Future Research Possibilities

However, this paper also has some limitations: on the one hand, the research sample of this paper is the data from the city level in China. In the future, when carbon emissions data at the corporate level becomes available, we can investigate the carbon effect of the city cluster policy from a micro-enterprise perspective. On the other hand, although this paper explores the carbon reduction effect of the city cluster policy, it does not examine its impact on human health. In the future, we will merge relevant micro-survey data to study the impact of city cluster policy on individual health.

Author Contributions: Conceptualization, X.L.; Data curation, C.H., S.Z. and Y.W.; Formal analysis, C.H.; Investigation, S.Z.; Methodology, X.L.; Project administration, X.L.; Supervision, Y.W.; Validation, S.Z.; Visualization, Y.W.; Writing—review & editing, C.H. and S.Z. All authors have read and agreed to the published version of the manuscript.

Funding: This research received no external funding.

Institutional Review Board Statement: Not applicable.

Informed Consent Statement: Not applicable.

Data Availability Statement: The data presented in this study are available on request from the corresponding author.

Conflicts of Interest: The authors declare no conflict of interest.

References

1. Mi, Z.; Wei, Y.M.; Wang, B.; Meng, J.; Liu, Z.; Shan, Y.; Liu, J.; Guan, D. Socioeconomic impact assessment of China's CO_2 emissions peak prior to 2030. *J. Clean. Prod.* **2017**, *142*, 2227–2236. [CrossRef]
2. Zhang, Y.J.; Peng, H.R.; Su, B. Energy rebound effect in China's Industry: An aggregate and disaggregate analysis. *Energy Econ.* **2017**, *61*, 199–208. [CrossRef]
3. Song, J.; Feng, Q.; Wang, X.; Fu, H.; Jiang, W.; Chen, B. Spatial association and effect evaluation of CO_2 emission in the Chengdu-Chongqing urban agglomeration: Quantitative evidence from social network analysis. *Sustainability* **2018**, *11*, 1. [CrossRef]
4. Li, X.; Wang, Y.; Zhou, H.; Shi, L. Has China's war on pollution reduced employment? Quasi-experimental evidence from the Clean Air Action. *J. Environ. Manag.* **2020**, *260*, 109851. [CrossRef]
5. Zhu, J.; Xu, J. Air pollution control and enterprise competitiveness–A re-examination based on China's Clean Air Action. *J. Environ. Manag.* **2022**, *312*, 114968. [CrossRef] [PubMed]
6. Wu, W.P.; Yan, J.J.; Chen, Y.H.; Wang, Z.J.; Lin, Y.R. Has environmental policy improved the job quality of migrant workers? A quasi-natural experiment on China's Clean Air Action. *J. Clean. Prod.* **2022**, *347*, 131231. [CrossRef]
7. Lin, B.; Ge, J. Valued forest carbon sinks: How much emissions abatement costs could be reduced in China. *J. Clean. Prod.* **2019**, *224*, 455–464. [CrossRef]
8. Zheng, Y.; Sun, X.; Zhang, C.; Wang, D.; Mao, J. Can emission trading scheme improve carbon emission performance? Evidence from China. *Front. Energy Res.* **2021**, *9*, 759572. [CrossRef]
9. Lin, B.; Wu, N. Will the China's carbon emissions market increase the risk-taking of its enterprises? *Int. Rev. Econ. Financ.* **2022**, *77*, 413–434. [CrossRef]
10. Yang, X.; Xu, J.; Zhu, M.; Yang, Y. Environmental regulation and corporate tax avoidance—Evidence from China. *PLoS ONE* **2022**, *17*, e0261037. [CrossRef]
11. Han, Y. Impact of environmental regulation policy on environmental regulation level: A quasi-natural experiment based on carbon emission trading pilot. *Environ. Sci. Pollut. Res.* **2020**, *27*, 23602–23615. [CrossRef] [PubMed]
12. Wang, H.; Zhang, R. Effects of environmental regulation on CO_2 emissions: An empirical analysis of 282 cities in China. *Sustain. Prod. Consum.* **2022**, *29*, 259–272. [CrossRef]

13. Borck, R.; Pflüger, M. Green cities? Urbanization, trade, and the environment. *J. Reg. Sci.* **2019**, *59*, 743–766. [CrossRef]
14. Song, M.; Wang, S.; Sun, J. Environmental regulations, staff quality, green technology, R&D efficiency, and profit in manufacturing. *Technol. Forecast. Soc. Change* **2018**, *133*, 1–14.
15. Xu, Q.; Dong, Y.X.; Yang, R.; Zhang, H.O.; Wang, C.J.; Du, Z.W. Temporal and spatial differences in carbon emissions in the Pearl River Delta based on multi-resolution emission inventory modeling. *J. Clean. Prod.* **2019**, *214*, 615–622. [CrossRef]
16. Chen, P.; Xie, R.; Lu, M. "Resource conservation" or "environmental friendliness": How do urban clusters affect total-factor ecological performance in China? *Int. J. Environ. Res. Public Health* **2020**, *17*, 4361. [CrossRef]
17. Glaeser, E.L.; Ponzetto, G.A.M.; Zou, Y. Urban networks: Connecting markets, people, and ideas. *Pap. Reg. Sci.* **2016**, *95*, 17–59. [CrossRef]
18. Huang, Y.; Li, L.; Yu, Y. Do urban agglomerations outperform non-agglomerations? A new perspective on exploring the eco-efficiency of Yangtze River Economic Belt in China. *J. Clean. Prod.* **2018**, *202*, 1056–1067. [CrossRef]
19. Boix, R.; Trullén, J. Knowledge, networks of cities and growth in regional urban systems. *Pap. Reg. Sci.* **2007**, *86*, 551–574. [CrossRef]
20. Bertinelli, L.; Black, D. Urbanization and growth. *J. Urban Econ.* **2004**, *56*, 80–96. [CrossRef]
21. Wu, F.; Liu, Z. Research on the mechanism of how city group drive economic growth: Empirical evidences from 16 cities of Yangtze River Delta. *Econ. Res. J.* **2008**, *11*, 126–136. (In Chinese)
22. Zhao, M.; Tan, L.; Zhang, W.; Ji, M.; Liu, Y.; Yu, L. Decomposing the influencing factors of industrial carbon emissions in Shanghai using the LMDI method. *Energy* **2010**, *35*, 2505–2510. [CrossRef]
23. Lu, Q.; Yang, H.; Huang, X.; Chuai, X.; Wu, C. Multi-sectoral decomposition in decoupling industrial growth from carbon emissions in the developed Jiangsu Province, China. *Energy* **2015**, *82*, 414–425. [CrossRef]
24. Duranton, G.; Puga, D. Nursery cities: Urban diversity, process innovation, and the life cycle of products. *Am. Econ. Rev.* **2001**, *91*, 1454–1477. [CrossRef]
25. Huallacháin, B.Ó.; Lee, D.S. Technological specialization and variety in urban invention. *Reg. Stud.* **2011**, *45*, 67–88. [CrossRef]
26. Lin, B.; Jia, H. Does the development of China's high-speed rail improve the total-factor carbon productivity of cities? *Transp. Res. Part D Transp. Environ.* **2022**, *105*, 103230. [CrossRef]
27. Li, P.; Lu, Y.; Wang, J. Does flattening government improve economic performance? Evidence from China. *J. Dev. Econ.* **2016**, *123*, 18–37. [CrossRef]
28. Heckman, J.J.; Ichimura, H.; Todd, P. Matching as an econometric evaluation estimator. *Rev. Econ. Stud.* **1998**, *65*, 261–294. [CrossRef]
29. Bertrand, M.; Duflo, E.; Mullainathan, S. How much should we trust differences-in-differences estimates? *Q. J. Econ.* **2004**, *119*, 249–275. [CrossRef]
30. Chen, F.; Zhao, T.; Wang, D. Research on China cities' total factor productivity of carbon emission: Based on decoupling effect. *Int. J. Environ. Res. Public Health* **2022**, *19*, 2007. [CrossRef]
31. Du, K.; Cheng, Y.; Yao, X. Environmental regulation, green technology innovation, and industrial structure upgrading: The road to the green transformation of Chinese cities. *Energy Econ.* **2021**, *98*, 105247. [CrossRef]
32. Lyu, L.; Sun, F.; Huang, R. Innovation-based urbanization: Evidence from 270 cities at the prefecture level or above in China. *J. Geogr. Sci.* **2019**, *29*, 1283–1299. [CrossRef]
33. Liu, J.; Hou, X.; Wang, Z.; Shen, Y. Study the effect of industrial structure optimization on urban land-use efficiency in China. *Land Use Policy* **2021**, *105*, 105390. [CrossRef]

energies

Article

Do Urbanization and Energy Consumption Change the Role in Environmental Degradation in the European Union Countries?

Bartosz Jóźwik *, Antonina-Victoria Gavryshkiv and Kinga Galewska

Department of International Economics, The John Paul II Catholic University of Lublin, Al. Racławickie 14, 20-950 Lublin, Poland

* Correspondence: bjozwik@kul.pl

Abstract: Nowadays, increased urbanization is visible in most European Union countries. At the same time, it can be noticed that in the studied period (2000–2018), GDP per capita increased, and CO_2 emissions per capita and energy consumption per capita decreased. These trends should be assessed in an unequivocally positive way. Considering these trends, especially with regard to economic development, our research goal is to answer the following questions: is there a long-run relationship between urbanization, energy consumption, economic growth, and carbon dioxide emissions, and what roles do urbanization and energy consumption play in the concept of the environmental Kuznets curve? This study aims to contribute to this growing area of research by exploring the European Union countries in the period covering the accession of new member states from Central Europe that needs intensifying European environmental policy. In order to test cointegration, we used Pedroni and Westerlund's panel tests. To estimate the long-run coefficients, we employed the FMOLS, MG, CCEMG, and AMG tests. Our findings confirmed the long-run relationship between variables. We find that urbanization has a high negative impact on carbon dioxide emissions per capita. Interestingly, our studies' results differ from those in most of the previously published articles about European countries. For this reason, our results provide a new insight for policymakers in European Union institutions.

Keywords: environmental Kuznets curve; carbon dioxide emissions; CO_2; urbanization; energy consumption; European Union

Citation: Jóźwik, B.; Gavryshkiv, A.-V.; Galewska, K. Do Urbanization and Energy Consumption Change the Role in Environmental Degradation in the European Union Countries? *Energies* **2022**, *15*, 6412. https://doi.org/10.3390/en15176412

Academic Editors: Junpeng Zhu and Xinlong Xu

Received: 26 July 2022
Accepted: 30 August 2022
Published: 2 September 2022

1. Introduction

Nowadays, over half of the world's inhabitants live in cities. The United Nations' prognoses point out that the total population in the world in 2050 will reach 9.31 billion, while the urban population will increase to 6.25 billion, and the urbanization index will be 67.2% [1]. It is largely the civilization advance, together with all the accompanying effects, which has made the population in cities grow dramatically. However, all these aspects and assumptions have consequences as far as the natural environment, the population growth, and the population distribution in particular areas of the globe, especially in cities, are concerned. In the European Union countries, urbanization is progressing continuously, extending into new regions. In the years 2000–2018, its increase was visible in most countries. The urbanization index dropped only in four countries: Slovak Republic, Austria, Cyprus, and Poland. Minor changes (less than 1%) in this respect occurred only in the Baltic countries (Estonia, Latvia, and Lithuania), the Czech Republic, and Belgium (Table A1 in Appendix A). Based on World Bank Statista data, in 2019, 75% of the population lived in cities and the suburbs of the European Union countries, while only 25% lived in rural areas. It is noteworthy that at the same time, a decrease in carbon dioxide emissions can be observed along with the process of growing urbanization.

In most countries, a decreasing carbon dioxide emissions tendency is seen in comparing the emissions in the years 2000 and 2018. For example, Luxembourg, the leader in this ranking in 2000, witnessed significant changes in the field of applied policy concerning emissions reduction. In this region, the estimated decrease in emissions was from 19.7 metric tons per capita in 2000 to 15.3 in 2018 (Table A1 in Appendix A). The visible changes (more than 1%), in this respect, proceed in another direction in the Baltic countries (Lithuania, Latvia, and Estonia). There, an increase in carbon dioxide emissions took place by 1.03, 1.13, and 1.50 metric tons per capita, respectively. However, it must be noted that in most countries, the ongoing changes are an element of the idea of climate neutrality, which is the aim of the European Union for the next decades.

The European Union treats the problem of climate change in a very emphatic way, and it undertakes activities in this direction. Prevention of those changes is one of its priority goals, reflected in the tasks designed for the decades to come, for example, through a reduction in greenhouse gas emissions [2]. The European Green Deal is a strategy for growth, transforming the economic and political union of 27 European democratic countries into places that are neutral to climate. The activities accompanying the major goal refer to significant aspects. Firstly, it is the establishment of a modern, resource-efficient, and competitive economy where there will be no net emissions of greenhouse gases in 2050. Secondly, it is a separation of economic growth from the use of resources. The third aspect refers to guaranteeing the protection and strengthening of neutral capital. Finally, the fourth, but nonetheless very important point, is to ensure citizens' health protection, security, and well-being, which is aimed at protecting them from the environmental effects of climate change.

Considering the formulated aims of climate neutrality, as well as the economic development and the progressing process of urbanization in the European Union countries, our main research goal is to answer the question, is there a long-run relationship between urbanization, energy consumption, economic growth, and the carbon dioxide emissions, and what roles do urbanization and energy consumption play in the concept of the environmental Kuznets curve? This study aims to contribute to this growing area of research by exploring the European Union countries in the period, which covers the accession of new member states from Central Europe. This enlargement needs intensifying cooperation between EU member states, especially in environmental policy. The relationship is tested using the concept of the environmental Kuznets curve, where, apart from carbon dioxide emissions and economic growth, urbanization and final energy consumption are considered. To this aim, the Pedroni and Westerlund panel cointegration tests are used. To estimate the long-run coefficients of the cointegration association, we employed the panel Fully Modified Ordinary Least Squares (FMOLS) test. To test the robustness of the estimation results, we used the Pesaran and Smith Mean Group (MG) estimator, the Pesaran Common Correlated Effects Mean Group (CCEMG), and Augmented Mean Group (AMG) estimators.

The remaining sections of this research are planned as follows. Section 2 presents a brief literature review on the relationship between urbanization, environmental degradation, and economic growth, analyzed within the environmental Kuznets curve (EKC) concept. Section 3 contains the data, model and empirical methodology. The research results and discussion are presented in Section 4. Section 5 concludes the research and provides policy recommendations.

2. Literature Review

Research on the effect of urbanization on the quality of the environment is frequently conducted using the concept of the environmental Kuznets curve, which appeared at the beginning of the 1990s in the work by Grossman and Krueger [3]. The Authors proved that the scale of environmental pollution is connected with the level of economic development of a given country. In the initial stages of economic development, an increase in the level of pollution related to the exploitation of natural resources also takes place, intending to create welfare. This tendency is reversed after a certain level of income (turning point) is trespassed. Then, the situation changes, and expenditures on environmental protection start to increase. The conclusions drawn by Grossman and Kruger became the basis for creating a model according to which the relationships between economic growth and the emissions of pollutants have an inverted U-shaped curve. In recent years, the popularity of the environmental Kuznets curve, which additionally used different variables, grew. A complex review of the literature in this area can be found, for instance, in Shahbaz and Sinha [4,5], Purcel [6], Koondhar et al. [7], and Xia et al. [8]. This article focuses on research that primarily considers the variables characterizing the urbanization process, urban population, and energy consumption.

It needs to be emphasized that this research was carried out in various regions and states with different levels of economic development, for instance, in emerging economies, developing countries, or developed countries. Most of those studies confirm the relationships defined by the environmental Kuznets curve, but the results of the effect of urbanization on the quality of the environment are not conclusive. For example, Maneejuk et al. [9] analyzed the relationship between GDP per capita, urbanization, financial development, the industrial sector, and the emissions of CO_2 for the Association of Southeast Asian Nations (ASEAN), the European Free Trade Association (EFTA), the European Union (EU), Group of Seven (G7), Gulf Cooperation Council (GCC), Mercosur, the North American Free Trade Agreement (NAFTA) and the Organization for Economic Co-operation and Development (OECD) in the years 2001–2016. The findings indicate that the EKC hypothesis is valid in only three out of eight international economic communities, namely, the EU, OECD, and G7. It follows from the research that urbanization, as well as financial development and the industrial sector, increase CO_2 emissions, while the use of renewable energy reduces degradation of the environment. In the case of urbanization, statistical significance and the highest positive effect were displayed by ASEAN (0.823), and then by GCC (0.563), Mercosur (0.553), UEU (0.123), and G7 (0.019). In the other groups, the effect of urbanization on CO_2 emissions was statistically insignificant.

Similar results were obtained by Wang et al. [1]. The Authors analyzed the effect of urbanization on economic growth and the quality of the environment in the period 1996–2015 based on data from 134 countries. Studies confirmed the occurrence of an inverted U-shaped relationship between economic growth and CO_2 emissions for the countries in the lower middle-income group, and a U-shaped relationship for the high-income group of countries. The Authors showed that the emissions of CO_2 increased together with increased urbanization. The same direction of the effect of urbanization on carbon dioxide emissions was defined by Sun Y. et al. [10], who conducted research on the Middle East and North African (MENA) economies.

However, it deserves to be pointed out that an increase in urbanization can increase the emissions of carbon dioxide only to a certain level, after which its further progress will reduce these emissions. Such relationships were confirmed in the studies by Gierałtowska et al. [11], who indicated that urbanization has an inverted U-shaped relationship with CO_2 emissions in the group covering 163 countries over the period from 2000 to 2016. This relationship can be confirmed by the results of studies obtained by Li and Haneklaus [12], where the Authors showed that increased urbanization decreases CO_2 emissions in the group of G7 countries in the years 1979 to 2019, and by Balsalobre-Lorente et al. [13] in the BRICS states in the years 1914–2014. Likewise, studies by Saidi and Mbarek [14], on the effect of urbanization, income, trade openness, and financial development on the carbon dioxide emissions in nineteen emerging economies during 1990–2013, indicate that urbanization decreases CO_2 emissions. According to the Authors, this is a powerful argument for politicians and city planners in shaping contemporary policies in those regions.

Previous studies conducted in European countries indicated the opposite results. Based on research carried out in 33 European countries and covering the period 1996–2017, Ali et al. [15] showed that urbanization together with economic growth, export, import, and energy consumption are the main factors that increase environmental degradation. The coefficients associated with urbanization are positive and statistically significant: 0.188 for model I, and 0.011 for model II. At the same time, the Authors point to energy innovation, which should help to reduce the rate of environmental degradation. A similar group of European countries (36) was studied by Wang et al. [16], who indicated a positive and significant effect of urbanization, as well as economic growth and foreign direct investment, on CO_2 emissions in the years 2000–2018. A slightly bigger group was examined by Khezri et al. [17]. The results of studies for 43 European countries between 1996 and 2018 also confirmed the relationship defined as the environmental Kuznets curve and urbanization's positive effect on carbon dioxide emissions (coefficients 0.659–0.760).

Comparable results, but for smaller groups of European countries, were obtained by Balsalobre-Lorente et al. [18]. The Authors studied the relationships between GDP per capita, urbanization, foreign direct investment, renewable energy consumption, and CO_2 emissions in Portugal, Ireland, Italy, Greece, and Spain, in the years 1990–2019. The study confirmed the relationship between economic growth and CO_2 emissions in the inverted U-shaped and N-shaped curves. The urbanization process increases the emissions of CO_2 in such a way that an increase in urbanization by 1% increases CO_2 emissions within the range from 0.44% to 6.36%, depending on the adopted model. Verbič et al. [19] conducted studies for the countries of South-Eastern Europe in the years 1997–2014. The evidence points to an inverted U-shaped relationship between GDP per capita and the emissions of carbon dioxide in the long run in the whole sample. Short-term estimates evidence the existence of EKC in the inverted U-shape only for Greece and Moldavia. The Authors pointed to a statistically significant positive influence of urbanization on CO_2 emissions (coefficient 1.057, FMOLS).

However, not all studies confirm the negative or positive effects of urbanization on the emissions of carbon dioxide. To give an example, no relation between urbanization and carbon dioxide emissions was indicated by Destek et al. [20]. Their research sample comprised Central European countries such as Albania, Bulgaria, Croatia, the Czech Republic, Macedonia, Hungary, Poland, Romania, Slovakia, and Slovenia. The main goal was to find the relationship between CO_2 emissions, urbanization, GDP per capita, energy consumption, and trade openness in the years 1991–2011. Studies confirmed the hypothesis of the environmental Kuznets curve in the sample. Results indicate a short-run two-directional causal relation between CO_2 and GDP per capita as well as between GDP per capita and energy consumption. There is no relation, however, between urbanization and carbon dioxide emissions. Similar results were obtained by Amin, et al. [21], who point out that urbanization in European countries does have a positive effect on environmental pollution, but it is statistically insignificant. Interestingly, the Authors saw a need to analyze the transport sector as a consequence of the process of urbanization. The Authors argue that transport significantly affects the air quality. They also point out that using renewable energy reduces carbon dioxide emissions from transportation. At the same time, they emphasize that necessary measures should be taken to increase ecological consciousness, especially among the urban population. In this process, it is important to promote environmentally friendly and energy-efficient means of transportation.

Although, the impact of urbanization on the environment in the European Union is related to the fact that some countries have undergone deindustrialization and offshored the environmental effects of their consumption to other countries. Research on industrialization's impact on carbon dioxide emissions mainly focuses on structural changes, where structural changes towards services, usually at higher levels of economic development, improve environmental quality [22–24]. Previous works, including Cherniwchan [25], and Raheem and Ogebe [26], have shown that industrialization is an important determinant in environmental quality changes. Another problem is offshoring the negative ecological impacts, which is often the result of differences in carbon prices in different regions. This phenomenon can lead to the production of energy-intensive goods into "carbon havens", thus creating a "carbon leakage". The observed industry relocation is a significant problem for the European Union and national policymakers [27].

3. Materials and Methods

We use the model that characterizes the relationships between economic development and the degree of environmental pollution. The first studies on these relationships included those by Grossman and Krueger [3,28], Shafik and Bandyopadhyay [29], Panayotou [30], and Selden and Song [31]. A fast increase in the number of studies led to the formulation of the concept of the environmental Kuznets curve, for example, see Gruszecki and Jóźwik [32]. It assumes a relationship between economic escalation (GDP per capita) and the level of nature contamination (e.g., due to carbon dioxide emissions), mostly in the inverted U-shaped curve. It happens because industrialization is followed by certain negative consequences (for example, pollution of man's natural environment), which grow to a certain point, after which they decrease, even though economic development proceeds. This, on the other hand, follows on from the fact that at a certain stage of advanced economic development, a change can be noticed in the mechanism of demands exhibited by consumers who then, to function, need more services and a cleaner environment. Technological progress also takes place, which does away with the negative effects of contamination of the surrounding world following economic development.

Considering the realization of our research goal, an important problem proves to be the aforementioned relationship described by the environmental Kuznets curve and the observed increase in urbanization and technological progress in the European Union countries. This relationship induces a search for the answer to the following question: is there a long-run relationship between urbanization, energy consumption, economic growth, and carbon dioxide emissions, and what roles do urbanization and energy consumption play in the concept of the environmental Kuznets curve? We use the econometric model with the urbanization variable to answer the questions. The model with the urbanization variable was employed, for example, by Kasman and Duman [33], Ozatac, Gokmenoglu, and Taspinar [34] as well as by Kirikkaleli and Kalmaz [35], Musa et al. [36], and Anwar et al. [37]. Our model will also consider final energy consumption.

The relationship between carbon dioxide emissions, GDP per capita, urban population (urbanization), and final energy consumption per capita is expressed in model I (Equation (1)). We also use model II (Equation (2)) for a robustness check where environmental degradation is proxied as greenhouse gas emissions, expressed in units of CO_2 equivalents. All variables are transformed into a natural logarithm format, to avoid multicollinearity issues, reduce the possible outliers from the dataset, as well as overcome the chances of data sharpness and normality [13].

$$\mathrm{LnCO_{2it}} = \beta_0 + \beta_1 \ln \mathrm{GDP} + \beta_2 (\ln \mathrm{GDP})^2 + \beta_3 \ln \mathrm{URB} + \beta_4 \ln \mathrm{ENC} + \mu_{it} \quad (1)$$

$$\mathrm{LnGHG_{it}} = \beta_0 + \beta_1 \ln \mathrm{GDP} + \beta_2 (\ln \mathrm{GDP})^2 + \beta_3 \ln \mathrm{URB} + \beta_4 \ln \mathrm{ENC} + \mu_{it} \quad (2)$$

where β—regression coefficients, CO_2—carbon dioxide emissions in metric tons per capita, GHG—greenhouse gas emissions per capita, GDP—gross domestic product per capita (constant 2015 USD), URB—urban population (% of total population), ENC—final energy consumption in tonnes of oil equivalent per capita, μ_{it}—error correction term. It should be pointed out that in many scientific studies, the standard measure of urbanization is the share of the population living in urban areas [38].

Before estimating the models, some preliminary tests need to be applied to the panel data. Figure 1 shows the entire research procedure. Initially, we test for cross-section dependence using the Pesaran CD-test [39]. Afterward, in order to discover whether the data of selected variables have stationarity or non-stationarity, we apply the Im–Pesaran–Shin panel unit root test [40] and the second-generation unit root test in the presence of cross-section dependence proposed by Pesaran [41]. Next, we test the long-run relationship (cointegration) among selected variables. To do this, we performed the Pedroni [42,43] and Westerlund [44] panel cointegration tests. These tests are recommended when inter-country convergence is confirmed [45]. The final step of the empirical analysis is estimating the long-run coefficients (elasticities) of the cointegration association concerning urbanization.

For that purpose, we employed the panel Fully Modified Ordinary Least Squares test, the Pesaran and Smith [46] Mean Group estimator, the Pesaran [47] Common Correlated Effects Mean Group, and Augmented Mean Group estimators.

Step 1	Cross-Sectional Dependence Test	Pesaran CD test (2004)
Step 2	Panel Unit Root Test	Im, Pesaran and Shin test (2003) Pesaran (2007)
Step 3	Panel Cointegration Tests	Pedroni (1999, 2004), Westerlund (2005) tests
Step 4	Long-Run Coefficients Estimation	FMOLS, MG, CCEMG, AMG

Figure 1. The model estimation method. Notes: FMOLS—Fully Modified Ordinary Least Squares test; MG—Pesaran and Smith Mean Group estimator; CCEMG—Pesaran Common Correlated Effects Mean Group estimator; AMG—Augmented Mean Group estimator.

Our study sample consists of 28 countries for which we have complete data for the 2000–2018 period (532 observations for each variable). All data were retrieved from the World Bank and Eurostat databases. Table 1 describes the variables and sources of data. Table 2 shows the summary statistics. It should be noted that the differences between the values of the variables are appreciable in our research sample. The CO_2 emissions range between 2.97 metric tons per capita in Latvia and 25.67 in Luxemburg, while GDP per capita ranges between USD 3668.65 in Bulgaria and 105,454.7 in Luxemburg. At the same time, we observe considerable differentiation in the urban population, where the smallest values occur in Slovenia (59.7), and the biggest in Belgium (98.0).

Table 1. Variables descriptions and sources of data.

Variable	Description	Data Source
	dependent variable	
CO_2	carbon dioxide emissions (metric tons per capita)	WDI
GHG	greenhouse gas emissions per capita. Indicator expressed in units of CO_2 equivalents in metric tons per capita	EEA
	Independent variable	
GDP	gross domestic product per capita (constant 2015 USD)	WDI
URB	urban population (% of total population)	WDI
ENC	tonnes of oil equivalent per capita	WDI

Notes: WDI—World Development Indicators; EEA—European Environment Agency.

Table 2. Summary statistics.

Variable	Mean	Std. Dev.	Min.	Max.	Variance	Kurt.
CO_2	8.104	3.975	2.927	25.669	15.799	6.260
GHG	10.426	4.157	4.3	30.8	17.278	8.304
GDP	29,163.460	19,691.830	3668.654	105,454.7	3.880	6.012
URB	72.140	12.444	50.754	98.001	154.861	2.129
ENC	2.487	1.372	0.930	9.630	1.883	12.648

Notes: CO_2—carbon dioxide emissions (metric tons per capita); GDP—GDP per capita (constant 2015 USD); URB—Urban population (% of total population); ENC—Energy final consumption (tonnes of oil equivalent) per capita.

Figure 2 presents changes in aggregated variables for the European Union countries in the years 2000–2018. In the examined period, a decrease in per capita CO_2 and greenhouse gas emissions, final energy consumption, and an increase in GDP per capita and urban population occurred. These trends should be assessed in a positive way. Another issue is a growing proportion of the population living in cities (urbanization), which is undoubtedly connected with the demographic changes, economic development, and technological advance observed in economically developed countries.

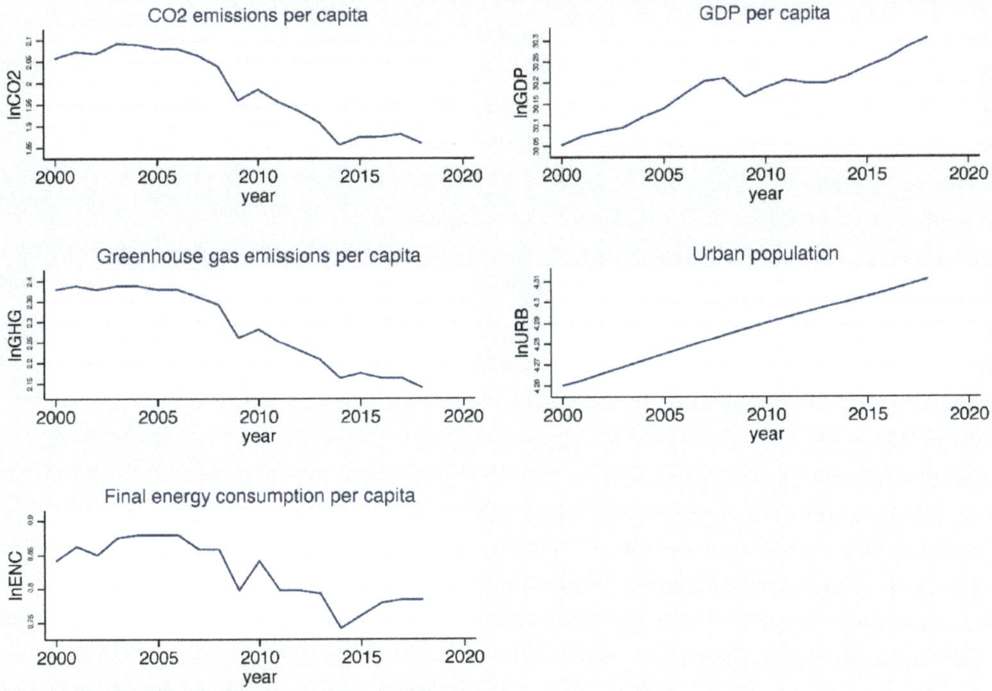

Figure 2. CO_2 emissions per capita, greenhouse gas emissions per capita, GDP per capita, urban population, and final energy consumption per capita in the European Union between 2000 and 2018.

4. Results and Discussion

As outlined above, the first step in our method is to observe whether series are generating common shocks in the long run. To this aim, we test for cross-section dependence in our panel time-series data. The outcomes of the Pesaran cross-sectional dependency test [39] are shown in Table 3. The test results rejected the null hypothesis and confirmed the presence of cross-country dependency, which is not unexpected because the European Union countries share a common market and economic policy. A number of the conducted studies point to systematic economic convergence between these countries in recent years, for example, see Jóźwik [48] or Bernardelli et al. [49]. Because of this convergence, one country's economic and environmental transformations can easily be transferred to its neighboring countries. Therefore, we need to use a proper stationarity approach to circumvent the common effect and provide reliable results [45].

In the second step, we identify the order of integration of the variables by employing the Im–Pesaran–Shin panel unit root test. We subtracted the cross-sectional averages from the series and requested that the number of lags of the series be chosen in such a way that the AIC for the regression is minimized (max AIC is four). The stationarity test results in Table 4 confirm that the data series is unstable at this level. However, after considering the first difference, the test confirmed that the series became stationary at the 1% significance.

Table 3. Results of cross-sectional dependency Pesaran test.

Variable	Cd-Test	p-Value	Corr	Abs (Corr)
$\ln CO_2$	45.12 ***	0.000	0.532	0.666
lnGHG	40.12 ***	0.000	0.473	0.711
lnGDP	58.27 ***	0.000	0.688	0.768
lnGDPsq	58.26 ***	0.000	0.687	0.768
lnURB	26.65 ***	0.000	0.314	0.843
lnENC	24.48 ***	0.000	0.289	0.565

Note: Under the null hypothesis of cross-section independence CD ~ $n(0,1)$. *** denotes statistical significance at the 1% level.

Table 4. Im–Pesaran–Shin Panel unit root test (W-t-bar statistics).

Variables	Drift and No Trend	p-Value	Drift and Trend	p-Value
		at level		
$\ln CO_2$	−0.8740	0.1910	−3.8365 ***	0.0001
lnGHG	−0.6831	0.2473	−4.6078 ***	0.0000
lnGDP	−0.0958	0.4618	−1.2311	0.1091
lnGDPsq	−0.2414	0.4046	−1.4185 **	0.0780
lnURB	−0.7913	0.2144	4.1977	1.0000
lnENC	−0.7915	0.2143	−4.6526 ***	0.0000
		at 1st difference		
$\ln CO_2$	−15.6464 ***	0.0000	−10.9659 ***	0.0000
lnGHG	−14.9536 ***	0.0000	−10.1475 ***	0.0000
lnGDP	−6.6989 ***	0.0000	−4.1460 ***	0.0000
lnGDPsq	−6.7379 ***	0.0000	−4.5406 ***	0.0000
lnURB	−2.4104 ***	0.0080	−6.2014 ***	0.0000
lnENC	−14.1526 ***	0.0000	−8.6068 ***	0.0000

Notes: H0: All panels contain unit roots. Ha: Some panels are stationary. Cross-sectional means removed. Max AIC is 4. The number of lags of the series is chosen in such a way that the AIC for the regression is minimized. **, *** denote statistical significance at the 5% and 1% levels, respectively.

In addition, we employed the panel second generation unit root test in the presence of cross-section dependence proposed by Pesaran [41]. We assume that the serial correlation order to be tasted with the Breusch–Godfrey Lagrange multiplier test in each regression is one, and the number of lags is four. Table 5 displays results for two deterministic models' specifications: with individual-specific intercepts and incidental linear trends. The test results confirm that the variables are stationary at the first difference, almost all at the 1% significance.

Table 5. Pesaran panel unit root test in the presence of cross-section dependence.

Variables	At Level		At 1st Difference	
	Individual-Specific Intercepts	Incidental Linear Trends	Individual-Specific Intercepts	Incidental Linear Trends
$\ln CO_2$	−2.065	−2.801 **	−4.291 ***	−4.334 ***
lnGHG	−2.202 **	−3.034 ***	−4.331 ***	−4.460 ***
lnGDP	−2.023 **	−2.183 **	−3.024 ***	−3.068 ***
lnGDPsq	−1.978	−2.160 **	−2.941 ***	−3.060 ***
lnURB	−0.903	−1.126	−2.174 **	−3.680 ***
lnENC	−1.436	−2.908 ***	−4.205 ***	−4.171 ***

Notes: critical values: at ***—1% significant level is −2.32 and at **—5% is −2.15; the serial correlation order to be tasted with the Breusch–Godfrey Lagrange multiplier test in each individual regression is 1; the number of lags is 4.

As noted previously, environmental degradation can be proxied in various ways. We selected CO_2 emissions as a proxy for environmental degradation in the model I, but for a robustness check, we also used the greenhouse gas emissions per capita variable. In this respect, we checked the cointegration (long-run relationship) between variables using the Pedroni and Westerlund tests. The tests have a common null hypothesis of no cointegration.

The alternative hypothesis of the Pedroni tests is that the variables are cointegrated in all panels. In the version of the Westerlund test in which the AR parameter is panel specific, the alternative hypothesis is that the variables are cointegrated in some of the panels. In the version of the Westerlund test in which the AR parameter is the same over the panels, the alternative hypothesis is that the variables are cointegrated in all the panels. In the Pedroni tests, we subtracted the cross-sectional averages from the series and requested that the number of lags of the series be chosen in such a way that the AIC for the regression is minimized (max AIC is four), as in the panel unit root tests calculations. Table 6 reports results for cointegration where six out of seven Pedroni tests confirm cointegration in Model I and Model II. The results of the Westerlund tests indicate that the variables are cointegrated in some of the panels.

Table 6. Pedroni and Westerlund panel cointegration tests results.

Tests	Model I	Model II
	Pedroni test AR parameter: Same	
Modified variance ratio	−3.2956 ***	−3.5891 ***
Modified Phillips–Perron t	0.9245	1.1932
Phillips–Perron t	−8.0501 ***	−6.8419 ***
Augmented Dickey–Fuller t	−10.4652 ***	−10.1033 ***
	Pedroni test AR parameter: Panel specific	
Modified Phillips–Perron t	3.0399 ***	3.1541 ***
Phillips–Perron t	−8.6913 ***	−7.3555 ***
Augmented Dickey–Fuller t	−12.4472 ***	−11.0408 ***
	Westerlund test AR parameter: Same	
Variance ratio	−1.2798	−1.3705 *
	Westerlund test AR parameter: Panel specific	
Variance ratio	−2.5575 ***	−2.6808 ***

Notes: Westerlund test AR parameter: Same. Ha: All panels are cointegrated; Panel specific. Ha: Some panels are cointegrated. *, *** denote statistical significance at the 10% and 1% levels, respectively.

In the final step, we estimated the coefficients of Equations (1) and (2). Table 7 provides the FMOLS test results. To test the robustness of the estimated results, we used the Pesaran and Smith Mean Group estimator, the Pesaran Common Correlated Effects Mean Group, and the Augmented Mean Group estimators. These tests, which concern with correlation across panel members (cross-section dependence), were introduced by Eberhardt and Teal [50] and Bond and Eberhardt [51]. The advantage of using these estimators is that they are designed for 'moderate-T and moderate-N' macro panels. These results are presented in Table 8.

Table 7. Panel FMOLS test results.

Variable	Coefficient	t-Stat	*p*-Value
	model I		
lnGDP	29.81 ***	813.83	$p < 0.00001$
lnGDPsq	−1.41 ***	−854.28	$p < 0.00001$
lnURB	−5.02 ***	−123.71	$p < 0.00001$
lnENC	1.05 ***	248.03	$p < 0.00001$
	model II		
lnGDP	17.65 ***	702.78	$p < 0.00001$
lnGDPsq	−0.82 ***	−752.76	$p < 0.00001$
lnURB	−6.54 ***	−144.88	$p < 0.00001$
lnENC	0.83 ***	209.09	$p < 0.00001$

Notes: The number of observations is 532. *p*-value for two-tailed hypothesis. *** denotes statistical significance at the 1% level.

Table 8. Mean Group (MG), Augmented Mean Group (AMG), and Common Correlated Effects Mean Group (CCEMG) estimation results.

Test	Coefficient				
	lnGDP	**lnGDPsq**	**lnURB**	**lnENC**	**Const.**
			model I		
MG	44.505 ** (0.023)	−2.038 ** (0.024)	−4.289 *** (0.002)	0.994 *** (0.000)	−223.513 ** (0.038)
AMG	25.251 * (0.076)	−1.176 * (0.083)	−3.688 (0.205)	0.907 *** (0.000)	−118.128 (0.121)
CCEMG	−10.037 (0.350)	0.511 (0.302)	−6.895 (0.159)	0.951 *** (0.000)	57.116 (0.330)
			model II		
MG	28.605 ** (0.041)	−1.290 ** (0.047)	−6.067 *** (0.004)	0.8170 *** (0.000)	−130.276 * (0.087)
AMG	29.452 ** (0.019)	−1.375 ** (0.021)	−10.449 (0.194)	0.672 *** (0.000)	−109.249 (0.076)
CCEMG	−10.037 (0.350)	0.511 (0.302)	−6.895 (0.159)	0.951 *** (0.000)	57.116 (0.330)

Notes: numbers in parentheses are *p*-value. ***, **, * denote statistical significance at the 1%, 5%, and 10% level, respectively.

As can be seen from Tables 7 and 8, in both models, the significant coefficients of the real GDP per capita are positive, whereas those of the squared GDP per capita are negative. It means that the long-run linkage between CO_2 emissions per capita and GDP per capita is an inverted U-shape implying that the environmental Kuznets curve concept is verified for the whole group of the European Union countries. The economic growth and development are supportive of carbon emissions in that region. Similar results of studies for the group of European countries were recently obtained by Destek et al. [20], Maneejuk et al. [9], and Verbič et al. [19], as well as by Balsalobre-Lorente et al. [18]. Despite significant technological advances in the European Union countries, energy consumption still positively influences carbon emissions per capita. Our results are similar to the papers mentioned earlier, as well as to those that have been published recently, namely, Khezri et al. [17], Kar [52], and Mohsin et al. [53].

Interestingly, all results from Tables 7 and 8 indicate that urbanization negatively impacts carbon emissions per capita. However, there are differences in the significance level of the coefficients depending on the method used. Nevertheless, this shows that urbanization has an essential effect on environmental protection in the European Union area, nowadays. For example, the results of FMOLS for model I show that a 1% increase in a share of the urban population decreases emissions per capita by 5.02% if all other variables remain the same. We want to highlight that our study on the urbanization process' effect on carbon dioxide emissions points to different results than many studies mentioned in the literature review section. As we remember, the significant results indicate that urbanization positively impacted the carbon dioxide emissions in different groups of European countries. To give an example, in an article by Ali et al. [15], coefficients are positive and equal to 0.188 and 0.011; in the study by Balsalobre-Lorente et al. [18], between 0.44 and 6.36; while in the studies by Destek et al. [20] and Amin et al. [21] there was no relationship. This difference probably results from a few reasons, some of which include research methods, samples, and periods. Another reason can be related to trespassing the threshold after which both increased income per capita and the coefficient of urbanization give rise to improved quality of the environment, which was indicated by Gierałtowska et al. [11]. This effect can be enhanced by the deindustrialization process we wrote about in the literature review. Dong et al. [54] highlighted that from the perspective of income level, industrialization contributes to the growth in carbon emissions. The effect of industrialization on CO_2 emissions gradually increases in the low- and intermediate-income levels. Azam et al. [55] also state that the industrialization process in OPEC economies increases environmental pollution, while the impacts on income are the opposite. However, the effect of industrialization begins to weaken at the high-income level according to research conducted by Dong et al. [54]. Probably this effect is observed in the European Union countries with a high-income level. Furthermore, economic development supports human capital, which significantly improves environmental quality [56]. Thus, our results indicate that studies in this area should be extended to different research models and methods.

5. Conclusions and Recommendations

In our research, we took into consideration two trends. First is the urbanization process, which increased the urban population in most European Union countries in years 2000–2018, and the second trend is a decrease in carbon dioxide emissions, which is indirectly the consequence of technological advances and the applied European climate policy. Considering these two trends, our research goal was to answer the following question: is there a long-run relationship between urbanization, energy consumption, economic growth, and carbon dioxide emissions, and what roles do urbanization and energy consumption play in the concept of the environmental Kuznets curve in European Union countries? We used the data from 28 European Union countries to assess the relationships. Our findings confirmed the long-run relationship between variables. We validated the environmental Kuznets curve hypothesis, indicating that economic growth has an inverted U-shaped effect on CO_2 emissions.

However, energy consumption still positively influences carbon emissions per capita, even though European Union countries have made significant economic and technological progress. At the same time, urbanization has a highly negative impact on carbon dioxide emissions per capita. If all other variables remain the same, a 1% increase in a share of the urban population decreases CO_2 emissions per capita by 5.02%. The result of our study is different from the results in the majority of earlier published articles. This difference probably arises from a few reasons. One of them may be the fact that the threshold after which both an increase in income per capita and urban population causes a decrease in carbon dioxide emissions in European Union countries has been trespassed in recent years.

Our results provide new insights for policymakers in European Union institutions. The findings suggest that the European policy should support the process of urbanization in a complex manner to fulfill the European Green Deal and achieve the Sustainable Development Goals related to improving environmental quality, especially by promoting urbanization with a low-carbon infrastructure and transport (smart technology and energy-efficient hybrid vehicles). A positive coefficient associated with energy consumption indicates that local authorities should support the development of home renewable energy infrastructure, for example, energy-efficient electric appliances and solar energy. Another important practical implication is related to human capital. The urban population can be motivated to adopt a sustainable lifestyle, including energy-saving, renewable energy sources, and public transportation [56]. It is very important in this context that urbanization be carried out according to environmental norms, possibly without social compromises in this respect. In addition, modern technological solutions enable the development of intelligent cities that are environmentally neutral.

However, we only conducted a preliminary empirical analysis of the relationship between environmental degradation and urbanization, and our study has a few limitations. The first limitation refers to sample size. The sample covers the period 2000–2018, this means we should be cautious in generalizing the findings. Second, although we have robust results using an alternative measure of environmental degradation, the two proxies (CO_2 and greenhouse emissions) might limit the ecological degradation effects. Additionally, it would be interesting to examine the consumption environmental impacts offshored to other countries and the deindustrialization processes. Third, we did not divide the European Union countries, for example, into less developed countries (Central European countries) and developed countries (Western European countries) to make a comparative analysis. These limitations could be addressed in future research.

Undoubtedly, in further research, we must also remember that climate neutrality is a global challenge. This means that it requires international dialogue and cooperation between the states. Although the pressure applied usually refers to particular countries and their economic structures, international activity is also an issue that plays a predominant role. It is especially important due to the necessity of creating a synergy between the European and international climate initiatives. For this reason, understanding that adaptation to climate changes is important; however, this is not in itself the aim, but rather a principle. It should, however, be a component of properly functioning and developing countries and societies.

Author Contributions: Conceptualization, B.J. and K.G.; methodology, B.J. and A.-V.G.; software, B.J.; formal analysis and investigation, B.J., K.G. and A.-V.G.; writing—original draft preparation, B.J., K.G. and A.-V.G.; writing—review and editing, B.J.; visualization, B.J.; project administration, B.J. All authors have read and agreed to the published version of the manuscript.

Funding: The APC was funded by the John Paul II Catholic University of Lublin.

Institutional Review Board Statement: Not applicable.

Informed Consent Statement: Not applicable.

Data Availability Statement: Publicly available datasets were analyzed in this study. This data can be found here: https://data.worldbank.org and https://ec.europa.eu/eurostat.

Conflicts of Interest: The authors declare no conflict of interest.

Appendix A

Table A1. CO_2 emissions per capita and urban population in the European Union countries.

Country	CO_2 Emissions (Metric Tons per Capita)		Urban Population (% of Total Population)	
	2000	2018	2000	2018
Austria	17,690	15,476	60,213	58,297
Belgium	11,441	8180	97,129	98,001
Bulgaria	5303	5855	68,899	75,008
Croatia	4040	4056	53,428	56,947
Cyprus	7495	6079	68,648	66,810
Czech Rep.	12,011	9641	73,988	73,792
Denmark	12,011	9641	85,100	87,874
Estonia	10,609	12,103	69,368	68,880
Finland	10,645	8043	82,183	85,382
France	6127	4619	75,871	80,444
Germany	10,097	8558	74,965	77,312
Greece	8742	6083	72,716	79,058
Hungary	5350	4746	64,575	71,351
Ireland	11,201	7624	59,155	63,170
Italy	7662	5376	67,222	70,438
Latvia	2927	3959	68,067	68,142
Lithuania	3003	4137	66,986	67,679
Luxembourg	19,665	15,330	84,216	90,981
Malta	5460	3198	92,368	94,612
Netherlands	10,191	8773	76,795	91,490
Poland	7729	8235	61,716	60,058
Portugal	5992	4841	54,399	65,211
Romania	3960	3845	53,004	53,998
Slovak Rep.	7065	6059	56,233	53,726
Slovenia	7310	6775	50,754	54,541
Spain	7230	5520	76,262	80,321
Sweden	6005	3538	84,026	87,431
United Kingdom	9001	5399	78,651	83,398

Source: World Development Indicators.

References

1. Wang, Q.; Wang, X.; Li, R. Does urbanization redefine the environmental Kuznets curve? An empirical analysis of 134 Countries. *Sustain. Cities Soc.* **2022**, *76*, 103382. [CrossRef]
2. Lapinskienė, G.; Peleckis, K.; Slavinskaitė, N. Energy consumption, economic growth and greenhouse gas emissions in the European Union countries. *J. Bus. Econ. Manag.* **2017**, *18*, 1082–1097. [CrossRef]
3. Grossman, G.M.; Krueger, A.B. Environmental Impacts of a North American Free Trade Agreement. NBER Working Papers. 1991. No. 3914. Available online: http://www.nber.org/papers/w3914 (accessed on 20 June 2022). [CrossRef]
4. Shahbaz, M.; Sinha, A. Environmental Kuznets Curve for CO_2 Emission: A Literature Survey. *J. Econ. Stud.* **2019**, *46*, 106–168. [CrossRef]
5. Sarkodie, S.A.; Strezov, V. A review on environmental Kuznets curve hypothesis using bibliometric and meta-analysis. *Sci. Total Environ.* **2019**, *649*, 128–145. [CrossRef] [PubMed]
6. Purcel, A.-A. New insights into the environmental Kuznets curve hypothesis in developing and transition economies: A literature survey. *Environ. Econ. Policy Stud.* **2020**, *22*, 585–631. [CrossRef]
7. Koondhar, M.A.; Shahbaz, M.; Memon, K.A.; Ozturk, I.; Kong, R. A visualization review analysis of the last two decades for environmental Kuznets curve "EKC" based on co-citation analysis theory and pathfinder network scaling algorithms. *Environ. Sci. Pollut. Res.* **2021**, *28*, 16690–16706. [CrossRef]
8. Xia, L.; Wang, Z.; Du, S.; Tian, D.; Zhao, S. New insight into the Bibliometric Analysis on the Topic of Environmental Kuznets Curve. *E3s Web Conf.* **2022**, *350*, 01003. [CrossRef]
9. Maneejuk, N.; Ratchakom, S.; Maneejuk, P.; Yamaka, W. Does the Environmental Kuznets Curve Exist? An International Study. *Sustainability* **2020**, *12*, 9117. [CrossRef]
10. Sun, Y.; Li, H.; Andlib, Z.; Genie, M.G. How do renewable energy and urbanization cause carbon emissions? Evidence from advanced panel estimation techniques. *Renew. Energ.* **2022**, *185*, 996–1005. [CrossRef]
11. Gierałtowska, U.; Asyngier, R.; Nakonieczny, J.; Salahodjaev, R. Renewable Energy, Urbanization, and CO_2 Emissions: A Global Test. *Energies* **2022**, *15*, 3390. [CrossRef]
12. Li, B.; Haneklaus, N. Reducing CO_2 emissions in G7 countries: The role of clean energy consumption, trade openness and urbanization. *Energy Rep.* **2022**, *8*, 704–713. [CrossRef]
13. Balsalobre-Lorente, D.; Driha, O.M.; Halkos, G.; Mishra, S. Influence of growth and urbanization on CO_2 emissions: The moderating effect of foreign direct investment on energy use in BRICS. *Sustain. Dev.* **2022**, *30*, 227–240. [CrossRef]
14. Saidi, K.; Mbarek, M.B. The impact of income, trade, urbanization, and financial development on CO_2 emissions in 19 emerging economies. *Environ. Sci. Pollut. Res.* **2017**, *24*, 12748–12757. [CrossRef] [PubMed]
15. Ali, M.; Raza, S.A.; Khamis, B. Environmental degradation, economic growth, and energy innovation: Evidence from European countries. *Environ. Sci. Pollut. Res.* **2020**, *27*, 28306–28315. [CrossRef] [PubMed]
16. Wang, J.; Mamkhezri, J.; Khezri, M.; Karimi, M.S.; Khan, Y.A. Insights from European nations on the spatial impacts of renewable energy sources on CO_2 emissions. *Energy Rep.* **2022**, *8*, 5620–5630. [CrossRef]
17. Khezri, M.; Karimi, M.S.; Khan, Y.A.; Khodaei, M. Environmental implications of regional financial development on air pollution: Evidence from European countries. *Environ. Dev. Sustain.* **2022**, 1–21. [CrossRef]
18. Balsalobre-Lorente, D.; Ibáñez-Luzón, L.; Usman, M.; Shahbaz, M. The environmental Kuznets curve, based on the economic complexity, and the pollution haven hypothesis in PIIGS countries. *Renew. Energy* **2022**, *185*, 1441–1455. [CrossRef]
19. Verbič, M.; Satrovic, E.; Muslija, A. Environmental Kuznets curve in Southeastern Europe: The role of urbanization and energy consumption. *Environ. Sci. Pollut. Res.* **2021**, *28*, 57807–57817. [CrossRef]
20. Destek, M.A.; Balli, E.; Manga, M. The relationship between CO_2 emission, energy consumption, urbanization and trade openness for selected CEECs. *Res. World Econ.* **2016**, *7*, 52–58. [CrossRef]
21. Amin, A.; Altinoz, B.; Dogan, E. Analyzing the determinants of carbon emissions from transportation in European countries: The role of renewable energy and urbanization. *Clean Technol. Environ. Policy* **2020**, *22*, 1725–1734. [CrossRef]
22. Duro, J.A.; Alcántara, V.; Padilla, E. International inequality in energy intensity levels and the role of production composition and energy efficiency: An analysis of OECD countries. *Ecol. Econ.* **2010**, *69*, 2468–2474. [CrossRef]
23. Mulder, P.; De Groot, H.L.; Pfeiffer, B. Dynamics and determinants of energy intensity in the service sector: A cross-country analysis, 1980–2005. *Ecol. Econ.* **2014**, *100*, 1–15. [CrossRef]
24. Du, X.; Xie, Z. Occurrence of turning point on environmental Kuznets curve in the process of (de) industrialization. *Struct. Chang. Econ. Dyn.* **2020**, *53*, 359–369. [CrossRef]
25. Cherniwchan, J. Economic growth, industrialization, and the environment. *Resour. Energy Econ.* **2012**, *34*, 442–467. [CrossRef]
26. Raheem, I.D.; Ogebe, J.O. CO_2 emissions, urbanization and industrialization: Evidence from a direct and indirect heterogeneous panel analysis. *Manag. Environ. Qual.* **2017**, *28*, 851–867. [CrossRef]
27. Aus dem Moore, N.; Grosskurth, P.; Themann, M. Multinational corporations and the EU Emissions Trading System: The specter of asset erosion and creeping deindustrialization. *J. Environ. Econ. Manag.* **2019**, *94*, 1–26. [CrossRef]
28. Grossman, G.M.; Krueger, A.B. Economic growth and the environment. *Q. J. Econ.* **1995**, *110*, 353–377. [CrossRef]
29. Shafik, N.; Bandyopadhyay, S. *Economic Growth and Environmental Quality Time-Series and Cross-Country Evidence*; Background Paper for the World Development Report 1992; The World Bank: Washington, DC, USA, 1992.
30. Panayotou, T. *Empirical Tests and Policy Analysis of Environmental Degradation at Different Stages of Economic Development*; World Employment Programme Research, Working Paper No 238; International Labor Office: Geneva, Switzerland, 1993; pp. 1–42.

31. Selden, T.; Song, D. Environmental quality and development: Is there a Kuznets curve for air pollution emissions. *J. Environ. Econ. Manag.* **1994**, *27*, 147–162. [CrossRef]
32. Gruszecki, L.; Jóźwik, B. Theoretical reconstruction of the environmental Kuznets curve. Gospodarka Narodowa. *Pol. J. Econ.* **2019**, *299*, 95–117. [CrossRef]
33. Kasman, A.; Duman, Y.S. CO_2 emissions, economic growth, energy consumption, trade and urbanization in new EU member and candidate countries: A panel data analysis. *Econ. Model.* **2015**, *44*, 97–103. [CrossRef]
34. Ozatac, N.; Gokmenoglu, K.K.; Taspinar, N. Testing the EKC hypothesis by considering trade openness, urbanization, and financial development: The case of Turkey. *Environ. Sci. Pollut. Res.* **2017**, *24*, 16690–16701. [CrossRef] [PubMed]
35. Kirikkaleli, D.; Kalmaz, D.B. Testing the moderating role of urbanization on the environmental Kuznets curve: Empirical evidence from an emerging market. *Environ. Sci. Pollut. Res.* **2020**, *27*, 38169–38180. [CrossRef] [PubMed]
36. Musa, K.S.; Maijama'a, R.; Yakubu, M. The causality between urbanization, industrialization and CO_2 emissions in Nigeria: Evidence from Toda and Yamamoto Approach. *Energy Econ. Lett.* **2021**, *8*, 1–14. [CrossRef]
37. Anwar, A.; Sinha, A.; Sharif, A.; Siddique, M.; Irshad, S.; Anwar, W.; Malik, S. The nexus between urbanization, renewable energy consumption, financial development, and CO_2 emissions: Evidence from selected Asian countries. *Environ. Dev. Sustain.* **2022**, *24*, 6556–6576. [CrossRef]
38. Aller, C.; Ductor, L.; Grechyna, D. Robust determinants of CO_2 emissions. *Energy Econ.* **2021**, *96*, 105154. [CrossRef]
39. Pesaran, M.H. General Diagnostic Tests for Cross Section Dependence in Panels' IZA; Discussion Paper No. 1240; 2004. Available online: https://www.iza.org/publications/dp/1240/general-diagnostic-tests-for-cross-section-dependence-in-panels (accessed on 5 July 2022).
40. Im, K.S.; Pesaran, M.H.; Shin, Y. Testing for unit roots in heterogeneous panels. *J. Econom.* **2003**, *115*, 53–74. [CrossRef]
41. Pesaran, M.H. A Simple Panel Unit Root Test in The Presence Of Cross-section Dependence. *J. Appl. Econom.* **2007**, *22*, 265–312. [CrossRef]
42. Pedroni, P. Critical values for cointegration tests in heterogeneous panels with multiple regressors. *Oxf. Bull. Econ. Stat.* **1999**, *61*, 653–670. [CrossRef]
43. Pedroni, P. Panel cointegration: Asymptotic and finite sample properties of pooled time series tests with an application to the PPP hypothesis. *Econ. Theory* **2004**, *20*, 597–625. [CrossRef]
44. Westerlund, J. New simple tests for panel cointegration. *Econom. Rev.* **2005**, *24*, 297–316. [CrossRef]
45. Sharma, R.; Shahbaz, M.; Kautish, P.; Vo, X.V. Analyzing the impact of export diversification and technological innovation on renewable energy consumption: Evidences from BRICS nations. *Renew. Energy* **2021**, *178*, 1034–1045. [CrossRef]
46. Pesaran, M.H.; Smith, R.P. Estimating long-run relationships from dynamic heterogeneous panels. *J. Econom.* **1995**, *68*, 79–113. [CrossRef]
47. Pesaran, M.H. Estimation and inference in large heterogeneous panels with a multifactor error structure. *Econometrica* **2006**, *74*, 967–1012. [CrossRef]
48. Jóźwik, B. *Realna Konwergencja Gospodarcza Państw Członkowskich Unii Europejskiej z Europy Środkowej i Wschodniej: Transformacja, Integracja i Polityka Spójności*; Wydawnictwo Naukowe PWN: Warszawa, Poland, 2017.
49. Bernardelli, M.; Próchniak, M.; Witkowski, B. Towards the Similarity of the Countries in Terms of Business Cycle Synchronization and Income Level Equalization–Empirical Analysis. *J. Adm. Bus. Stud.* **2019**, *5*, 291–302. [CrossRef]
50. Teal, F.; Eberhardt, M. *Productivity Analysis in Global Manufacturing Production*; Economics Series Working Papers, 515; University of Oxford: Oxford, UK, 2010.
51. Eberhardt, M.; Bond, S. *Cross-Section Dependence in Nonstationary Panel Models: A Novel Estimator*; MPRA Paper, No. 17692; University Library of Munich: Munich, Germany, 2009. Available online: https://mpra.ub.uni-muenchen.de/id/eprint/17692 (accessed on 5 July 2022).
52. Kar, A.K. Environmental Kuznets curve for CO_2 emissions in Baltic countries: An empirical investigation. *Environ. Sci. Pollut Res.* **2022**, *29*, 47189–47208. [CrossRef] [PubMed]
53. Mohsin, M.; Naseem, S.; Sarfraz, M.; Azam, T. Assessing the effects of fuel energy consumption, foreign direct investment and GDP on CO_2 emission: New data science evidence from Europe & Central Asia. *Fuel* **2022**, *314*, 123098. [CrossRef]
54. Dong, F.; Wang, Y.; Su, B.; Hua, Y.; Zhang, Y. The process of peak CO_2 emissions in developed economies: A perspective of industrialization and urbanization. *Resour. Conserv. Recyc.* **2019**, *141*, 61–75. [CrossRef]
55. Azam, M.; Rehman, Z.U.; Ibrahim, Y. Causal nexus in industrialization, urbanization, trade openness, and carbon emissions: Empirical evidence from OPEC economies. *Environ. Dev. Sustain.* **2022**, 1–21. [CrossRef]
56. Ahmed, Z.; Zafar, M.W.; Ali, S. Linking urbanization, human capital, and the ecological footprint in G7 countries: An empirical analysis. *Sustain. Cities Soc.* **2020**, *55*, 102064. [CrossRef]

energies

MDPI

Article

The Interaction Mechanism of Tourism Carbon Emission Efficiency and Tourism Economy High-Quality Development in the Yellow River Basin

Shuxiao Li [1], Zhanhong Cheng [2], Yun Tong [1] and Biao He [1,3,4,*]

1 School of Tourism, Hainan University, Haikou 570228, China
2 College of Culture Tourism, Shanxi University of Finance and Economics, Taiyuan 030031, China
3 Hainan Province Holistic Tourism Research Base, Haikou 570228, China
4 Academician Desheng Wu Station of Hainan Province, Haikou 570228, China
* Correspondence: hebiao@hainanu.edu.cn

Abstract: Exploring the relationship between the tourism carbon environment and high-quality economic development in the Yellow River Basin is a national strategy to meet the realistic requirements of the goal of "Carbon Peak and Carbon Neutral". It is also conducive to the realization of "Ecological Protection and High-quality Development Strategy in the Yellow River Basin". Therefore, based on the calculation of tourism's carbon emission efficiency and the evaluation of the tourism economy's high-quality development, the interaction mechanism between them was observed. The results showed that, firstly, the tourism carbon emission efficiency of the Yellow River Basin increased slightly from 2010 to 2019, with an average of 0.9782, which was at a medium efficiency level. Secondly, the tourism economy's high-quality development level is rising, and the speed of development is fast, especially in western provinces. Thirdly, there is a parasitic relationship between the two, but in each province, there is a positive or negative asymmetric symbiotic relationship. The tourism economy's high-quality development has a greater impact on the efficiency of tourism's carbon emissions. Fourthly, energy and capital input, as well as coordination and innovation factors, are important driving factors of the symbiosis between the two, among which the role of labor input was gradually revealed, and the impact factor experienced the changing process of "sharing-coordination-innovation". This study provides a theoretical framework and evaluation methods for evaluating and analyzing the relationship between tourism's carbon emission efficiency and the tourism economy's high-quality development, and it provides data support and policy suggestions for the real development.

Keywords: tourism carbon emission efficiency; tourism economy high-quality development; interactive mechanism; the Yellow River Basin

Citation: Li, S.; Cheng, Z.; Tong, Y.; He, B. The Interaction Mechanism of Tourism Carbon Emission Efficiency and Tourism Economy High-Quality Development in the Yellow River Basin. *Energies* **2022**, *15*, 6975. https://doi.org/10.3390/en15196975

Academic Editor: Attilio Converti

Received: 29 August 2022
Accepted: 20 September 2022
Published: 23 September 2022

1. Introduction

China is the world's largest emitter of greenhouse gases, and its low-carbon development faces huge challenges. In September 2020, China proposed the climate goal of "carbon peak by 2030 and carbon neutral by 2060" for the first time at the UN General Assembly. The Central Economic Work Conference in December 2020, the Government Work Report in March 2021, and the 14th Five-Year Plan have repeatedly reaffirmed this climate goal.

Tourism has become one of the major sources of global climate change [1]. The carbon emission of tourism accounts for 5% of the total global carbon emission, and the greenhouse effect formed by the carbon emission of tourism accounts for about 14% of the total global effect. By 2035, tourism's carbon emissions are expected to increase by 152%, and its contribution to the greenhouse effect is expected to increase by 188% [2]. In response to this problem, more attention should be given to key issues such as sustainability and adaptation to climate change [3]. Tourism should strive to reduce carbon emissions and improve the efficiency of tourism's carbon emissions, and some actions based on anticipatory action planning are needed in the tourism sector [4]. The availability and sharing of knowledge and information related to tourism's carbon emissions is a basic requirement for the successful planning of the tourism sector regarding this phenomenon [5].

Global tourist arrivals are expected to maintain an annual growth rate of 3.3% between 2010 and 2030 to reach 1.8 billion arrivals [6], generating rapid economic growth for the tourism industry as well as huge challenges. As a pillar industry of China's national economy, the economic development of tourism is also an important issue that requires attention. The China Tourism Economy Blue Book (No. 13) put forward the tourism economy's high-quality development. The development of the tourism economy is closely related to social, economic, and ecological environments [7,8]. Studies have shown that, with the passage of time, tourism consumption and total tourism emissions are roughly 2:1 in direct proportion [9]. Therefore, the next step is to realize how "ecological benefits" and "economic benefits" go hand in hand. It is important to study the interaction mechanism between tourism's carbon emission efficiency and the tourism economy. It helps to realize low-carbon tourism and the high-quality development of tourism.

The Yellow River Basin is an important economic belt and ecological barrier in China. The general secretary's speech in 2019 at the Symposium on Ecological Protection and High-quality Development in the Yellow River Basin, as well as the proposal made at the 6th Meeting of the Financial and Economic Commission of the CPC Central Committee in 2020 to make overall planning and coordinated progress based on the whole Basin and the ecosystem, all illustrate the importance of achieving a win-win situation between economic development and environmental protection in the Yellow River Basin. Therefore, this paper analyzed tourism's carbon emission efficiency (TCEE) and the tourism economy's high-quality development (TEHQD) in nine provinces in the Yellow River Basin from 2010 to 2019. Then, we explored the symbiotic interaction mechanism between TCEE and TEHQD. This is part of the tourism industry's active response to climate change. It provides a basis for the tourism economy's high-quality development, and it also provides ideas and paths for the realization of low-carbon tourism and high-quality economic development in the Yellow River Basin.

There are three main contributions of this paper. Firstly, from a theoretical perspective, this study is conducive to deepening the research on low-carbon tourism, exploring the tourism economy's high-quality development (TEHQD) from a low-carbon perspective, improving the research on the relationship between tourism's carbon emission efficiency (TCEE) and the tourism economy from the symbiotic perspective, enriching the research framework of the interaction mechanism. Secondly, this study adapted to the goal of "Carbon Peak and Carbon Neutral" and the requirements of the tourism economy's high-quality development. Research on the carbon emission efficiency of tourism can guide the development of tourism to better assume corresponding responsibilities for carbon reduction. The evaluation system for the tourism economy's high-quality development (TEHQD) was established, which can help to evaluate the development level of the tourism economy comprehensively and provide ideas for the tourism economy's high-quality development. Thirdly, this study contributes to the realization of the Ecological Protection and High-quality Development strategy in the Yellow River Basin. Taking the nine provinces in the Yellow River Basin as the study area, the study of tourism's carbon emission efficiency (TCEE) was used to connect with ecological protection, and the study of the tourism economy's high-quality development was used to correspond with a high-quality development strategy, which can not only promote the development of the two but also contribute to the integration of the two.

The rest of this paper is organized as follows. The second part reviews the relevant literature of this paper. The third part introduces the model, the data source, the specific evaluation index system, and the formula. The fourth part is the empirical analysis of this paper, including the evaluation of TCEE, the evaluation of TEHQD, the evaluation of the symbiosis between TCEE and TEHQD, and the construction of the symbiosis interaction mechanism between TCEE and TEHQD. Following that, the research conclusion, the countermeasure suggestion, the shortage, and the prospect are given in the fifth part.

2. Literature Review

Tourism's carbon emission efficiency (TCEE) refers to tourism's ecological efficiency based on carbon emissions from the perspective of low carbon, which is used to observe the value that can be achieved by the cost of carbon emissions. At present, research on tourism's ecological efficiency has covered concept [10], mechanism analysis [11], countermeasures and suggestions [12], model building [13,14], and efficiency measurement [15,16]. In terms of the specific content of the research, it mainly includes research on the input and output effect of tourism resources. For example, Jiang (2022) took tourism's CO_2 emission efficiency as undesired output, established an index system based on the input and output of tourism's CO_2 emission efficiency, and measured the tourism CO_2 emission efficiency of Chinese provinces [17]. The application of tourism's ecological efficiency to destination management, as well as the research on energy consumption and carbon emission intensity generated in the process of tourism are also receiving attention. Reilly (2010) studies have shown that tourism traffic is the most important part of energy consumption, and promoting the efficiency of transportation energy will help to enhance the efficiency of tourism ecology [18]. From the perspective of research methods, the current evaluation of tourism's ecological efficiency mostly adopts the single ratio method [19], carbon footprint model [20], life cycle assessment [21], carrying capacity of the low-carbon tourism environment model [22], DEA model [23], and SBM-DEA model [24]. Some researchers who use the single ratio method to measure tourism's carbon emission efficiency usually choose the two indexes of tourism's carbon emissions and tourism's income for accounting [25]. However, most researchers choose to use the "input-output" index and calculate the tourism carbon emission efficiency by establishing a model. From the perspective of the "input" index, it mainly focuses on capital input, labor input, resource input, and energy input [24]. From the perspective of "output" indicators, desirable output indicators such as tourism's income and number of tourists are mainly used [26]. In terms of influencing factors, urbanization, economic development level, government regulation, and tourism development level have an impact on tourism's ecological efficiency [27]. Other researchers analyzed the impact of foreign direct investment [28] and technology embedding [29] on tourism eco-efficiency. It can be seen that, at present, research on the influencing factors of tourism's carbon emission efficiency are mostly focused on the single level of economy, industry, and technology, while the influencing factors of humanities, society, and environment are relatively rare. the exploration of multi-faceted influencing factors has not yet received attention. Researchers mostly use exponential decomposition [24], the regression model [28], or the spatial econometric model [24] to study the influencing factors and analyze the linear relationship between each influencing factor and tourism's ecological efficiency, but they rarely consider the dynamic relationship and interaction mechanism between each influencing factor and tourism's ecological efficiency.

Tourism development is closely related to economy, culture, and ecological environment [30]. With China's requirements of high-quality development, the tourism economy's high-quality development (TEHQD) has also entered the horizon of researchers. Research on the influencing factors of the development quality of the tourism economy has always been the focus of scholars [31]. It has been found that resource conservation, ecological environmental protection, and sustainable development are the important goals of tourism's economic development and the important content of quality improvement [32]. Efficiency improvement, structural optimization, and environmental coordination are the core contents and important ways to promote the development of the tourism economy [33]. Scientific and reasonable arrangements should be made to maximize the adjustment of tourism's resource development and ecological environment protection to improve the sustainable development capacity of the tourism economy [34].

The relationship between the development of the tourism economy and other factors has always been an important issue, such as the relationship between carbon emissions and international tourism growth [35], between tourism investment and energy innovation on carbon dioxide emissions [36], between tourism economy and regional integration [37], and between tourism economic development and government policy [38,39]. With the development of tourism and the improvement of the quality requirements of the tourism economy, the relationship between tourism and the ecological environment is increasingly concerned. Tourism development not only brings economic benefits to the local area, but it also increases the pressure on the local ecological environment [40]. Therefore, it is necessary to take certain measures to promote the coordination between tourism and the environment to realize the sustainable development of tourism [41]. In terms of research content, studies on the tourism economy and ecological environment mainly include their interaction [42], their coupling and coordinated development evaluation [43], influencing factors [44], policies and paths to promote their joint development [45], tourism's ecological footprint measurement [46,47], tourism's ecological efficiency [24], tourism's environmental capacity [48,49], and tourism's ecological security [50]. From the perspective of research methods, in addition to macro qualitative description, spatial data analysis [24], and econometric analysis [28], the coupled coordination model is the most common quantitative study [51].

To sum up, the evaluation of tourism's efficiency from the perspective of ecology is the basis and premise for the realization of the high-quality and sustainable development of tourism, when low carbon tourism has become the goal and mode of tourism development. However, the current research on tourism's carbon emission efficiency usually focuses on the form of tourism's ecological efficiency and less directly considers the more detailed tourism carbon emission efficiency. In the calculation of tourism's carbon emission efficiency, the single ratio method is mainly used, and the research method needs to be expanded urgently. The construction of the "input-output" evaluation index of tourism's carbon emission efficiency rarely considers the undesired output. The few evaluation systems that include an undesired output rarely take tourism's carbon emissions as a specific index. The influencing factors of tourism's carbon efficiency focus on a single aspect, such as economy, industry, technology, society, and environment factors, and multifaceted influence factors are uncommon. Research on the relationship between various influencing factors and tourism's ecological efficiency that gives priority to a linear relationship between the interaction mechanism and dynamic relationship is relatively lacking. The consideration of the development of the tourism economy is mainly based on a single factor, such as industry or technology, and lacks the comprehensive consideration of the tourism economy's high-quality development. Studies on the relationship between the tourism economy and other factors are mainly about social and industrial factors; the relationship between the tourism economy and the ecological environment needs more attention. Most of the studies on tourism's economic development related to the ecological environment are focused on the environment of the whole society, and few of them are detailed towards tourism or even tourism's carbon emission efficiency. The research methods are mainly coupled and coordinated, while other methods should be applied.

3. Research Methods and Data Sources

3.1. Modeling and Data Sources

3.1.1. Modeling

The empirical analysis in this paper was based on the following analysis framework, as shown in Figure 1. This paper first used the Super-SBM model to calculate tourism's carbon emission efficiency (TCEE) of the Yellow River Basin from 2010 to 2019. Secondly, the entropy method and linear weighting were used to comprehensively evaluate the tourism economy's high-quality development (TEHQD). Finally, taking tourism's carbon emission efficiency (TCEE) and the tourism economy's high-quality development (TEHQD) as the symbiosis unit, the symbiosis degree model was used to reflect the correlation degree of their mutual influence. The symbiosis coefficient was used to measure the mutual influence degree between the two, and the geographical detector was used to explore the driving factors and obstructive factors affecting the symbiosis development of the two to construct an interactive mechanism model of tourism's carbon emission efficiency (TCEE) and the tourism economy's high-quality development (TEHQD).

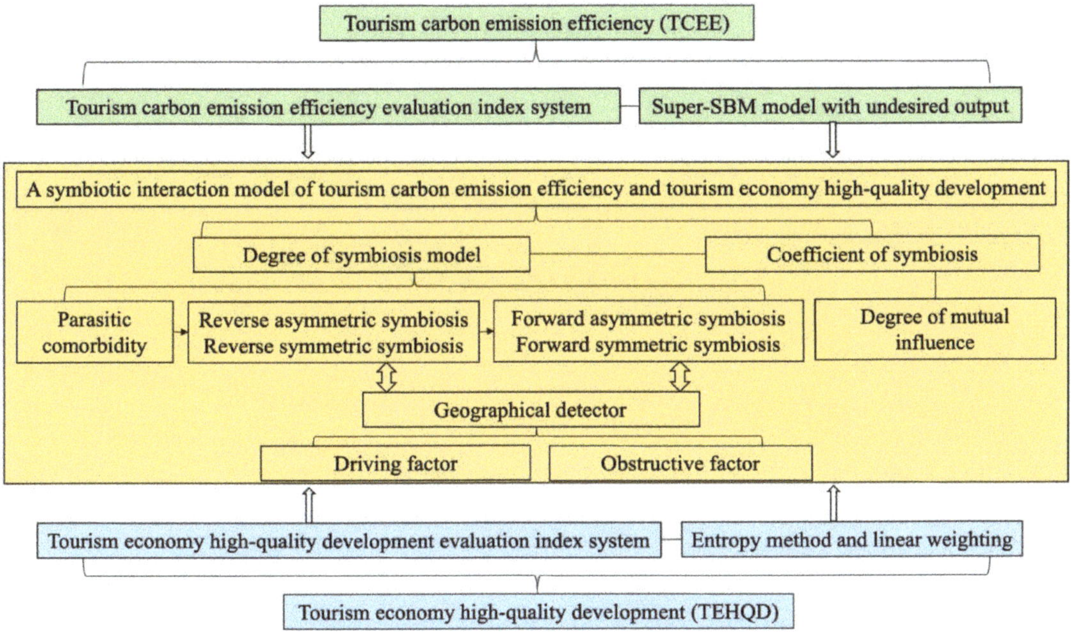

Figure 1. Analysis framework of the symbiotic interaction model between tourism's carbon emission efficiency (TCEE) and the tourism economy's high-quality development (TEHQD).

3.1.2. Data Sources

This study covered nine provinces in the Yellow River Basin from 2010 to 2019. In the establishment of the index system of this paper, the actual operability and feasibility were considered. Thus, the statistical data from government departments were selected. The original data were from the "China Tourism Statistical Yearbook", the "China Energy Statistical Yearbook", the "China Transportation Statistical Yearbook", the EPS database, the provincial statistical yearbook, and the social and economic development bulletin. In this paper, data with inconsistent accounting ranges in different years were processed, and some missing data were supplemented and improved by third-order moving average.

The "bottom-up" method was used to calculate the carbon emissions in advance for the undesired output data of tourism's carbon emission efficiency (TCEE). The "bottom-up" carbon emission calculation method is based on the six elements of tourism, such as tourism transportation, tourism accommodation, and tourism activities. It is actually the best choice for the carbon emission calculation, according to the actual situation in China, and this method has been widely used in the tourism field in China. Because China has yet to establish a dedicated database of carbon accounts for tourism satellites, much of the data are not available. Therefore, the "bottom-up" carbon emission calculation is more reasonable and effective. The data measured by the symbiotic interaction model come from the evaluation results of tourism's carbon emission efficiency (TCEE) and the tourism economy's high-quality development (TEHQD).

3.2. Tourism's Carbon Emission Efficiency (TCEE) Evaluation Index System and Evaluation Method

3.2.1. Evaluation Index System of Tourism's Carbon Emission Efficiency (TCEE)

The input-output index system of TCEE was established as shown in Table 1. Among them, energy input was expressed by the ratio of the tertiary industry energy consumption to the total energy consumption. The desirable output included tourism income and tourist reception. Tourism income was measured by the ratio of the sum of the income of star hotels, travel agencies, and tourist attractions to the total tourism income, and tourist reception was measured by the ratio of the total number of tourists in tourist attractions to the total number of tourists.

Table 1. Evaluation index system of tourism's carbon emission efficiency (TCEE).

TCEE System	Indicator Type	Indicators	Indicators Direction
Input	A1 Capital input	A11 Original value of fixed assets of star hotels/thousand yuan	+
		A12 Original value of fixed assets of travel agency/thousand yuan	+
	B1 Resource input	B11 Number of star-rated hotels	+
		B12 Number of travel agencies	+
		B13 The scenic area number	+
	C1 Labor input	C11 Number of hotel employees	+
		C12 Number of travel agency employees	+
		C13 Number of employees in scenic spots	+
	D1 Energy input	D11 Total energy consumption/tons of standard coal	+
		D12 Energy consumption in the tertiary industry/tons of standard coal	+
Output	E1 Desirable output	E11 Tourism revenue/100 million yuan	+
		E12 Star hotel operating income/100 million yuan	+
		E13 Travel agency revenue/100 million yuan	+
		E14 Tourist attractions operating income/100 million yuan	+
		E15 Total number of visitors in tourist attractions/100 million yuan	+
		E16 Total number of visits/100 million yuan	+
	F1 Undesirable output	F11 Tourism transport carbon emissions/tone	−
		F12 Tourism accommodation carbon emissions/tone	−
		F13 Tourism activity carbon emission/tone	−

3.2.2. Super-SBM Model

Since the traditional DEA model has the deviation of efficiency value caused by the relaxation of input and output, the undesirable output was incorporated into the evaluation system [52]. The non-radial and non-directional Super-SBM model based on relaxation variables was used to achieve the effective ordering of decision-making units [53]. Suppose there are n DMU (decision units), and each DMU has m input indicators, s_1 desirable output indicators, s_2 undesirable output indicators, and x, y^e, and y^u are the elements of the corresponding input matrix, desirable output matrix, and undesirable output matrix, respectively. Input matrix $X = [x_1, x_2, x_3, \cdots, x_n] \in R^{m\times n}$, and desirable output matrix $Y^e = [y_1^e, y_2^e, y_3^e, \cdots, y_n^e] \in R^{s_1\times n}$. The undesirable output matrix $Y^u = [y_1^u, y_2^u, y_3^u, \cdots, y_n^u] \in R^{s_2\times n}$. The Super-SBM model containing the undesired outputs is:

$$\begin{cases} min\rho = \dfrac{\frac{1}{m}\sum_{i=1}^{m}\frac{\overline{x}}{x_{ik}}}{\frac{1}{s_1+s_2}\left(\frac{\sum_{r=1}^{s_1}\overline{y^e}}{y^e_{rk}}+\frac{\sum_{t=1}^{s_2}\overline{y^u}}{y^u_{rk}}\right)} \\ \overline{x} \geq \sum_{j=1,\neq k}^{n} x_{ij}\lambda_j;\ \overline{y^e} \leq \sum_{j=1,\neq k}^{n} y^e_{rj}\lambda_j;\ \overline{y^u} \geq \sum_{j=1,\neq k}^{n} y^d_{tj}\lambda_j; \\ \overline{x} \geq x_k;\ \overline{y^e} \leq y^e_k;\ \overline{y^u} \geq y^u_k \\ \lambda_j \geq 0,\ i = 1,2,\cdots,m; j = 1,2,\cdots,n,\ j \neq 0; \\ r = 1,2,\cdots,s_1; t = 1,2,\cdots,s_2 \end{cases} \quad (1)$$

$\overline{x}$, $\overline{y^e}$, and $\overline{y^u}$ represent the input, desirable output, and undesirable output vectors considering the slack variables, respectively, j represents the decision unit, n is the number of decision-making units, k is the production period, and λ_j is the weight vector of decision-making units. ρ is the efficiency value, $\rho \geq 1$ is a relatively effective decision unit, and $0 < \rho < 1$ is a relatively invalid decision unit.

3.3. Evaluation Index System and Evaluation Method for the Tourism Economy's High-Quality Development (TEHQD)

3.3.1. Evaluation Index System of the Tourism Economy's High-Quality Development (TEHQD)

The evaluation index system of TEHQD was established from five dimensions of "innovation, coordination, green, openness and sharing", as shown in Table 2. A21 tourism R&D expenditure is represented by "the whole society R&D expenditure" multiplied by "the ratio of tourism production value to the gross national economic product". A22 is represented by "R&D personnel in the whole society" multiplied by "ratio of tourism employees to total employment in the region". A23 is represented by "total social fixed asset investment" multiplied by "ratio of tourism output value to GDP". B24 is represented by the difference between "turnover of local passengers" and "total turnover of national passengers". C22 is represented by the ratio of "garden green space area" to "total urban area". D22 is represented by the ratio of "international tourists per 10,000 people" to "tourism employees". E22 is expressed by the ratio of "park area" to "total population".

Table 2. Evaluation index system of the tourism economy's high-quality development (TEHQD).

TEHQD System	Indicator Type	Indicators	Indicators Direction
Tourism economy high-quality development	A2 Innovation	A21 Tourism R&D expenditure/yuan	+
		A22 Tourism R&D personnel	+
		A23 Investment in fixed assets of tourism/thousand yuan	+
	B2 Coordination	B21 Proportion of the primary industry in tourism economy/%	+
		B22 Proportion of the secondary industry in tourism economy/%	+
		B23 Proportion of the tertiary industry in tourism economy/%	+
		B24 Regional difference in passenger turnover/100 million passenger-km	+
	C2 Green	C21 Green coverage rate of built-up area/%	+
		C22 Tourism greening contribution/%	+
		C23 Per capita green area of park/ square meters	+
		C24 Proportion of investment in environmental governance in GDP/%	+
	D2 openness	D21 Proportion of foreign tourists in inbound tourists/%	+
		D22 Number of international tourism employees per 10,000 people	+
		D23 Foreign exchange income from tourism/100 million dollars	+
		D24 Foreign investment in tourism/100 million dollars	+
	E2 Sharing	E21 Tourism employment contribution/%	+
		E22 Per capita public recreation area m_2/person	+
		E23 Capita disposable income of households/yuan	+
		E24 Capita GDP/yuan	+

3.3.2. Entropy Value Method

The method of assigning weight to entropy can avoid subjective judgment and ensure a scientific and effective index score [54]. First of all, standardized treatment should be carried out according to the basic indicators, and the formula is as follows:

$$x_{ij} = \begin{cases} \frac{X_{ij} - X_{j,\min}}{X_{j,max} - X_{j,min}} \text{ positive indicators} \\ \frac{X_{j,max} - X_{ij}}{X_{j,max} - X_{j,min}} \text{ negative indicators} \end{cases} \quad (2)$$

In the formula, X_{ij} is the original value of the j index of the i sample, x_{ij} is the normalized value of X_{ij}. $X_{j,max}$ and $X_{j,min}$ are the maximum and minimum values of the

j index, respectively, and there are m samples and n indexes. Since there is a value of 0 after normalization, x_{ij} is shifted to the right by 1 unit to obtain x'_{ij} prime for logarithmic operation in the information entropy.

Determine the entropy value of item j:

$$H_j = -\frac{1}{lnm}\sum_{i=1}^{m}(P_{ij} \times InP_{ij}),\ P_{ij} = x'_{ij} / \sum_{i=1}^{m} x'_{ij} \tag{3}$$

Determine the weight of item j:

$$w_j = (1 - H_j) / \sum_{j=1}^{n}(1 - H_j) \tag{4}$$

The linear weighted model was adopted to measure the comprehensive development level of the tourism economy's high-quality development (TEHQD). The formula is as follows:

$$v_E = \sum_{j=1}^{n} w_{je} e_j \tag{5}$$

v_E is the value of TEHQD, w_{je} is the weight of each index of TEHQD, and e_j is the standardized value of each index of TEHQD.

3.4. Symbiotic Interaction Model between Tourism's Carbon Emission Efficiency (TCEE) and the Tourism Economy's High-Quality Development (TEHQD)

3.4.1. Symbiosis Model

Symbiosis can describe the correlation degree of the variation of quality parameters between two symbiosis units or systems and reflect the correlation degree of their mutual influence [55]. This paper took tourism's carbon emission efficiency (TCEE) and the tourism economy's high-quality development (TEHQD) as symbiotic units and selected the added value of the comprehensive score of TCEE and TEHQD as the main quality parameters. Then, the symbiotic degree of TCEE and TEHQD is:

$$\delta_{CE} = \frac{dv_C / v_C}{dv_E / v_E} = \frac{v_E}{v_C}\frac{dv_C}{dv_E} \tag{6}$$

Similarly, the symbiosis degree between TEHQD and TCEE is:

$$\delta_{EC} = \frac{dv_E / v_E}{dv_C / v_C} = \frac{v_C}{v_E}\frac{dv_E}{dv_C} \tag{7}$$

If $\delta_{CE} = \delta_{EC} > 0$, it indicates that TCEE and TEHQD are in a positive symbiotic state. If $\delta_{CE} \neq \delta_{EC} > 0$, then the two parties are in a positive asymmetric symbiosis. If $\delta_{CE} = \delta_{EC} < 0$, it indicates that TCEE and TEHQD are in a state of reverse symmetry symbiosis. If $\delta_{CE} \neq \delta_{EC} < 0$, it indicates that both parties are in a state of reverse asymmetric symbiosis. If $\delta_{CE} > 0$ (=0, <0), $\delta_{EC} < 0$ (=0, >0), it indicates that the two parties are in the parasitic, coexisting, and parasitic states, respectively.

3.4.2. Symbiosis Coefficient

The symbiosis coefficient is usually used to measure the degree of mutual influence between symbiosis units. The symbiosis coefficient of the main quality parameters of tourism's carbon emission efficiency (TCEE) and the tourism economy's high-quality development (TEHQD) can be expressed as follows:

$$\theta_C^M = \frac{|\delta_{CE}^m|}{|\delta_{CE}^m| + |\delta_{EC}^m|} \tag{8}$$

$$\theta_E^M = \frac{|\delta_{EC}^m|}{|\delta_{CE}^m| + |\delta_{EC}^m|} \tag{9}$$

$$\theta_C^M + \theta_E^M = 1 \quad (10)$$

If $\theta_C^M = 0$, it indicates that TCEE has no influence on the TEHQD. If $\theta_C^M = 1$, it indicates that the TEHQD has no impact on the TCEE, but only the TCEE has an impact on the TEHQD. If $0 < \theta_C^M < 0.5$, it indicates that the TEHQD has a relatively large impact on TCEE. If $\theta_C^M = 0.5$, the interaction between TCEE and TEHQD is the same. If $0.5 < \theta_C^M < 1$, it indicates that TCEE has a relatively large impact on the TEHQD.

3.4.3. Geographical Detector

The spatial differentiation of tourism's carbon emission efficiency (TCEE) and the tourism economy's high-quality symbiotic development (TEHQD) in the Yellow River Basin is explored by using geographic detectors [56]. The driving factors behind it were revealed, and the interactive mechanism of the two was explored.

Factor detection was used to analyze the spatial differentiation of dependent variable Y and the explanatory power of independent variable X_i to the dependent variable, which is measured by the q value and expressed as follows:

$$q = 1 - \frac{1}{N\sigma}\sum_{h=1}^{L} N_h\sigma_h^2 = 1 - \frac{SSW}{SST},\ SSW = \sum_{h=1}^{L} N_h\sigma_h^2, SST = N\sigma^2 \quad (11)$$

where q represents the explanatory power of the influencing factor X_i, $q \in [0,1]$. The larger the q value is, the stronger the explanatory power of the independent variable X to attribute Y is, and vice versa. N is the total number of provincial units, and N_h is the total number of units in the province of the layer h divided by the variable factor. σ^2 is the total variance of Y value, and σ_h^2 is the variance of the h layer. SSW and SST are the sum of variances and total variances within layers, respectively.

4. Empirical Analysis

4.1. Calculation of Tourism's Carbon Emission Efficiency (TCEE)

Under the condition that tourism's carbon emissions are taken as an undesirable output, according to Formula (1), the Super-SBM model was used to calculate the tourism carbon emission efficiency (TCEE) of nine provinces in the Yellow River Basin from 2010 to 2019. The results are shown in Table 3. Tourism's carbon efficiency in the Yellow River Basin has been fluctuating, rising slightly in 2010 compared to 2019. The average TCEE was 0.9782, in the medium level of efficiency, with the frontier still having room for improvement. Obviously, there is a big waste and diseconomy in tourism resources, and tourism's carbon efficiency has great development potential.

Table 3. Tourism's carbon emission efficiency (TCEE) in the nine provinces in the Yellow River Basin.

Province	Year										Average	Rank
	2010	2011	2012	2013	2014	2015	2016	2017	2018	2019		
Qinghai	0.1047	0.1345	1.2473	0.5643	1.3090	1.0319	1.0355	1.8984	1.7581	1.5516	1.0635	4
Sichuan	1.3264	1.2662	1.1335	1.5001	1.4894	1.2915	0.3882	0.2755	1.2859	1.2586	1.1215	2
Gansu	0.1967	1.1340	1.0840	0.2268	1.1552	1.0129	1.4472	0.3416	1.2239	1.0352	0.8858	7
Ningxia	1.3979	0.3129	1.2439	1.0934	1.1844	1.2531	1.3906	1.3607	1.1325	1.1486	1.1518	1
Inner Mongolia	1.1012	1.2510	1.2799	1.9010	0.6844	1.7658	0.4246	1.2377	0.3222	1.1193	1.1087	3
Shaanxi	0.0759	0.2193	0.4543	0.4235	0.3780	1.5434	1.0347	1.1039	0.4875	1.3300	0.7051	8
Shanxi	0.3560	1.3070	1.4161	1.5774	0.6610	0.2497	1.0480	1.9854	0.3656	1.2464	1.0213	6
Henan	1.1705	0.1923	1.1614	0.5534	1.3091	0.2008	2.1081	0.3329	1.6483	1.8764	1.0553	5
Shandong	1.4482	1.3876	0.3019	1.1720	1.2028	1.0578	0.1059	0.0629	0.0443	0.1241	0.6908	9
Yellow River Basin	0.7975	0.8005	1.0358	1.0013	1.0415	1.0452	0.9981	0.9555	0.9187	1.1878	0.9782	—

From the perspective of spatial distribution, there were six provinces whose average TCEE exceeded 1 and whose TCEE was effective, which were, respectively, Ningxia, Inner Mongolia, Sichuan, Qinghai, Henan, and Shanxi. Among them, Ningxia ranked first in tourism's carbon emission efficiency (1.1518). The last three provinces, Gansu, Shaanxi, and Shandong, had relatively low efficiency of tourism's carbon emissions. Among them, the efficiency of Shandong was at the bottom of the Yellow River Basin (0.6908). The difference of TCEE between Shandong and Ningxia was 0.461, and the latter was 1.67 times of the former, indicating that there is a large inter-provincial difference in TCEE in the Yellow River Basin.

From the perspective of time distribution, the inter-annual change rates of TCEE in the Yellow River Basin from 2010 to 2019 were higher in Henan, Gansu, and Qinghai, while the inter-annual change rates were lower in Ningxia, Sichuan, and Shandong. In 2019, the TCEE of all provinces was effective, except for Shandong. Compared to 2010, the TCEE increased significantly in Qinghai, Gansu, Shaanxi, Shanxi, and Henan provinces, and it decreased significantly in Ningxia and Shandong provinces. It can be seen that the TCEE in the Yellow River Basin also had a large time difference.

From the perspective of the weight of each resource index of TCEE (Table 4), the weight of the output index was higher than that of the input, indicating that input factors need to be strengthened in order to achieve a higher efficiency of tourism carbon.

Table 4. Weight of tourism's carbon emission efficiency (TCEE) evaluation index of the nine provinces in the Yellow River Basin.

TCEE System	Indicators	Qinghai	Sichuan	Gansu	Ningxia	Inner Mongolia	Shaanxi	Shanxi	Henan	Shandong
A1 (0.0981)	A11 (0.0198)	0.0117	0.0132	0.0276	0.0138	0.0346	0.0198	0.0184	0.0286	0.0102
	A12 (0.0156)	0.0147	0.0141	0.0136	0.0092	0.0169	0.0173	0.0139	0.0141	0.0262
B1 (0.0792)	B11 (0.0221)	0.0184	0.0186	0.0253	0.0297	0.0271	0.0179	0.0212	0.0187	0.0222
	B12 (0.0192)	0.0192	0.0179	0.0155	0.0168	0.0228	0.0179	0.0140	0.0338	0.0149
	B13 (0.0220)	0.0114	0.0175	0.0206	0.0154	0.0331	0.0331	0.0210	0.0216	0.0247
C1 (0.0899)	C11 (0.0258)	0.0188	0.0259	0.0235	0.0229	0.0372	0.0359	0.0235	0.0135	0.0309
	C12 (0.0169)	0.0139	0.0117	0.0124	0.0290	0.0098	0.0359	0.0109	0.0168	0.0118
	C13 (0.0166)	0.0183	0.0138	0.0201	0.0098	0.0099	0.0241	0.0113	0.0160	0.0259
D1 (0.1136)	D11 (0.0289)	0.0406	0.0138	0.0393	0.0153	0.0345	0.0283	0.0226	0.0351	0.0304
	D12 (0.0202)	0.0260	0.0142	0.0137	0.0153	0.0221	0.0205	0.0290	0.0222	0.0191
E1 (0.2879)	E11 (0.0330)	0.0376	0.0322	0.0349	0.0265	0.0326	0.0370	0.0350	0.0378	0.0231
	E12 (0.0269)	0.0256	0.0542	0.0123	0.0096	0.0237	0.0240	0.0364	0.0188	0.0378
	E13 (0.0162)	0.0185	0.0118	0.0117	0.0160	0.0116	0.0186	0.0258	0.0200	0.0118
	E14 (0.0346)	0.0248	0.0490	0.0429	0.1001	0.0225	0.0240	0.0216	0.0160	0.0106
	E15 (0.0329)	0.0320	0.0224	0.0433	0.0319	0.0285	0.0263	0.0411	0.0279	0.0425
	E16 (0.0431)	0.0417	0.0339	0.0295	0.0252	0.0346	0.0313	0.0359	0.0361	0.1196
F1 (0.1356)	F11 (0.0183)	0.0217	0.0205	0.0292	0.0172	0.0154	0.0132	0.0134	0.0193	0.0151
	F12 (0.0264)	0.0188	0.0141	0.0385	0.0322	0.0616	0.0111	0.0241	0.0183	0.0185
	F13 (0.0218)	0.0187	0.0489	0.0161	0.0133	0.0124	0.0237	0.0179	0.0182	0.0273

In terms of input index, energy input had the highest weight (0.1136), while resource input had the lowest weight (0.0792). Among them, total energy consumption (0.0289) had the highest weight. It can be seen that this index plays a significant role in TCEE. It is especially significant for the provinces with sparse population and abundant energy resources, such as Qinghai, Gansu, and Inner Mongolia, and the provinces with developed economy and large energy consumption, such as Henan and Shandong. The weight of the index of fixed asset investment of travel agencies (0.0156) was low, which shows that the existing capital investment of travel agencies cannot provide enough development space for tourism and needs to be strengthened, especially for Shanxi, Gansu, and Ningxia.

In terms of output index, the weight of desirable output (0.2879) was significantly higher than that of the undesirable output (0.1356), among which the weight of total number of visits (0.0431) was the highest, which is more significant for Qinghai, Sichuan, Inner Mongolia, Shanxi, and Henan. While the effect of total number of visits on Shandong is not obvious, the number of visitors plays a more important role in TCEE. The income of travel agencies (0.0162) had a low weight, and the contribution rate in Sichuan, Gansu, Inner Mongolia, and Shandong provinces was low, which is more dependent on the income of star hotels and scenic spots.

4.2. Evaluation of the Tourism Economy's High-Quality Development (TEHQD)

According to Formulas (2)–(5), the comprehensive evaluation value of the tourism economy's high-quality development (TEHQD) in the Yellow River Basin can be obtained, as shown in Table 5. From 2010 to 2019, the TEHQD in the Yellow River Basin showed a positive upward trend, except Shanxi, where the level of TEHQD fluctuated greatly. Other provinces had a gentle growth.

Table 5. Comprehensive evaluation of the tourism economy's high-quality development (TEHQD) of the nine provinces in the Yellow River Basin.

Province	Year										Average	Rank
	2010	2011	2012	2013	2014	2015	2016	2017	2018	2019		
Qinghai	0.1006	0.1488	0.1429	0.1849	0.2030	0.2317	0.3208	0.3672	0.3411	0.3849	0.2426	2
Sichuan	0.1102	0.1293	0.1500	0.1824	0.1978	0.2195	0.2688	0.2961	0.3300	0.4475	0.2332	3
Gansu	0.1101	0.0930	0.1603	0.2071	0.1945	0.2084	0.2522	0.2611	0.3122	0.4094	0.2208	5
Ningxia	0.0926	0.0988	0.1157	0.1451	0.1547	0.1733	0.2322	0.3333	0.3441	0.4503	0.2140	8
Inner Mongolia	0.1096	0.1191	0.1263	0.1675	0.1971	0.2387	0.2460	0.2908	0.3097	0.3963	0.2201	7
Shaanxi	0.1245	0.1116	0.1545	0.1881	0.2491	0.2667	0.3051	0.3431	0.3679	0.4224	0.2533	1
Shanxi	0.1528	0.1889	0.2312	0.2738	0.1992	0.1811	0.2351	0.2228	0.2771	0.3425	0.2304	4
Henan	0.1573	0.1332	0.1295	0.1388	0.1472	0.1818	0.2471	0.3053	0.3408	0.4212	0.2202	6
Shandong	0.1213	0.1221	0.1592	0.1711	0.2053	0.2162	0.2429	0.2750	0.2853	0.3021	0.2100	9
Yellow River Basin	0.1199	0.1272	0.1522	0.1843	0.1942	0.2131	0.2611	0.2994	0.3231	0.3974	0.2272	—

From the provincial TEHQD, it had the following three characteristics. First of all, the evaluation of TEHQD increased significantly, with an average growth rate of 3.3 times. Ningxia, Sichuan, and other western regions are developing faster because the economic development and policy dividend brought a series of advantages, such as industrial structure adjustment and tourism development, while the development speed of Shandong, Henan, and Shanxi are lower than the average level due to the eastern region as a whole entering the stage of slow development. The central region, such as Shanxi, should strive to adjust the industrial structure, change the current development mode, and pay more attention to the tourism economy's high-quality development. Secondly, the TEHQD in different provinces is more and more obvious. In 2010, there was a difference of 0.0646 between Henan (0.1573), which ranked first, and Ningxia (0.0926), which ranked last. By 2019, there was a difference of 0.1482 between Ningxia (0.4503), which ranked first, and Shandong (0.3021), which ranked last, in the comprehensive evaluation value of TEHQD. The difference in 2019 was 2.3 times of that in 2010. Thirdly, from the average value of comprehensive evaluation, there was a certain similarity between the TEHQD and the evaluation of TCEE; thus, there was a co-existing relationship to some extent.

From the perspective of the weight of each resource index of TEHQD (Table 6), the open index (0.1363) and the sharing index (0.1194) had a large weight, indicating that these two indexes play a more obvious role in promoting TEHQD in the Yellow River Basin than other indexes. However, the coordination index (0.0928) had the smallest weight; thus, in order to realize the high-quality development of the regional tourism economy in the Yellow River basin, attention should be paid to the coordinated development of regions, industries, and other aspects.

Table 6. Weight of the tourism economy's high-quality development (TEHQD) evaluation index of the nine provinces in the Yellow River Basin.

TEHQD System	Indicators	Qinghai	Sichuan	Gansu	Ningxia	Inner Mongolia	Shaanxi	Shanxi	Henan	Shandong
A2 (0.0959)	A21 (0.0363)	0.0432	0.0413	0.0353	0.0347	0.0349	0.0429	0.0339	0.0357	0.0245
	A22 (0.0277)	0.0231	0.0390	0.0184	0.0234	0.0210	0.0292	0.0296	0.0365	0.0292
	A23 (0.0319)	0.0401	0.0324	0.0306	0.0259	0.0330	0.0350	0.0309	0.0352	0.0244
B2 (0.0928)	B21 (0.0214)	0.0270	0.0275	0.0173	0.0230	0.0207	0.0209	0.0208	0.0156	0.0194
	B22 (0.0191)	0.0169	0.0351	0.0153	0.0200	0.0180	0.0180	0.0221	0.0148	0.0120
	B23 (0.0318)	0.0269	0.0153	0.0307	0.0202	0.0485	0.0269	0.0320	0.0591	0.0270
	B24 (0.0205)	0.0235	0.0219	0.0201	0.0179	0.0197	0.0231	0.0212	0.0185	0.0183
C2 (0.0953)	C21 (0.0251)	0.0274	0.0316	0.0204	0.0251	0.0192	0.0223	0.0244	0.0318	0.0236
	C22 (0.0248)	0.0183	0.0108	0.0171	0.0490	0.0268	0.0193	0.0285	0.0295	0.0236
	C23 (0.0209)	0.0174	0.0266	0.0231	0.0246	0.0140	0.0161	0.0149	0.0313	0.0196
	C24 (0.0245)	0.0415	0.0151	0.0314	0.0122	0.0121	0.0302	0.0301	0.0256	0.0226
D2 (0.1363)	D21 (0.0254)	0.0196	0.0164	0.0265	0.0161	0.0299	0.0185	0.0533	0.0178	0.0308
	D22 (0.0283)	0.0171	0.0154	0.0315	0.0396	0.0118	0.0224	0.0438	0.0333	0.0401
	D23 (0.0299)	0.0335	0.0244	0.0366	0.0408	0.0210	0.0270	0.0431	0.0260	0.0168
	D24 (0.0527)	0.0499	0.0574	0.0672	0.0640	0.0541	0.0550	0.0409	0.0433	0.0427
E2 (0.1194)	E21 (0.0371)	0.0415	0.0548	0.0283	0.0183	0.0641	0.0371	0.0225	0.0344	0.0325
	E22 (0.0286)	0.0415	0.0241	0.0297	0.0484	0.0149	0.0335	0.0181	0.0261	0.0208
	E23 (0.0360)	0.0410	0.0400	0.0346	0.0326	0.0340	0.0411	0.0350	0.0327	0.0330
	E24 (0.0177)	0.0183	0.0229	0.0159	0.0148	0.0112	0.0216	0.0179	0.0203	0.0164

In terms of the innovation index, the weight of tourism R&D expenditure was the highest (0.0363), indicating that this index plays a significant role, especially for the central and western provinces such as Qinghai, Sichuan, and Shaanxi. For the economically developed eastern regions, such as Shandong, tourism R&D personnel are more important. In terms of the coordination index, the proportion of the tertiary industry in the tourism economy had the highest weight (0.0318), and the index weight of Inner Mongolia and Henan was higher than the average level of the Yellow River basin, indicating that the industrial coordination degree has brought great dividends to the high-quality development of the local tourism economy. However, the proportion of the tertiary industry of the tourism economy in Sichuan and Ningxia is very low, indicating that this index has not brought the advantage of the high-quality development of the tourism economy to the local area. The proportion of the secondary industry of the tourism economy in Sichuan and the proportion of the primary industry of the tourism economy in Ningxia play a more important role. In terms of green indicators, the weight of each indicator is above 0.2, indicating that all indicators of green development play an important role. Among them, the weight of green coverage in built-up areas is the highest (0.0251), especially for cities with a high urbanization rate and a relatively developed economy, such as Sichuan, Henan, and Shandong. However, central and western provinces such as Qinghai, Gansu, Shaanxi, and Shanxi, with slower tourism and economic development and more serious environmental pollution, had the highest proportion of investment in environmental governance in GDP. For areas with excellent tourism development and sparse population such as Ningxia and Inner Mongolia, the contribution of tourism greening has become an important green index in TEHQD. In terms of the opening index, the foreign investment in tourism (0.0527) had the highest weight, but the foreign investment of tourism in Shanxi is relatively insufficient. The proportion of foreign tourists in the number of inbound tourists plays a more significant role in Shanxi (0.533), indicating that the proportion of foreign tourists in Shanxi is relatively large, which brings advantages to TEHQD. In terms of the sharing index, tourism per capita GDP (0.0177) had the lowest weight and needs to be strengthened.

4.3. Symbiotic Interaction between Tourism's Carbon Emission Efficiency (TCEE) and the Tourism Economy's High-Quality Development (TEHQD)

4.3.1. Calculation of Symbiosis Degree

According to Formulas (6) and (7), the symbiosis degree of tourism's carbon emission efficiency (TCEE) and the tourism economy's high-quality development (TEHQD) in the Yellow River Basin can be obtained, as shown in Table 7. In general, the average symbiosis degree of the Yellow River Basin is $\delta_{CE} = 0.2617 > 0$, $\delta_{EC} = -2.1192 < 0$, which indicates that there is a parasitic relationship between TCEE and TEHQD.

Table 7. Symbiosis between tourism's carbon emission efficiency (TCEE) and the tourism economy's high-quality development (TEHQD).

Province		2011	2012	2013	2014	2015	2016	2017	2018	2019	Average
Qinghai	δ_{CE}	0.0677	−4.4416	−0.6215	−0.5463	−0.3869	0.1920	−1.0987	1.2779	−0.5807	−0.6820
	δ_{EC}	14.7786	−0.2251	−1.6090	−1.8306	−2.5847	5.2080	−0.9102	0.7825	−1.7220	1.3208
Sichuan	δ_{CE}	−1.4406	0.0702	−0.8545	0.5205	0.5550	0.7392	−0.6277	−1.4235	0.8648	−0.1774
	δ_{EC}	−0.6942	14.2513	−1.1703	1.9212	1.8018	1.3528	−1.5932	−0.7025	1.1564	1.8137
Gansu	δ_{CE}	−0.8155	−0.6781	−0.2908	1.1998	1.4569	−0.8283	2.8167	−1.2152	1.5376	0.3537
	δ_{EC}	−1.2263	−1.4747	−3.4391	0.8335	0.6864	−1.2073	0.3550	−0.8229	0.6504	−0.6272
Ningxia	δ_{CE}	−12.7345	0.5669	−0.5746	−0.4786	−0.1739	−0.0509	−0.0745	5.8721	−0.0968	−0.8605
	δ_{EC}	−0.0785	1.7639	−1.7405	−2.0894	−5.7503	−19.6466	−13.4288	0.1703	−10.3330	−5.6814
Inner Mongolia	δ_{CE}	0.0498	0.6651	−1.0111	−0.4304	−0.1089	12.9749	−2.9512	2.5985	0.5747	1.3735
	δ_{EC}	20.0948	1.5036	−0.9891	−2.3235	−9.1804	0.0771	−0.3388	0.3848	1.7400	1.2187
Shaanxi	δ_{CE}	0.4897	0.1655	−0.7004	−0.5425	−2.7104	1.2113	0.1498	3.2747	−2.2085	−0.0968
	δ_{EC}	2.0420	6.0428	−1.4278	−1.8434	−0.3689	0.8256	6.6734	0.3054	−0.4528	1.3107
Shanxi	δ_{CE}	−1.0298	0.0404	−0.0081	−0.2984	−1.0413	0.2521	1.7477	0.4055	0.3821	0.0500
	δ_{EC}	−0.9711	24.7507	−123.6111	−3.3511	−0.9603	3.9674	0.5722	2.4664	2.6174	−10.5022
Henan	δ_{CE}	0.4666	2.0937	−0.0325	1.5051	−0.0237	−0.1591	0.5901	−2.0234	0.0757	0.2769
	δ_{EC}	2.1432	0.4776	−30.7391	0.6644	−42.1233	−6.2838	1.6946	−0.4942	13.2040	−6.8285
Shandong	δ_{CE}	23.9285	−2.2827	−2.3001	−1.0972	−0.6460	1.2108	−0.1528	0.9130	−0.5104	2.1181
	δ_{EC}	0.0418	−0.4381	−0.4348	−0.9114	−1.5481	0.8259	−6.5463	1.0953	−1.9593	−1.0972
Yellow River Basin	δ_{CE}	0.9980	−0.4223	−0.7104	−0.0187	−0.3421	1.7269	0.0444	1.0755	0.0043	0.2617
	δ_{EC}	4.0145	5.1835	−18.3512	−0.9923	−6.6698	−1.6534	−1.5024	0.3539	0.5446	−2.1192

From a provincial perspective, the results show that, firstly, the average symbiosis of Qinghai, Sichuan, Gansu, Shaanxi, Shanxi, Henan, and Shandong provinces are $\delta_{CE} > 0$, $\delta_{EC} < 0$ or $\delta_{CE} < 0$, $\delta_{EC} > 0$, indicating that the TCEE and TEHQD in these regions are parasitic, but from the point of view of each year, it is positive or negative asymmetric symbiosis. The difference of the symbiosis coefficient in each year resulted in the deviation of the overall mean coefficient. Secondly, the average symbiosis degree of Ningxia was $\delta_{CE} = -0.8605 \neq \delta_{EC} = -5.6814 < 0$, indicating that the TCEE and the TEHQD show a reverse asymmetric symbiosis. Thirdly, as the average symbiosis degree δ_{CE} and δ_{EC} were close, it can be considered that the TCEE and the TEHQD in Inner Mongolia have approximately presented a positive symbiosis and experienced a transition from positive asymmetric symbiosis to reverse asymmetric symbiosis to positive asymmetric symbiosis from 2010 to 2019.

4.3.2. Calculation of Symbiosis Coefficient

According to Formulas (8) and (9), the symbiosis coefficient of tourism's carbon emission efficiency (TCEE) and the tourism economy's high-quality development (TEHQD) in the Yellow River Basin was calculated, as shown in Table 8. The results all meet the feature that the sum of the symbiosis coefficients is 1 in Formula (10). In general, the average symbiosis coefficient of the Yellow River Basin is $0 < \theta_C^M = 0.3885 < 0.5$, $0.5 < \theta_E^M = 0.6115 < 1$, indicating that TEHQD has a relatively large impact on TCEE.

Table 8. Symbiosis coefficient of tourism's carbon emission efficiency (TCEE) and the tourism economy's high-quality development (TEHQD).

Province		2011	2012	2013	2014	2015	2016	2017	2018	2019	Average
Qinghai	θ_C^M	0.0046	0.9518	0.2786	0.2298	0.1302	0.0356	0.5469	0.6202	0.2522	0.3389
	θ_E^M	0.9954	0.0482	0.7214	0.7702	0.8698	0.9644	0.4531	0.3798	0.7478	0.6611
Sichuan	θ_C^M	0.6748	0.0049	0.4220	0.2132	0.2355	0.3533	0.2826	0.6696	0.4279	0.3649
	θ_E^M	0.3252	0.9951	0.5780	0.7868	0.7645	0.6467	0.7174	0.3304	0.5721	0.6351
Gansu	θ_C^M	0.3994	0.3150	0.0780	0.5901	0.6797	0.4069	0.8881	0.5962	0.7028	0.5174
	θ_E^M	0.6006	0.6850	0.9220	0.4099	0.3203	0.5931	0.1119	0.4038	0.2972	0.4826
Ningxia	θ_C^M	0.9939	0.2432	0.2482	0.1864	0.0294	0.0026	0.0055	0.9718	0.0093	0.2989
	θ_E^M	0.0061	0.7568	0.7518	0.8136	0.9706	0.9974	0.9945	0.0282	0.9907	0.7011
Inner Mongolia	θ_C^M	0.0025	0.3067	0.5055	0.1563	0.0117	0.9941	0.8970	0.8710	0.2483	0.4437
	θ_E^M	0.9975	0.6933	0.4945	0.8437	0.9883	0.0059	0.1030	0.1290	0.7517	0.5563
Shaanxi	θ_C^M	0.1934	0.0267	0.3291	0.2274	0.8802	0.5947	0.0220	0.9147	0.8299	0.4464
	θ_E^M	0.8066	0.9733	0.6709	0.7726	0.1198	0.4053	0.9780	0.0853	0.1701	0.5536
Shanxi	θ_C^M	0.5147	0.0016	0.0001	0.0818	0.5202	0.0597	0.7534	0.1412	0.1274	0.2444
	θ_E^M	0.4853	0.9984	0.9999	0.9182	0.4798	0.9403	0.2466	0.8588	0.8726	0.7556
Henan	θ_C^M	0.1788	0.8143	0.0011	0.6938	0.0006	0.0247	0.2583	0.8037	0.0057	0.3090
	θ_E^M	0.8212	0.1857	0.9989	0.3062	0.9994	0.9753	0.7417	0.1963	0.9943	0.6910
Shandong	θ_C^M	0.9983	0.8390	0.8410	0.5463	0.2944	0.5945	0.0228	0.4546	0.2067	0.5331
	θ_E^M	0.0017	0.1610	0.1590	0.4537	0.7056	0.4055	0.9772	0.5454	0.7933	0.4669
Yellow River Basin	θ_C^M	0.4400	0.3892	0.3004	0.3250	0.3091	0.3407	0.4085	0.6715	0.3122	0.3885
	θ_E^M	0.5600	0.6108	0.6996	0.6750	0.6909	0.6593	0.5915	0.3285	0.6878	0.6115

From a provincial perspective, the results show that, firstly, the symbiosis coefficient (θ_C^M) of Qinghai, Sichuan, Ningxia, Inner Mongolia, Shaanxi, Shanxi, and Henan is between 0 and 0.5, indicating that TEHQD has a greater impact on TCEE. That means TEHQD can play a positive impact on TCEE. Its development process fluctuated from 2010 to 2019, which is related to multiple factors such as provincial tourism development, economic development speed, and environmental background. Secondly, the symbiosis coefficient (θ_C^M) of Gansu and Shandong was between 0.5 and 1, indicating that TCEE has a relatively large impact on TEHQD. That is to say, the improvement of TCEE can promote the TEHQD to some extent. Among them, in recent years, Shandong province gradually changed into a TEHQD with a greater impact on TCEE, while Gansu province was in the promotion stage of TEHQD before 2014 ($0 < \theta_C^M < 0.5$) and began to change into a TCEE with a greater impact after 2014. Thirdly, the symbiosis coefficient (θ_C^M) of Inner Mongolia, Shaanxi, Gansu, and

Shandong was close to 0.5, which can be approximated as the state of interaction between TCEE and TEHQD, forming a relatively benign symbiosis state.

In conclusion, with the continuous development of economy, society, the demand of the nation, and people changing and under the influence of the national policies about the Yellow River Basin, tourism's carbon efficiency has improved, and the tourism economy is changing to high quality development. In this procession, the TEHQD has undoubtedly contributed to the local economy, environment, and other aspects. At the same time, the improvement of TCEE can not only benefit the carbon environment, but it can also play an important role in the efficient and high-quality development of tourism, thus promoting the TEHQD. It can be seen that TCEE and the tourism economy in the Yellow River Basin form a corresponding symbiotic interface in the symbiotic environment, influencing, promoting, and developing each other in the symbiotic state.

4.4. *Symbiotic Interaction Mechanism between Tourism's Carbon Emission Efficiency (TCEE) and the Tourism Economy's High-Quality Development (TEHQD)*

4.4.1. Research on the Influencing Factors of Symbiotic Interaction

From the symbiotic interaction between TCEE and TEHQD, through the analysis of the explanatory power q of each effective factor, it was found that (Table 9), firstly, on the whole, energy input and capital input are the most important factors affecting the symbiosis between the TEHQD and the TCEE. Among them, D11 (the total energy consumption (0.9726)), C11 (the number of employees in hotels (0.9112)), and A11 (the original value of fixed assets of star hotels (0.8983)) have q values greater than 0.89, which are the key factors affecting symbiosis. Secondly, in 2011, energy factors were important factors affecting the symbiosis between the TEHQD and TCEE, and there was a large gap in explanatory power q with other factors. Among them, D11 (the total energy consumption (0.9612)) and D12 (the energy consumption of tertiary industry (0.9599)) were the key factors affecting symbiosis. Thirdly, in 2015, the effect of energy input factors was still significant, but the desirable output became the most important factor affecting the symbiosis between the TEHQD and TCEE. The gap between the explanatory power q of each factor gradually narrowed, and the explanatory power gap of other factors was very small except for the resource input, which all become relatively important influencing factors. Among them, E15 (the total number of visitors at scenic spot (0.9942)), E16 (the total number of visitors (0.9928)), and F13 (the carbon emission from tourism activity (0.9928)) are the most critical factors affecting symbiosis. Fourthly, compared to 2015, the explanatory power q of each factor decreased in 2019, and the gap widened. Capital input and labor input are important factors affecting the symbiosis between TEHQD and TCEE. C11 (the number of employees in hotels (0.9918)), A11 (the original value of fixed assets in star hotels (0.9854)), and D11 (the total energy consumption (0.9736)) are the key influencing factors. To sum up, with the development of time, the role of capital and labor input in TCEE gradually emerges, experiencing a development process from "energy" to "capital". Among them, D12 (the effect intensity of tertiary industry energy consumption) and C12 (travel agency employees) decreased significantly, while the effect of C13 (scenic area employees), E12 (star hotel operating income), E15 (tourist attractions total number of reception), A11 (star hotel fixed assets), and other indicators gradually increased and changed significantly. Indicators such as D11 (total energy consumption), E14 (operating income of tourist attractions), and C11 (number of hotel employees) were important influencing factors in recent years.

Table 9. Factor detection of symbiosis between tourism's carbon emission efficiency (TCEE) and the tourism economy's high-quality development (TEHQD).

Year	2011		2015		2019		Average
Variable	q	p	q	p	q	p	q
A11	0.7484	0.0401 *	0.9610	0.0000 *	0.9854	0.0000 *	0.8983
A12	0.7620	0.0314 *	0.9470	0.0000 *	0.7445	0.0036 *	0.8178
B11	0.7659	0.0281 *	0.9628	0.0000 *	0.7178	0.0130 *	0.8155
B12	0.7567	0.0343 *	0.9806	0.0000 *	0.8544	0.0010 *	0.8639
B13	0.7551	0.0351 *	0.4231	0.2745	0.7473	0.0189 *	0.6418
C11	0.7691	0.0148 *	0.9727	0.0000 *	0.9918	0.0000 *	0.9112
C12	0.7702	0.0264 *	0.9565	0.0000 *	0.6250	0.0391 *	0.7839
C13	0.7327	0.0285 *	0.9628	0.0000 *	0.8507	0.0012 *	0.8487
D11	0.9612	0.0000 *	0.9830	0.0000 *	0.9736	0.0000 *	0.9726
D12	0.9599	0.0000 *	0.9724	0.0000 *	0.4939	0.1655	0.8087
E11	0.7604	0.0323 *	0.9806	0.0000 *	0.8410	0.0017 *	0.8607
E12	0.7484	0.0401 *	0.9628	0.0000 *	0.8507	0.0012 *	0.8540
E13	0.7406	0.0258 *	0.9780	0.0000 *	0.7593	0.0139 *	0.8260
E14	0.7739	0.0246 *	0.9876	0.0000 *	0.8567	0.0009 *	0.8727
E15	0.7659	0.0281 *	0.9942	0.0000 *	0.8559	0.0009 *	0.8720
E16	0.7567	0.0342 *	0.9928	0.0000 *	0.6785	0.0253 *	0.8093
F11	0.3164	0.5999	0.9705	0.0000 *	0.7170	0.0299 *	0.6680
F12	0.7692	0.0281 *	0.9628	0.0000 *	0.8559	0.0009 *	0.8626
F13	0.7659	0.0281 *	0.9928	0.0000 *	0.4819	0.1888 *	0.7469
A21	0.8874	0.0008 *	0.9400	0.0000 *	0.9450	0.0000 *	0.9241
A22	0.8980	0.0001 *	0.7125	0.0459 *	0.7165	0.0561 *	0.7757
A23	0.8847	0.0009 *	0.9791	0.0000 *	0.9450	0.0000 *	0.9363
B21	0.9021	0.0004 *	0.8199	0.0084 *	0.9333	0.0000 *	0.8851
B22	0.8846	0.0004 *	0.9584	0.0000 *	0.8525	0.0011 *	0.8985
B23	0.8859	0.0009 *	0.8633	0.0027 *	0.8353	0.0022 *	0.8615
B24	0.8868	0.0008 *	0.9375	0.0000 *	0.9661	0.0000 *	0.9301
C21	0.9977	0.0000 *	0.7020	0.0365 *	0.6456	0.0347 *	0.7818
C22	0.9039	0.0003 *	0.8520	0.0032 *	0.9669	0.0000 *	0.9076
C23	0.6082	0.1367	0.9439	0.0000 *	0.7845	0.0013 *	0.7789
C24	0.8850	0.0009 *	0.8284	0.0068 *	0.9342	0.0000 *	0.8825
D21	0.8696	0.0007 *	0.8087	0.0054 *	0.7171	0.0326 *	0.7985
D22	0.9004	0.0001 *	0.8488	0.0032 *	0.8935	0.0001 *	0.8809
D23	0.9010	0.0004 *	0.8306	0.0058 *	0.6382	0.0107 *	0.7899
D24	0.8851	0.0009 *	0.8061	0.0059 *	0.8240	0.0028 *	0.8384
E21	0.9036	0.0003 *	0.7917	0.0165 *	0.7234	0.0517 *	0.8062
E22	0.9959	0.0000 *	0.9439	0.0000 *	0.4467	0.4316	0.7955
E23	0.9961	0.0000 *	0.8087	0.0115 *	0.7333	0.0492 *	0.8460
E24	0.8337	0.0060 *	0.8559	0.0003 *	0.8144	0.0037 *	0.8347

* Indicates that the factor is valid after passing the significance test at 0.05 level.

From the symbiotic interaction between TEHQD and TCEE (Table 9), through the analysis of the explanatory power q of each effective factor, it can be found that, firstly, overall, coordination and innovation were the most important factors affecting the symbiosis between TCEE and TEHQD. The q values of A23 (tourism fixed asset investment (0.9363)), B24 (tourist turnover (0.9301)), A21 (tourism R&D expenditure (0.9241)), and C22 (tourism greening contribution (0.9076)) were all greater than 0.9, and these are the key factors affecting symbiosis. Secondly, in 2011, sharing factors were the most important factors affecting the symbiosis between TCEE and TEHQD. Among them, C21 (the green coverage rate of built-up area (0.9977)), E23 (the per capita disposable income of residents (0.9961)), and E22 (the per capita public recreation area (0.9959)) were the key factors affecting the symbiosis. Thirdly, in 2015 and 2019, coordination became the most important factor affecting the symbiosis between TCEE and TEHQD, followed by innovation. In 2015, A23 (tourism fixed asset investment (0.9791)), B22 (tourism economy secondary industry proportion (0.9584)), C23 (per capita park green area (0.9439)), E22 (per capita public recreation area (0.9439)), and A21 (tourism R&D expenditure (0.94)) were the key factors affecting symbiosis. By 2019, compared to 2015, in addition to A21 (tourism R&D expenditure (0.945)) and A23 (tourism fixed asset investment (0.945)), C22 (tourism greening contribution (0.9669)) and B24 (tourist turnover regional differences (0.9661)) became important influencing factors. To sum up, with the change of time, the role of innovation and coordination factors in TEHQD gradually came into prominence. It experienced a development procession of "sharing-coordination-innovation". C21 (the effect intensity of built-up area green coverage rate), E23 (residents per capita disposable income), E22 (per capita public recreation area), and D23 (tourism foreign exchange income) decreased significantly. A23 (tourism fixed asset investment), C23 (per capita park green area), A21 (tourism R&D expenditure), B24 (regional differences in tourist turnover), and C24 (environmental governance investment in GDP) gradually increased and changed significantly. C22 (contribution to tourism greening) and D22 (number of international tourism employees per 10,000 people) have been important influencing factors in recent years. However, the gap between the explanatory power q of each influencing factor gradually widened.

4.4.2. Interactive Mechanism Model

1. Based on the symbiotic system theory

Based on the symbiotic system theory, a symbiotic interaction mechanism model of tourism's carbon emission efficiency (TCEE) and the tourism economy's high-quality development (TEHQD) was established by taking TCEE and TEHQD as two main symbiotic units (Figure 2). TCEE can evaluate the development quality of tourism from the dual perspectives of economy and ecology, and the TEHQD is based on a tourism economic development concept of "innovation, coordination, green, open and sharing", which also involves many aspects such as economy, ecology, and tourism's development quality. Therefore, there is a quality parameter compatibility between the two symbiotic units. That is, some attributes of the symbiotic units are related, so it can be judged that there is a symbiotic interaction between the two units. The symbiotic interface is the energy transmission channel between symbiotic units, and the transmission of different types of energy requires different symbiotic interfaces. Because the symbiosis between TCEE and TEHQD is relatively complex, involving multiple symbiotic internal and external environments, such as politics, economy, culture, society, and ecology, the tourism market, as the result of the comprehensive effect of TCEE and TEHQD, can represent the comprehensive interaction between the two symbiotic units and, thus, can serve as the symbiotic interface of the symbiotic interaction mechanism.

2. Influencing factors based on interaction mechanism

From the perspective of the influencing factors of the interaction mechanism, symbiosis is a phenomenon of mutual influence and interaction between the two symbiotic units of TCEE and TEHQD (Figure 3). Through various interactions, the symbiotic relationship between the two symbiotic units can generate a new energy or achieve energy transformation, namely symbiosis energy generation. The specific performance is the symbiosis unit's ability to improve. From the interactive influence of TCEE on TEHQD, energy input and the capital input factor are the main factors driving the two symbiotic coordination interaction. From the interactive impact of TEHQD on TCEE, coordination and innovation are the main factors promoting the symbiotic coordination and interaction between the two. It can be found that the total energy consumption, the number of hotel employees, and the original value of fixed assets of star hotels are the key attraction factors affecting the symbiosis from the perspective of the symbiotic interaction between TCEE and TEHQD. From the perspective of the symbiotic interaction between TEHQD and TCEE, the investment in fixed assets of tourism, the R&D expenditure of tourism, regional differences in passenger turnover, and the tourism greening contribution are the key attraction factors affecting the symbiosis.

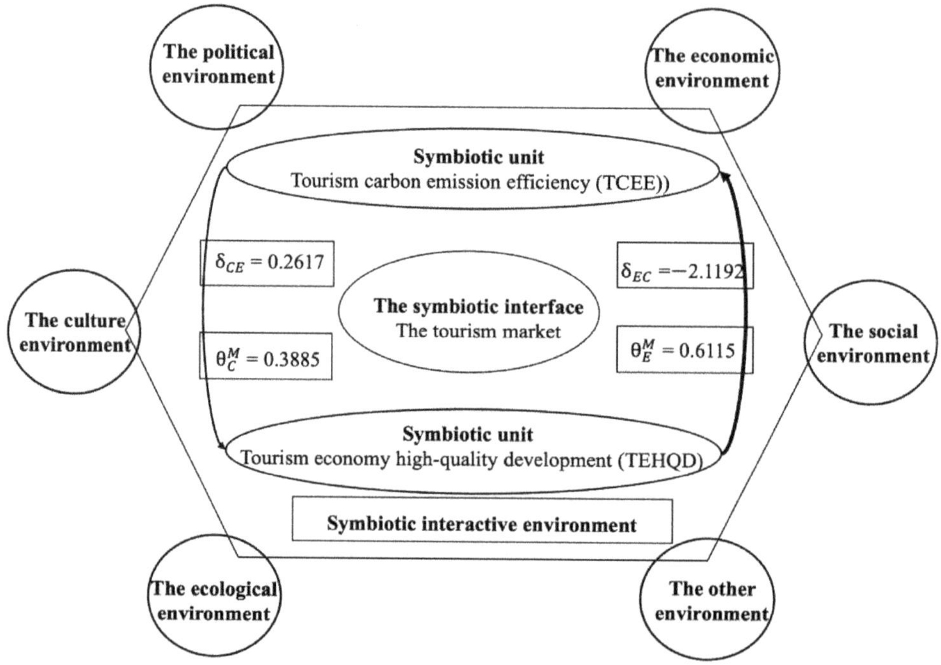

Figure 2. Symbiotic interaction mechanism model of tourism's carbon emission efficiency (TCEE) and the tourism economy's high-quality development (TEHQD) in the Yellow River Basin.

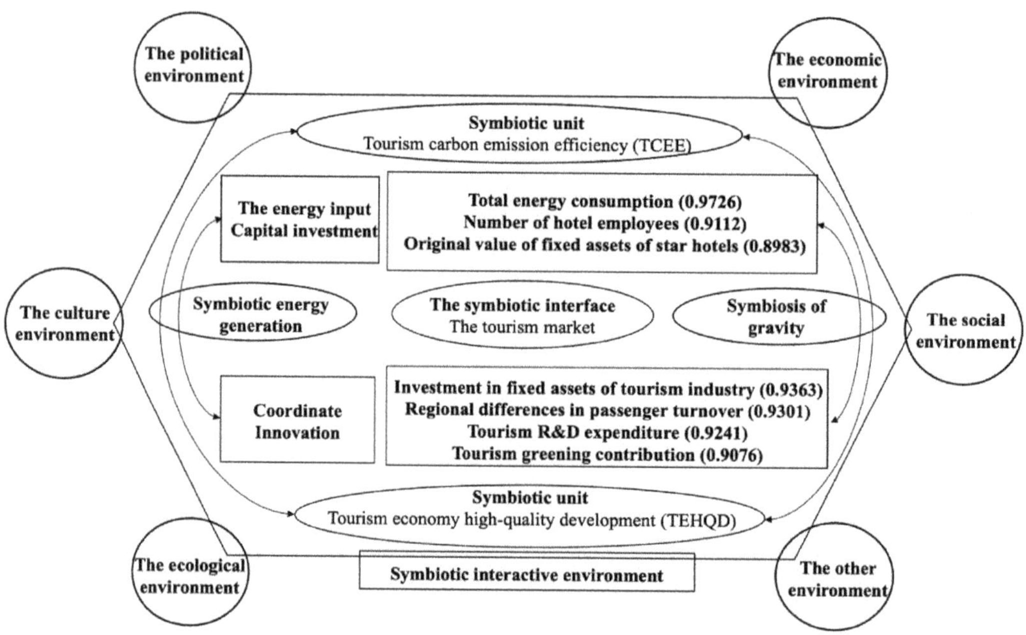

Figure 3. Interactive mechanism model of tourism's carbon emission efficiency (TCEE) and the tourism economy's high-quality development (TEHQD) in the Yellow River Basin.

5. Conclusions and Suggestions

5.1. Research Conclusions

Based on the calculation of tourism's carbon emission efficiency (TCEE) and the evaluation of the tourism economy's high-quality development (TEHQD), this paper discussed the symbiotic interaction mechanism between TCEE and TEHQD in nine provinces of the Yellow River Basin. The main conclusions are as follows:

- From the perspective of TCEE, the TCEE of the Yellow River Basin was in a state of fluctuation from 2010 to 2019, with a large time difference. The average value of TCEE in the Yellow River Basin was 0.9782, which is in the middle efficiency level. However, there was a large spatial difference in the TCEE of each province.
- From the perspective of TEHQD, the evaluation of TEHQD in the Yellow River Basin increased from 2010 to 2019, and the speed of development was fast, especially in western provinces. The inter-provincial differences in the TEHQD gradually widened.
- From the perspective of the symbiotic interaction between TCEE and TEHQD, on the whole, there was a parasitic relationship between TCEE and TEHQD in the Yellow River Basin. However, from the perspective of each year, all provinces showed positive or negative asymmetric symbiosis. The TEHQD in the Yellow River Basin has a greater impact on the TCEE. The TCEE and the TEHQD in Inner Mongolia, Shaanxi, Gansu, and Shandong provinces showed mutual influence and interaction (θ_C^M is close to 0.5), forming a relatively harmonious symbiotic state.
- From the influencing factors of symbiotic interaction between TCEE and TEHQD, energy input and capital input were the most important influence factors, but as time changed, the role of the energy input factor significantly reduced, and the role of labor input gradually emerged. Capital investment is always the key factor of symbiotic interaction between TCEE and TEHQD. Coordination and innovation are two important factors that affect the symbiosis between TCEE and TEHQD. With the change of time, the main influencing factors experienced a process of "sharing-coordination-innovation".

5.2. Research Suggestions

Based on the above conclusions, the following suggestions and countermeasures are proposed:

- In terms of tourism's carbon emission in the Yellow River Basin, especially in Shandong, Henan, Sichuan, and other provinces with large carbon emissions, tourism transportation carbon emissions should be taken as the main body of emission reduction, focusing on the rail infrastructure and related supporting construction. The construction of central and western provinces especially need to strengthen the transport network and expand the advantages of rail transport to reduce high carbon emissions from air and road transport. Secondly, tourism's energy consumption should be changed from the internal source. For example, the government can regulate the high energy consumption behavior of tourism enterprises and individuals by carbon emission tax, subsidy, and other ways. Tourism enterprises can also provide corresponding incentives and compensation measures for tourists.
- In terms of tourism's carbon emission efficiency (TCEE), firstly, on the basis of strengthening desirable output, input and undesirable output should be continuously reduced. Secondly, the Yellow River Basin provinces should make efforts to break the administrative regional barriers and promote experience exchange among provinces. Full play should be given to the leading role of provinces with high efficiency in tourism carbon emissions and the intercommunication of technology, concept, management, and other aspects among provincial administrative regions should be realized in the Yellow River Basin. In particular, the eastern region needs to focus on the issue of tourism's carbon emission efficiency to solve the existing large regional differences. Thirdly, the creation of structural (tangible) measures are fundamental [5]. Governments and municipal councils should issue guidelines and establish participatory networks to involve various stakeholders related to tourism and planning.

- In terms of the tourism economy's high-quality development (TEHQD), firstly, we should focus on the "coordinated" development of regions, industries, and other aspects. Secondly, giving full play to the advantages of the provinces, the construction of the tourism economy's high-quality development should be promoted in the Yellow River Basin. For example, Shandong and Henan rely on their good location conditions and economic advantages; thus, they should enhance the level of innovation and open up, promoting the development of green. At the same time, two-way interactions with the central and western regions should be strengthened to narrow the gap in the quality development of the tourism economy in different regions. Some provinces, such as Inner Mongolia, Shaanxi, and Shanxi should, on the basis of maintaining their current development status, give full play to their geographical advantages to link the east with the west and play the role of a bridge for regional connection. Qinghai, Sichuan, Ningxia, and Gansu should make full use of the advantages of good tourism resources, increase the input and output of tourism efficiency, and promote the tourism economy's high-quality development.
- In terms of the symbiotic interaction between tourism's carbon emission efficiency (TCEE) and the tourism economy's high-quality development (TEHQD), the efficiency of tourism's carbon emissions should be improved to meet the tourism economy's high-quality development, from the parasitic development to the symbiotic development. Secondly, each province should put forward countermeasures for its own problems according to local conditions. For example, Ningxia should focus on improving the efficiency of tourism's carbon emissions; at the same time, they should strengthen the construction of the high-quality tourism economy, so that the reverse asymmetric symbiosis can gradually change to a positive symbiosis. Inner Mongolia should continue to maintain the approximate positive symbiosis relationship between tourism's carbon emission efficiency and the tourism economy's high-quality development and should promote its transition to positive symbiosis. Thirdly, the Yellow River Basin should pay more attention to the improvement of tourism's carbon emission efficiency, while Gansu and Shandong should focus on improving the high-quality development level of the tourism economy.
- In terms of the influencing factors of symbiotic interaction between tourism's carbon emission efficiency (TCEE) and the tourism economy's high-quality development (TEHQD). Firstly, energy input and capital input in tourism's carbon emission efficiency should be strengthened. Energy utilization efficiency should be improved, and the green development level of tourism's carbon emission efficiency should be promoted. Tourism capital input should be increased and the development of tourism boosted. Importance should be attached to the role of influencing factors such as total energy consumption, the number of hotel employees, and the original value of fixed assets of star hotels in tourism's carbon emission efficiency in particular. Secondly, importance should be attached to the role of "coordination" and "innovation" in the tourism economy's high-quality development. Increasing investment in innovation, increasing research and development funds, and cultivating innovative talents should all be priorities. Striving to narrow the gap between regions, industries, and departments, and realizing the coordinated and sustainable development of them should be considered as well. In particular, attention should be paid to the effects of fixed assets investment, R&D expenditure, regional differences in tourist turnover, and sharing degrees of tourism greening.

5.3. Limitations and Future Research Directions

This paper presented an evaluation method and constructed a symbiotic interaction mechanism between tourism's carbon emission efficiency and the tourism economy's high-quality development in the Yellow River Basin, which provides theoretical support for the subsequent evaluation and practical basis for subsequent policy practice. However, due to the availability of data and the inadequacy of previous relevant studies, the selection and establishment of the index system may have deficiencies. At the same time, the selection of indicators are quantitative indicators without qualitative evaluation, which has limitations. To select a case in this paper, on the other hand, there is a limit too. It can only reflect the status of the Yellow River basin, a single specific area. At the same time, the influencing factors of the interaction mechanism were not further subdivided, such as the different effects of political, economic, cultural, social, and ecological factors on the interaction mechanism.

Thus, in the future, the measurement of tourism's carbon emission and its efficiency should be more precise and specific. For example, a variety of methods can be used to compare to obtain more accurate results, or the carbon emissions of the ecosystem in terms of technogenic pollution can be taken into account. Further exploration will be carried out from the aspects of technology and energy efficiency, such as studying the impact of existing technologies on tourism's carbon emission efficiency and further on tourism economy or the impact of innovative technologies on energy efficiency in tourism. Secondly, the symbiotic interaction mechanism between tourism's carbon emission efficiency and the tourism economy's high-quality development can be explored from multiple regions and multiple levels. On the basis of expanding the research area, different research scales were explored from different levels, such as region, city, county, or village to build a complete interactive mechanism model. Thirdly, the symbiotic interaction mechanism between tourism carbon emission efficiency and tourism economy high-quality development was studied from multiple perspectives. To improve the selection of influencing factors and the establishment of the index system, different mechanisms of action can be studied from the internal and external aspects of the symbiotic interaction system. On this basis, different types of influencing factors can be explored separately to find the commonality and individuality of each type of factor.

Author Contributions: Conceptualization, S.L. and Z.C.; methodology, S.L. and Z.C.; funding acquisition, Y.T. and B.H.; investigation, S.L., Z.C., Y.T. and B.H.; resources, Z.C.,Y.T. and B.H.; software, S.L. and Z.C.; formal analysis, S.L. and Z.C.; data curation, S.L.; writing—original draft preparation, S.L. and Z.C.; writing—review and editing, S.L. and B.H.; visualization, S.L.; supervision, B.H.; project administration, B.H. and Y.T. All authors have read and agreed to the published version of the manuscript.

Funding: This research was supported by the Project of Hainan Academician Innovation Platform Scientific Research (YSPTZX202210), the Project of National Social Science Foundation of China (21BJY194).

Institutional Review Board Statement: Not applicable.

Informed Consent Statement: Not applicable.

Data Availability Statement: The sources of relevant data acquisition have been described in the text.

Conflicts of Interest: The authors declare no conflict of interest.

References

1. Gössling, S. Global Environmental Consequences of Tourism. *Glob. Environ. Chang.* **2002**, *12*, 283–302. [CrossRef]
2. Scott, D.; Amelung, B.; Ceron, J.P.; Dubois, G.; Gössling, S.; Peeters, P.; Simpson, M.C. *Climate Change and Tourism: Responding to Global Challenges. Climate Change & Tourism Responding to Global Challenges*; UNWTO: Madrid, Spain, 2007; pp. 168–181.
3. Lopes, H.; Remoaldo, P.; Silva, M.; Ribeiro, V.; Vide, J.M. Climate in Tourism's Research Agenda: Future Directions Based on Literature Review. *Geogr. Bull.* **2021**, *90*, 1–58. [CrossRef]
4. Scott, D.; Hall, C.M.; Gossling, S. Global Tourism Vulnerability to Climate Change. *Ann. Tour. Res.* **2019**, *77*, 49–61. [CrossRef]
5. Lopes, H.; Remoaldo, P.; Ribeiro, V.; Javier, M. Pathways for Adapting Tourism to Climate Change in an Urban Destination—Evidences Based on Thermal Conditions for the Porto Metropolitan Area (Portugal). *J. Environ. Manag.* **2022**, *315*, 115161. [CrossRef]
6. UNWTO. *Tourism Highlights*, 2014th ed.; World Tourism Organization: Madrid, Spain, 2014.
7. Stryzhak, O. Assessment of the Relationship Between the Tourism Sector Development and Other Sectors of Economy. *Econ. Dev.* **2019**, *18*, 10–18. [CrossRef]
8. Zhang, F.; Peng, H.; Sun, X.; Song, T. Influence of Tourism Economy on Air Quality-An Empirical Analysis Based on Panel Data of 102 Cities in China. *Int. J. Environ. Res. Public Health* **2022**, *19*, 4393. [CrossRef] [PubMed]

9. Sun, Y.Y. Decomposition of Tourism Greenhouse Gas Emissions: Revealing the Dynamics between Tourism Economic Growth, Technological Efficiency, and Carbon Emissions. *Tour. Manag.* **2016**, *55*, 326–336. [CrossRef]
10. William, D.J. Changing Course: A Global Business Perspective on Development and the Environment. *Thromb. Haemost.* **1992**, *111*, 575–582.
11. Ruan, W.; Li, Y.; Zhang, S.; Liu, C.H. Evaluation and Drive Mechanism of Tourism Ecological Security Based on the DPSIR-DEA Model. *Tour. Manag.* **2019**, *75*, 609–625. [CrossRef]
12. Surugiu, C.; Surugiu, M.R.; Breda, Z.; Dinca, A.I. An Input-output Approach of CO2 Emissions in Tourism Sector in Post-communist Romania. *Procedia Econ. Financ.* **2012**, *3*, 987–992. [CrossRef]
13. Liu, G.; Shi, P.; Hai, F.; Zhang, Y.; Li, X. Study on Measurement of Green Productivity of Tourism in the Yangtze River Economic Zone, China. *Sustainability* **2018**, *10*, 2786. [CrossRef]
14. Kytzia, S.; Walz, A.; Wegmann, M. How Can Tourism Use Land More Efficiently? A Model-based Approach to Land-use Efficiency for Tourist Destinations. *Tour. Manag.* **2011**, *32*, 629–640. [CrossRef]
15. Perch-Nielsen, S.; Sesartic, A.; Stucki, M. The Greenhouse Gas Intensity of the Tourism Sector: The Case of Switzerland. *Environ. Sci. Policy* **2010**, *13*, 131–140. [CrossRef]
16. Zha, J.; He, L.; Liu, Y.; Shao, Y. Evaluation on Development Efficiency of Low-carbon Tourism Economy: A Case Study of Hubei Province, China. *Socio-Econ. Plan. Sci.* **2019**, *66*, 47–57. [CrossRef]
17. Jiang, G.; Zhu, A.; Li, J. Measurement and Impactors of Tourism Carbon Dioxide Emission Efficiency in China. *J. Environ. Public Health* **2022**, *14*, 9161845. [CrossRef] [PubMed]
18. Reilly, J.; Williams, P.; Haider, W. Moving Towards More Eco-efficient Tourist Transportation to a Resort Destination: The Case of Whistler, British Columbia. *Res. Transp. Econ.* **2010**, *26*, 66–73. [CrossRef]
19. Li, P.; Yang, L.G. Ecological Footprint Study on Tourism Itinerary Products in Shangri-La, Yunnan Province, China. *Acta Ecol. Sin.* **2007**, *27*, 2954–2963. [CrossRef]
20. Li, Y.; Zhang, L. Ecological Efficiency Management of Tourism Scenic Spots Based on Carbon Footprint Analysis. *Int. J. Low-Carbon Technol.* **2020**, *15*, 550–554. [CrossRef]
21. Kuo, N.; Chen, P. Quantifying Energy Use, Carbon Dioxide Emission, and Other Environmental Loads from Island Tourism Based on a Life Cycle Assessment Approach. *J. Clean. Prod.* **2009**, *17*, 1324–1330. [CrossRef]
22. Han, J. Carrying Capacity of Low Carbon Tourism Environment in Coastal Areas from the Perspective of Ecological Efficiency. *J. Coast. Res.* **2018**, *83*, 199–203. [CrossRef]
23. Zha, J.; Tan, T.; Yuan, W.; Yang, X.; Zhu, Y. Decomposition Analysis of Tourism CO2 Emissions for Sustainable Development: A Case Study of China. *Sustain. Dev.* **2020**, *28*, 169–186. [CrossRef]
24. Haibo, C.; Ke, D.; Fangfang, W.; Ayamba, E.C. The Spatial Effect of Tourism Economic Development on Regional Ecological Efficiency. *Environ. Sci. Pollut. Res. Int.* **2020**, *27*, 1–18. [CrossRef] [PubMed]
25. Qiu, X.; Fang, Y.; Yang, X.; Zhu, F. Tourism Eco-Efficiency Measurement, Characteristics, and Its Influence Factors in China. *Sustainability* **2017**, *9*, 1634. [CrossRef]
26. Guo, L.; Li, P.; Zhang, J.; Xiao, X.; Peng, H. Do Socio-economic Factors Matter? A Comprehensive Evaluation of Tourism Eco-efficiency Determinants in China Based on the Geographical Detector Model. *J. Environ. Manag.* **2022**, *320*, 115812. [CrossRef]
27. Xiong, G.; Deng, J.; Ding, B. Characteristics, Decoupling Effect, and Driving Factors of Regional Tourism's Carbon Emissions in China. *Environ. Sci. Pollut. Res. Int.* **2022**, *29*, 47082–47093. [CrossRef]
28. Tong, Y.; Zhou, H.; Jiang, L. Exploring the Transition Effects of Foreign Direct Investment on the Eco-efficiency of Chinese Cities: Based on Multi-source Data and Panel Smooth Transition Regression Models. *Ecol. Indic.* **2020**, *121*, 107073. [CrossRef]
29. Nitti, M.; Pilloni, V.; Giusto, D.; Popescu, V. IoT Architecture for a Sustainable Tourism Application in a Smart City Environment. *Mob. Inf. Syst.* **2017**, *2017*, 9201640. [CrossRef]
30. Cortés-Jiménez, I. Which Type of Tourism Matters to the Regional Economic Growth? The Cases of Spain and Italy. *Int. J. Tour. Res.* **2010**, *10*, 127–139. [CrossRef]
31. Hassan, T.H.; Salem, A.E.; Abdelmoaty, M.A. Impact of Rural Tourism Development on Residents' Satisfaction with the Local Environment, Socio-Economy and Quality of Life in Al-Ahsa Region, Saudi Arabia. *Int. J. Environ. Res. Public Health* **2022**, *19*, 4410. [CrossRef]
32. Shi, C.; Peng, J. Construction of Low-carbon Tourist Attractions Based on Low-carbon Economy. *Energy Procedia* **2011**, *5*, 759–762. [CrossRef]
33. Zhang, Y.; Tian, L. The Sustainable Development of Circular Economy under the Perspective of Ecological Tourism. *Adv. Mater. Res.* **2014**, *1010–1012*, 2090–2093. [CrossRef]
34. Tang, Z.; Shi, C.B.; Liu, Z. Sustainable Development of Tourism Industry in China under the Low-carbon Economy. *Energy Procedia* **2011**, *5*, 1303–1307. [CrossRef]
35. Akadiri, S.S.; Lasisi, T.T.; Uzuner, G.; Akadiri, A.C. Examining the Impact of Globalization in the Environmental Kuznets Curve Hypothesis: The Case of Tourist Destination States. *Environ. Sci. Pollut. Res. Int.* **2019**, *26*, 12605–12615. [CrossRef]
36. Balsalobre-Lorente, D.; Driha, O.M.; Leitão, N.C.; Murshed, M. The Carbon Dioxide Neutralizing Effect of Energy Innovation on International Tourism in EU-5 Countries Under the Prism of The EKC Hypothesis. *J. Environ. Manag.* **2021**, *298*, 113513. [CrossRef] [PubMed]

37. Andraz, J.M.; Norte, N.M.; Goncalves, H.S. Effects of Tourism on Regional Asymmetries: Empirical Evidence for Portugal. *Tour. Manag.* **2015**, *50*, 257–267. [CrossRef]
38. Araujo, L.M.D.; Bramwell, B. Partnership and Regional Tourism in Brazil. *Ann. Tour. Res.* **2002**, *29*, 1138–1164. [CrossRef]
39. Baidal, J. Regional Development Policies: An Assessment of Their Evolution and Effects on the Spanish Tourist Model. *Tour. Manag.* **2003**, *24*, 655–663. [CrossRef]
40. Morri, E.; Pruscini, F.; Scolozzi, R.; Santolini, R. A Forest Ecosystem Services Evaluation at the River Basin Scale: Supply and Demand Between Coastal Areas and Upstream Lands (Italy). *Ecol. Indic.* **2014**, *37*, 210–219. [CrossRef]
41. Mbaiwa, J.E. The Socio-economic and Environmental Impacts of Tourism Development on the Okavango Delta, North-western Botswana. *J. Arid Environ.* **2003**, *54*, 447–467. [CrossRef]
42. Wang, Z.W. Interaction Between the Tourism Industry and Ecological Environment Based on the Complicated Adaptation System (CAS) Theory: A Case Study on Henan Province, China. *Nat. Environ. Pollut. Technol.* **2020**, *19*, 1039–1045. [CrossRef]
43. Wang, Z.W.; Han, H.L. Analysis on Tourism Environmental Pollution and Tourism Economy-Ecological Environmental Coordination Degree: A Case Study from China. *Nat. Environ. Pollut. Technol.* **2021**, *20*, 1353–1361. [CrossRef]
44. Pan, Y.; Weng, G.; Li, C.; Li, J. Coupling Coordination and Influencing Factors among Tourism Carbon Emission, Tourism Economic and Tourism Innovation. *Int. J. Environ. Res. Public Health* **2021**, *18*, 1601. [CrossRef]
45. Liu, Y.; Suk, S. Coupling and Coordinating Relationship between Tourism Economy and Ecological Environment-A Case Study of Nagasaki Prefecture, Japan. *Int. J. Environ. Res. Public Health* **2021**, *18*, 12818. [CrossRef] [PubMed]
46. Hunter, C.; Shaw, J. The Ecological Footprint as a Key Indicator of Sustainable Tourism. *Tourism Management* **2007**, *28*, 46–57. [CrossRef]
47. Gössling, S.; Hansson, C.B.; Horstmeier, O.; Saggel, S. Ecological Footprint Analysis as A Tool to Assess Tourism Sustainability. *Ecol. Econ.* **2002**, *43*, 199–211. [CrossRef]
48. Lawson, S.R.; Manning, R.E.; Valliere, W.A.; Wang, B. Proactive Monitoring and Adaptive Management of Social Carrying Capacity in Arches National Park: An Application of Computer Simulation Modeling. *J. Environ. Manag.* **2003**, *68*, 305–313. [CrossRef]
49. Zhang, X.H.; Lin, M.; Lin, S. The Method of Tourism Environmental Carrying Capacity. *Ecosyst. Assess. Fuzzy Syst. Manag.* **2014**, *254*, 87–92.
50. Xu, M.; Liu, C.; Li, D.; Zhong, X.L. Tourism Ecological Security Early Warning of Zhangjiajie, China Based on the Improved TOPSIS Method and the Grey GM (1,1) Model. *J. Appl. Ecol.* **2017**, *28*, 3731–3739.
51. Lai, Z.; Ge, D.; Xia, H.; Yue, Y.; Wang, Z. Coupling Coordination between Environment, Economy and Tourism: A Case Study of China. *PLoS ONE* **2020**, *15*, e0228426. [CrossRef]
52. Zahedi-Seresht, M.; Khosravi, S.; Jablonsky, J.; Zykova, P. A Data Envelopment Analysis Model for Performance Evaluation and Ranking of DMUs with Alternative Scenarios. *Comput. Ind. Eng.* **2020**, *152*, 107002. [CrossRef]
53. Zhao, P.J.; Zeng, L.E.; Lu, H.Y.; Zhou, Y.; Hu, H.Y.; Wei, X.Y. Green Economic Efficiency and its Influencing Factors in China From 2008 to 2017: Based on the Super-SBM Model with Undesirable Outputs and Spatial Dubin Model. *Sci. Total Environ.* **2020**, *741*, 140026. [CrossRef]
54. Wang, F.; Liu, P.; Wang, P. An Evaluation Study of Rural Scientific and Technological Talents Based on TODIM Method with Hybrid Indicator. *J. Intell. Fuzzy Syst.* **2021**, *40*, 1–14. [CrossRef]
55. Lan, Q.; Liu, C.; Ling, S. Research on Measurement of Symbiosis Degree Between National Fitness and the Sports Industry from the Perspective of Collaborative Development. *Int. J. Environ. Res. Public Health* **2019**, *16*, 2191. [CrossRef]
56. Zhou, C.S.; Chen, J.; Wang, S.J. Examining the Effects of Socioeconomic Development on Fine Particulate Matter (PM2.5) in China's Cities Using Spatial Regression and the Geographical Detector Technique. *Sci. Total Environ.* **2018**, *619–620*, 436–445. [CrossRef]

energies

MDPI

Article

Tourism Development, Carbon Emission Intensity and Urban Green Economic Efficiency from the Perspective of Spatial Effects

Xiaorong He [1], Jizhi Shi [1], Haichao Xu [1,*], Chaoyue Cai [1,*] and Qiangsheng Hu [2]

1 Tourism College, Hunan Normal University, Changsha 410081, China
2 College of Humanities and Arts, Hunan University of Finance and Economics, Changsha 410205, China
* Correspondence: xuhaic@hunnu.edu.cn (H.X.); caichaoyue2021@hunnu.edu.cn (C.C.)

Abstract: In recent years, China has increasingly emphasized green development. Therefore, it is of theoretical and practical significance to study the green economic effect and carbon reduction effect of tourism development for the transformation of economic development. Using the superefficient EBM to measure the green economic efficiency of 280 cities from 2007–2019, we rely on the spatial Durbin model to explore the spatial spillover utility and nonlinear characteristic relationship of tourism development on green economic efficiency and carbon emission intensity and test the mediating effect of carbon emission intensity. The findings are as follows: (1) Under the exogenous shock test of the "low-carbon city" pilot policy, the spatial spillover effect of tourism development on urban green economic efficiency and carbon emission intensity is robust to spatial heterogeneity. (2) The spatial spillover effects of tourism development on the green economic efficiency and carbon emission intensity of cities show a nonlinear characteristic relationship of "U" and "M" shapes. After tourism development reaches a certain high level, the green economy effect and carbon emission reduction effect are significantly increased. (3) Carbon emission intensity has a significant mediating effect on the impact of tourism development on urban green economic efficiency.

Keywords: tourism development; green economy; carbon emissions; spillover effect; superefficient EBM

Citation: He, X.; Shi, J.; Xu, H.; Cai, C.; Hu, Q. Tourism Development, Carbon Emission Intensity and Urban Green Economic Efficiency from the Perspective of Spatial Effects. *Energies* **2022**, *15*, 7729. https://doi.org/10.3390/en15207729

Academic Editors: Junpeng Zhu and Xinlong Xu

Received: 16 September 2022
Accepted: 17 October 2022
Published: 19 October 2022

1. Introduction

Anthropogenic activities such as deforestation and burning of fossil fuels produce a large amount of greenhouse gases, and the continued increase in carbon dioxide emissions is expected to have a catastrophic impact on the global climate system. Therefore, it has become the consensus of various countries around the world to mitigate the negative impacts of climate change and ensure sustainable development by continuously reducing carbon emissions [1]. The United Nations Environment Programme (UNEP) proposed in 2011 that a green economy can improve human well-being and social justice and that it can serve as a pattern for building a resource-efficient and environmentally friendly society. It is also an important way to promote sustainable development and aid in poverty eradication. Since its reform and opening up, China's economy has grown at a rapid rate of 9.7% per year [2], rapidly becoming the world's second largest economy, ranking among the upper middle-income countries and lifting 800 million people out of poverty [3]. However, China's economic development still has serious problems due to the widespread use of energy that exploits the environment to promote economic growth. According to the World Bank, China has become the world's largest energy consumer. China accounts for 27.6% of the world's CO_2 emissions [4], and the Chinese government has committed to ensure that carbon emissions peak by 2030. Thus, the Chinese government has recognized that current and potential environmental degradation poses a serious threat to China's sustainable development and that the previous aggressive development model is no longer appropriate. Solving China's problems will not only benefit the quality of China's economic development but also provide a model in balancing economic growth and the environment that could be used by other developing countries around the world. However, based on economic stability, ensuring the achievement of this goal is an important challenge for the Chinese government, and improving economic efficiency is certainly an effective way of achieving it [5]. For European countries, tourism, as a key sector of the European economy, is an important source of income and employment in Europe [6]. However, carbon intensity in Europe is growing steadily. Therefore, these countries must also mitigate carbon emissions through tourism reforms [6]. It can be seen that both developing and developed countries are actively seeking a balance between energy consumption and economic growth, exploring the carbon reduction effect of the tourism industry, the green economy effect and its interaction, so that the research results can become a guide to the reality of the dilemma between tourism and environmental protection.

In the economic sense, economic efficiency usually refers to how to obtain as much output of economic goods from as few factor inputs as possible, mainly considering the input–output ratio of labor and capital inputs [7]. In 2010, Yang and Hu first proposed green economic efficiency as a key indicator to measure the level of the green economy, and green economic efficiency further addresses the issues of energy constraints and undesired output [8]. Therefore, green economic efficiency can be defined as an economic production system that can achieve greater economic output or less environmental pollution with constant or reduced factor inputs, taking into account the constraints of resources and the environment [5,9,10]. For studies of green economy efficiency, data envelopment analysis (DEA) is a common method used by most scholars [11–16]. However, DEA is either input- or output-oriented and cannot consider both output and input, which is a limitation of the DEA measurement method. To avoid such problems, some scholars use the slack-based measurement (SBM) method to measure green economic efficiency [5,17–23]. Although the SBM model achieves a balance of inputs and outputs, it also ensures different proportional changes in inputs and outputs, which are closer to the true values [5]. However, the SBM model cannot solve the problem of undesired outputs. In recent years, the EBM method has been increasingly used by scholars to study energy efficiency or environmental efficiency [24–28]. Therefore, this paper adopts the EBM model to measure the green economy efficiency level of 280 prefecture-level cities in China.

It is widely acknowledged that the tourism industry makes a significant contribution to economic development in terms of income generation, tax revenue and employment [29]. After reform and opening up, tourism has developed along with China's economic take-off, and it has become a pillar industry of China's economy. According to data from the China Tourism Research Institute, the number of domestic tourists rose from 2.13 billion to 6.006 billion from 2010 to 2019, and the total revenue from tourism rose from 1.57 trillion RMB to 6.63 trillion RMB. This rise in tourism is not only the case in China; the tourism industry is also a key sector of the European economy, prioritized by the EU as an important source of income, employment and economic growth [6]. As the contribution of tourism agglomeration increases, scholars are increasingly looking at the relationship between tourism development and economic growth [6]. During the continuous development of tourism, the carbon emissions generated by the activities of tourism itself, such as transportation, accommodation and catering [30], as well as tourism-related industries [31], have gradually increased. For countries or regions with large populations or developed tourism industries, the relationship between tourism and carbon emissions has received much attention [32], for example, in China [33], the European Union [6] and Southeast Asia [32].

Compared with other industries, tourism is less polluting, less ecologically damaging and less energy-consuming, which are hallmarks of a typical green industry and demonstrate that it contributes to the development of the green economy in cities [34]. Therefore, based on the uniqueness of tourism, the relationship between tourism development, carbon emissions and economic growth has been a hot topic of research [6,32,35–42]. After the introduction of the concept of the green economy, what is the relationship between tourism as a green industry and carbon emissions and the green economy? Is there a necessary link between tourism development, carbon emissions and a green economy? As cities become more frequently and closely connected, the development of neighboring cities is mutually linked and influenced, so it is necessary and important to consider spatial effects. Furthermore, with the global advocacy for green development, it is crucial to understand the relationship between urban tourism development, carbon emission intensity and green economic efficiency. This is not only important for China to achieve the goal of reducing carbon emissions but also has important practical significance for the global exploration of green economic development paths.

The contribution of this paper may be as follows. First, the article defines the relationship among tourism development, carbon emission intensity and green economic efficiency. Supported by the data of 280 prefecture-level cities in China from 2007–2019, this paper confirms that urban tourism development has a significant positive effect on the reduction in carbon emission intensity and the improvement of green economic efficiency. Meanwhile, it confirms that carbon emission intensity plays a significant mediating role in the promotion of urban green economic efficiency by tourism development. Second, this paper compares the spatial relationship between tourism development on carbon emission intensity and green economic efficiency. By using the spatial Durbin model and introducing spatial effects, this paper confirms that tourism development has a significant spatial spillover effect on the mitigation of carbon emission intensity and the improvement of green economic efficiency. On this basis, this paper tests the robustness of the spatial spillover effect of tourism development on carbon emission intensity and green economic efficiency by introducing the "low carbon city" pilot policy as an exogenous shock. Third, the paper compares the nonlinear characteristic relationship of tourism development on carbon emission intensity and green economic efficiency. Based on the previous influence relationship, this paper further explores the nonlinear characteristic relationship of tourism development on the reduction in carbon emission intensity and the improvement of green economic efficiency and elaborates the influence of tourism development on carbon emission intensity and green economic efficiency.

In order to introduce the research of this paper more clearly, the main contents of each chapter are introduced in the form of flowcharts in the order of the research catalog of this paper, as shown in Figure 1. The literature review and research hypotheses are given in the next section. The econometric methodology and model utilized in this study are provided in section three. Section four addresses the empirical results, and the last section provides a discussion of the findings together with policy implications.

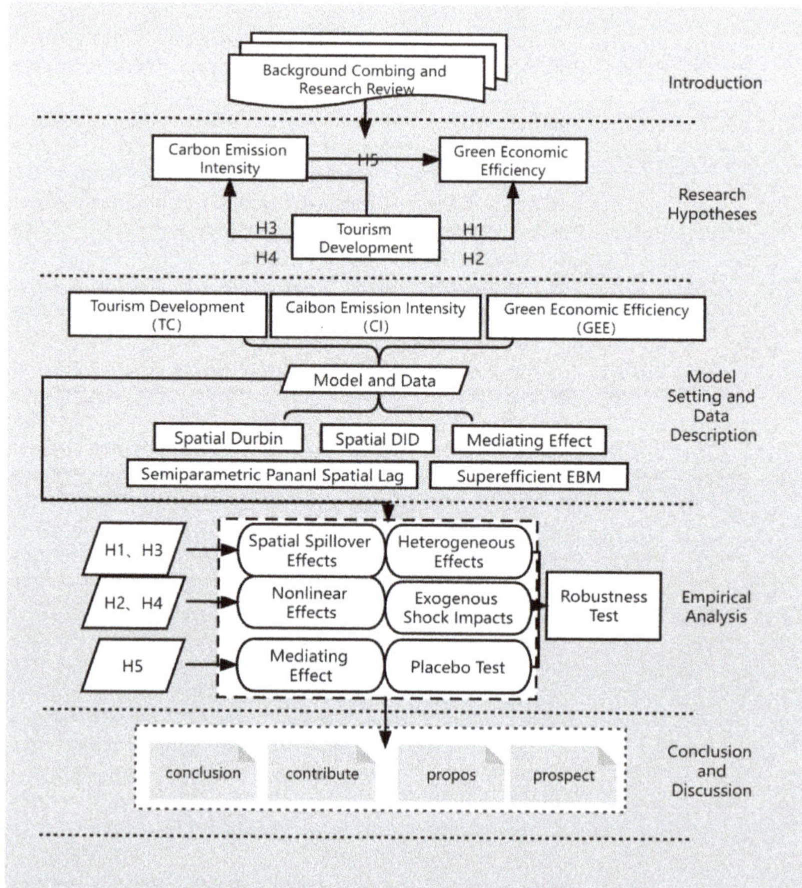

Figure 1. Flow chart.

2. Theoretical Foundations and Research Hypotheses

Although there is no unified academic definition of the green economy, scholars have recognized that achieving a positive interaction between economic growth and the ecological environment is the core connotation of green economic development [14,43]. The tourism industry is inextricably linked to the regional economy, both in terms of GDP contribution and employment contribution [6,36], and plays an important role in promoting economic development. In addition, the tourism industry is both dependent on the destination's ecological environment and protective of it. A good ecological environment improves the quality of the elements of tourism development, while tourism development further protects the local ecological environment and realizes the sustainable development of the regional ecology. From the core connotation of the green economy, the impact of tourism development on the green economy is mainly reflected in two aspects [44,45]: ecological and environmental effects and economic growth effects.

In the process of tourism development, it will have direct and indirect effects on the local economy and environment, which will affect the level of local green economic efficiency. As shown in Figure 2, tourism development and environmental pollution have an "EKC" effect, i.e., the environmental Kuznets curve is introduced into tourism development, and the two are found to have an "inverted U" curve relationship [32]. At the same time, tourism development has both positive and negative environmental externalities. The positive environmental externality refers to tourism as a friendly industry that does not involve industrial pollution [46] and has a significant carbon reduction effect [6]. The negative environmental externality, on the contrary, refers to the high-carbon nature of the tourism industry, which is a major contributor to greenhouse gas emissions [47]. Although it is recognized that the tourism industry itself has significant economic driving power [6], Corden et al. found that tourism development has a "Dutch disease" effect, i.e., tourism development promotes economic growth in the short term, but depresses the economy in the long term [48]. Based on the "Dutch disease" effect, scholars have shown that tourism development has a non-linear effect. More specifically, different levels of tourism specialization can have differential effects on economic growth [49]. These studies demonstrate that tourism development has a direct effect on economic efficiency and thus on green economic efficiency.

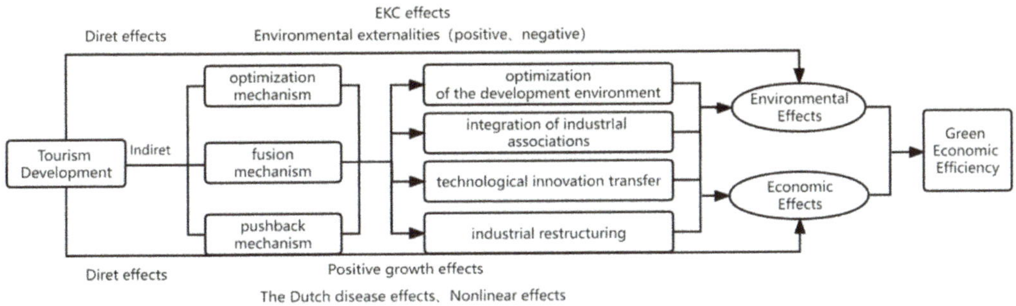

Figure 2. The mechanism of the impact of tourism development on green economy efficiency.

In addition to the direct impact, tourism development can also indirectly affect the efficiency of the green economy by influencing environmental and economic factors. Tourism development can optimize the ecological environment, industry linkage and integration, technological innovation and industrial structure. Tourism needs an excellent natural environment, so the development of tourism will inevitably require the relocation and withdrawal of highly polluting enterprises, performing a passive adjustment of industrial structure. In conclusion, environmental friendliness and economic growth are the core concepts of the green economy, tourism development in the process directly and indirectly affecting the local ecological environment and economic growth, with practical benefits and results to illustrate the existence of overlap between tourism development and green economic efficiency.

In terms of environmental benefits, the "environmental Kuznets" effect of tourism development is obvious [50], as the tourism industry produces environmental pollution such as waste gas and wastewater in the process of development, causing certain negative impacts on the ecological environment, but the economic growth effect brought by tourism development directly "compensates" by allowing environmental protection. In terms of indirect effects, tourism development can force the original industrial structure to be adjusted and optimized through the "crowding-out effect". In addition, the "dependency" of the local economy on the tourism industry raises the environmental awareness of the government and residents, creating formal and informal monitoring of the ecological environment and further optimizing the development environment. At the same time, tourism development also brings economic growth. For example, tourism development can directly lead to increased GDP and employment in cities and contribute to poverty alleviation in the destination. Moreover, the increasing integration of tourism with other industries not only facilitates the innovation and transfer of technology but also has a positive impact on the optimization of the industrial structure and market environment [21].

Hypothesis H1: *The impact of tourism development on the green economic efficiency of cities has a spillover effect.*

Hypothesis H2: *The impact of tourism development on urban green economic efficiency has a nonlinear relationship.*

The impact of tourism development on urban carbon emission intensity is mainly studied from the perspective of tourism industry agglomeration, and the environmental effect of tourism carbon emission reduction is mainly realized through externalities, which is a proven consensus [6,31,32,51]. The research on the relationship between tourism development and carbon emissions is mainly based on the environmental Kuznets curve hypothesis (EKC) proposed by American economists Grossman and Krueger. This hypothesis suggests that environmental quality deteriorates and then improves as the level of economic development increases, and there is an inverted "U" shaped relationship between environmental quality and economic development. The environmental effect brought by the development of the tourism industry is mainly studied from the two perspectives of the tourism economy and industrial agglomeration. On the one hand, the growth of the tourism economy follows the environmental Kuznets curve hypothesis, with a nonlinear characteristic relationship between the two [32], but at the same time, some scholars question this and test the relationship between the tourism economy and the decoupling of carbon emissions [33]. On the other hand, in the early stage of tourism industry agglomeration, the industry development method is relatively crude, and large-scale enterprises may be concentrated in the same area. Although this increases the regional GDP, as a process with fierce competition, it can also easily waste resources and energy and lead to a negative impact on the ecological environment [52]. As the level of agglomeration increases, the agglomeration of the tourism industry brings about a reduction in the cost of raw material transportation and transaction, a saving in energy and resources, and the agglomeration of enterprises with backward and forward linkage makes the exchange of knowledge and technology more convenient. The exchange and cooperation between tourism and other industries are likely to collide with new technologies, expand knowledge and technology spillover effects, and achieve synergistic innovation in technology, which in turn have a suppressive effect on carbon emissions and improve environmental pollution [6].

Hypothesis H3: *The impact of tourism development on urban carbon emission intensity has a spillover effect.*

Hypothesis H4: *The impact of tourism development on urban carbon emission intensity has a nonlinear relationship.*

Tourism development must both reduce carbon emissions and increase the GDP: on the one hand, tourism development needs to pay attention to reducing carbon dioxide emissions, both from tourism itself and from the manufacturing sector. According to existing research, tourism is not only highly correlated with manufacturing but also has a certain "crowding-out effect" on regional manufacturing [46]. Therefore, while developing tourism, clean energy should be used to reduce air pollution, thus attracting more tourists and generating more tourism revenue. This is also a mutually reinforcing process, as the development of tourism to a certain extent squeezes out the high-pollution manufacturing industry, which also alleviates regional carbon intensity. The ability of cities to reduce carbon emissions in turn directly affects the development of the urban green economy, and carbon emission intensity has a direct negative impact on the improvement of the urban green economy efficiency level [5,53,54]. Tourism development tends to consider the effect of its carbon emissions while influencing the efficiency of the urban green economy, which shows that carbon emissions have a very important role in the green development of tourism.

Hypothesis H5: *Carbon emission intensity has a mediating effect on the relationship between tourism development and urban green economic efficiency.*

3. Model Setting and Data Description

3.1. Model Construction

3.1.1. Superefficient EBM Model

The hybrid distance function model (EBM) can be compatible with the radial ratio of input frontier values to actual values and realize the effective combination of radial and nonradial methods in data envelopment analysis. The model makes up for the deficiencies of DEA and SBM, giving more consideration to the efficiency level [24–28]. The following superefficient EBM model with undesired outputs and nondirectional and constant payoffs of scale is used to measure the green economic efficiency of 280 cities across China, and the obtained combined efficiency value (GEE) is used as the core explanatory variable of the spatial econometric model.

$$r^* = \min \frac{\theta - \varepsilon_x \sum_{i=1}^{m} w_i{}^- s_i{}^- / x_{i0}}{\varphi + \varepsilon_y \sum_{r=1}^{s} w_r{}^+ s_r{}^+ / y_{r0} + \varepsilon_z \sum_{p=1}^{q} w_p{}^{z-} s_p{}^{z-} / z_{p0}} \quad \text{s.t.} \begin{cases} \sum_j^n x_{ij}\lambda_j + s_i{}^- = \theta x_{i0} \ (i = 1, 2, \cdots, m) \\ \sum_{j=1}^n y_{rj}\lambda_j - s_r{}^+ = \varphi y_{r0} \ (r = 1, 2, \cdots, s) \\ \sum_{j=1}^n z_{pj}\lambda_j + s_p^- = \varphi z_{p0} \ (p = 1, 2, \cdots, q) \\ \lambda_j \geqslant 0, s_i^-, Sr^+, s_p{}^- \geqslant 0 \end{cases} \tag{1}$$

Regarding the specific meaning of these variables, r* denotes the combined efficiency value, and x, y and z denote the input, desired output and undesired output elements, respectively. m, s and q denote their quantities. λ denotes the relative importance of the reference unit, ε is the core parameter representing the importance of the nonradial component, and θ is the efficiency value in the radial condition. w_i, w_r and w_p denote the i-th input, r-th desired output and p-th nonweights of the expected output indicators.

3.1.2. Spatial Durbin Model

This study adopts the spatial Durbin model (SDM) to test the spatial spillover effect of the tourism development level on urban green economic efficiency and carbon emission intensity and explores the nonlinear characteristic relationship on this basis. The acceleration of regional economic integration makes it possible for green economic efficiency to interact spatially among different cities. The spatial econometric model makes up for the deficiency that traditional measurement cannot introduce spatial factors, the spatial Durbin model contains the spatial dependence of both dependent and independent variables [14,20,21], and models (1)–(4) are as follows:

$$\begin{aligned} \ln GEE_{it} &= \rho W\ln GEE + \beta_1 \ln TC_{it} + \beta_2 \ln CON_{it} + \theta_1 W\ln TC_{it} + \theta_2 W\ln CON_{it} + \delta_i + \mu_t + \varepsilon_{it} \\ \ln GEE_{it} &= \rho W\ln GEE_{it} + \beta_1 \ln TC_{it} + \beta_2 \ln^2 TC_{it} + \beta_3 \ln CON_{it} + \theta_1 W\ln TC_{it} + \theta_2 W\ln^2 TC_{it} + \theta_3 W\ln CON_{it} + \delta_i + \mu_t + \varepsilon_{it} \\ \ln CI_{it} &= \rho W\ln CI_{it} + \beta_1 \ln TC_{it} + \beta_2 \ln CON_{it} + \theta_1 W\ln TC_{it} + \theta_2 W\ln CON_{it} + \delta_i + \mu_t + \varepsilon_{it} \\ \ln CI_{it} &= \rho W\ln CI_{it} + \beta_1 \ln TC_{it} + \beta_2 \ln^2 TC_{it} + \beta_3 \ln CON_{it} + \theta_1 W\ln TC + \theta_2 W\ln^2 TC_{it} + \theta_3 W\ln CON_{it} + \delta_i + \mu_t + \varepsilon_{it} \end{aligned} \tag{2}$$

where i and t denote region and time, respectively, W represents the spatial weight matrix, and β_i and θ_i are the parameter vectors to be estimated and the spatial regression coefficients of the tourism development level. "Green economic efficiency" is GEE, CI is "carbon emission intensity", TC is "tourism development", CON is the control variable, and ρ is the spatial regression coefficient. "Tourism development" is TC, CON is the control variable, and ρ is the spatial regression coefficient. δ_i is the individual fixed effect, μ_t is the time fixed effect, and ε_{it} is the random error term.

3.1.3. Semiparametric Panel Spatial Lag Model

The semiparametric panel spatial lag model not only analyzes the influence of spatial factors but can also test the spatial nonlinear relationship between variables. To further analyze the spatial nonlinear effects of tourism development on urban green economic efficiency and carbon emission intensity, this paper draws on the related research [55] to further construct a semiparametric panel spatial lag model.

$$\begin{aligned} \ln GEE_{it} &= \alpha_i + \rho W \ln GEE_{it} + \beta_1 \ln CON_{it} + \theta_1 W \ln CON_{it} + G(\ln TC_{it}) + \varepsilon_{it} \\ \ln CI_{it} &= \alpha_i + \rho W \ln CI_{it} + \beta_1 \ln CON_{it} + \theta_1 W \ln CON_{it} + G(\ln TC_{it}) + u_{it} \end{aligned} \tag{3}$$

where $G(\ln TC_{it})$ represents the nonparametric part of the unknown function, α_i represents the individual effect, ε_{it} and u_{it} represent the random perturbation term.

3.1.4. Mediating Effect Model

To test whether tourism development affects urban green economic efficiency through carbon emission intensity, a mediating effect model is constructed by drawing on the related research [14,21].

$$\begin{aligned} \ln GEE_{it} &= \alpha_0 + \alpha_1 \ln TC_{it} + \textstyle\sum_{j=2}^{n} \alpha_j \ln CON_{it} + \eta_i + \varepsilon_{it} \\ \ln CI_{it} &= \beta_0 + \beta_1 \ln TC_{it} + \textstyle\sum_{j=2}^{n} \beta_j \ln CON_{it} + \eta_i + \varepsilon_{it} \\ \ln GEE_{it} &= \delta_0 + \delta_1 \ln TC_{it} + \delta_2 \ln CI_{it} + \textstyle\sum_{j=3}^{n} \delta_j \ln CON_{it} + \eta_i + \varepsilon_{it} \end{aligned} \tag{4}$$

η_i is the individual fixed effect.

3.1.5. Exogenous Shock Testing Model of the Low-Carbon City Pilot Policy

1. Endogenous Relationship between Low-carbon Cities and Tourism Development

Low-carbon cities achieve green development by adjusting the industrial structure, reducing disposable energy use, using renewable resources as much as possible, and developing low-carbon transportation systems. Tourism can influence industrial optimization, clean energy use and low-carbon transportation. Under the "double carbon" strategy, low-carbon tourism is undoubtedly one of the paths for low-carbon city reform, and in addition to the low consumption of tourism itself and the low carbonization of the tourism process, the strong correlation of tourism itself with a reduction in carbon emissions can lead to the low-carbon transformation of more industries. Tourism development is in line with the essence and connotation of low-carbon city development, and low-carbon tourism can also be an effective path for green economic development in low-carbon cities [45]. Therefore, this paper uses the exogenous shock of low-carbon city pilots to evaluate the existing model and test the robustness of the spillover effect of tourism development on the green economic efficiency and carbon emission intensity of cities.

2. Model Setting and Testing

The dummy variable is constructed as to whether the city is a "low-carbon city" pilot or not, and takes the value of 1 if the city is already a "low-carbon city" pilot at the end of the year and 0 otherwise [56].

Based on the spatial Durbin model, a multitemporal spatial DID expansion estimation was constructed based on relevant studies [56], and the coefficients of the variables were estimated by randomly selecting cities and any year as the sample size and rerunning the model test. Comparing whether there is a significant difference between the true value and the estimated interval can indicate whether the model estimation is biased by omitting city-time-level variables.

3.2. *Description of Variables*

3.2.1. Tourism Development

Compared with a single indicator, this paper adopts tourism development indicators [57], including the tourism economy and tourism scale, to measure the level of tourism development more comprehensively. Based on the domestic tourism income and number of people, inbound tourism income and the number of people under the city scale, the entropy weight TOPSIS method is used to obtain the tourism development level (TC) of 280 cities nationwide, which is used as the core explanatory variable of this spatial econometric model.

3.2.2. Carbon Emission Intensity

Urban carbon emissions include carbon emissions from direct energy consumption as well as carbon emissions from electric energy and thermal energy consumption. Drawing on the related research [58], carbon emissions from liquefied petroleum gas, transportation, electric energy and thermal energy consumption are summed to obtain the total urban carbon emissions. The ratio of the total carbon emissions of a city to its GDP is used to measure the intensity of carbon emissions of the city.

3.2.3. Green Economy Efficiency

The green economy is an economic development model that maximizes resource utilization by improving development efficiency and reducing environmental pressure. According to the current situation of social, environmental and resource problems, taking into account the availability and consistency of data, the green economy efficiency index system of cities is constructed from the perspective of resource input-economic output-pollution output regarding relevant research. Based on the existing studies [14,20–22,59], we add "the number of industrial enterprises above the scale" to the labor factor level and change the previous consideration from the number of employees to the number of employees and enterprises to improve the index system. Table 1 shows the green economy efficiency measurement indicators.

Table 1. Urban green economy efficiency measurement index system.

Guideline Layer		Indicator Layer	Guideline Layer		Indicator Layer
Input metrics	Labor	The number of employees in the city at the end of the year	Output indicators	Expected output	Urban green area
		Number of industrial enterprises above designated size			Real GDP
	Capital	Local financial expenditure on science and technology			Total retail consumption per capita
		Investment in fixed assets		Undesirable output	Industrial wastewater discharge
		Area of urban construction land			Industrial sulfur dioxide emissions
	Energy	Total air supply			Industrial soot emissions
		Total water supply			
		Total electricity consumption			

3.2.4. Control Variables

Based on previous studies, such as Tong Yun [60], the current studies mostly analyze the effects of the economic level, environmental regulation, technological innovation and foreign investment on urban green economic efficiency and carbon emission intensity, while variables such as industrial structure, government intervention and financial development are gradually added as the research progresses. To reduce the estimation bias, the following control variables are chosen in this paper: industrial structure (is), using the share of secondary industry value added in GDP as a proxy variable [21,61]; economic development (eco), using the GDP per capita representation [59]; foreign investment (fdi), using the share of foreign direct investment in urban GDP as a proxy variable [14]; technological innovation (ino), using the share of science and technology expenditure in urban GDP [14]; environmental regulation (env), characterized by the comprehensive utilization rate of the industrial fixed waste [25].

3.3. Data Description

Two hundred and eighty prefecture-level cities in China are the subject of the data in this paper, which involve a total of 21 variables in the fields of energy, tourism, environment, and economy; the data collection was challenging; most city statistics were not updated in a timely manner, leading to a significant amount of missing data in 2020; and artificially completing the data would interfere with the validity of the research findings. Likewise, the pace of China's tourism development has slowed down dramatically as a result of the new crown pandemic, taking into account that force majeure circumstances will compromise the validity of the findings and have a negative influence on tourism's contribution to the green economy. Given the aforementioned justifications, based on the availability and consistency of data, this paper selects the relevant data of 280 cities in 30 provinces from 2007 to 2019, excluding the data of Tibet, Hong Kong, Macao and Taiwan. Due to the long period and the change in some city data, this paper takes 2019 prefecture-level cities as the benchmark, excludes the data of merged cities, and retains the data related to the removal of counties and promotion of cities. The relevant data are obtained from *The China City Statistical Yearbook, China City Construction Statistical Yearbook, China Energy Statistical Yearbook* and *Tourism Statistical Yearbook* of each city. Carbon emission data were calculated based on county-level data from the literature. The list of pilot low-carbon cities is based on the list announced by the National Development and Reform Commission. To avoid the interference of multicollinearity and pseudo-regression, the VIF test and unit variance test were conducted on the panel data, and the results showed that the variance inflation factor of the panel data was less than three, and they all passed the LLC test and Fisher-ADF test at the 1% significance level, so there was no multicollinearity problem, and the data were smooth. In addition, the effect of heteroskedasticity was eliminated by taking the logarithms of all variables.

4. Results

4.1. Spatial Autocorrelation Test and Spatial Econometric Model Selection

Urban carbon emission intensity and green economic efficiency are spatially correlated, which is a prerequisite for spatial econometric modeling [60]. Stata.23 is used to calculate the global Moran index to explore the spatial agglomeration characteristics. As shown in Table 2, the GEE Moran index is positive for all the years from 2007 to 2019, and the spatial agglomeration strengthens year by year except for 2013–2015, all of which pass the 1% significance level test. During the observation period, the Moran index of Carin is positive in all cases, and the degree of spatial agglomeration increases in fluctuation. Overall, both urban green economic efficiency and carbon emission intensity have significant spatial agglomeration characteristics.

Table 2. Global Moran Index test results for green economy efficiency and carbon emission intensity.

Year	GEE	CI
2007	0.011 ***	0.043 ***
2008	0.007 **	0.041 ***
2009	0.010 ***	0.039 ***
2010	0.014 ***	0.045 ***
2011	0.013 ***	0.041 ***
2012	0.017 ***	0.039 ***
2013	0.016 ***	0.042 ***
2014	0.012 ***	0.045 ***
2015	0.014 ***	0.057 ***
2016	0.020 ***	0.071 ***
2017	0.021 ***	0.074 ***
2018	0.034 ***	0.092 ***
2019	0.035 ***	0.083 ***

Note: ****, **, and * represent significance levels of 1%, 5%, and 10%, respectively.

Based on passing Moran's I test, the spatial effect econometric model can be selected. Additionally, to measure the nonlinear characteristics of tourism development on urban carbon emission intensity and green economic efficiency, models (3) and (4) containing quadratic terms of tourism development variables were introduced. First, the LM test, LR test and Wald test were performed to identify the spatial econometric models. Both the LM spatial lag test and LM spatial error test showed high significance, so both the SAR model and SEM were suitable for this study, and we chose the SDM model that combined both. Then, the LR test and Wald test were applied to further determine whether the SDM could be degraded to the SAR model or SEM. The comprehensive test results showed that the SDM was better than the SAR and SEM, so the SDM was selected as the baseline regression model in this paper. Based on the SDM model [62], all matrices passed the Hausman test at a 1% significance level except for model (1), which had a negative value, so the fixed-effects model was selected. As shown in Table 3, the overall R2 of the individual fixed-effects model is significantly better than that of the time-point fixed-effects and double fixed-effects models. Finally, the individual fixed-effects spatial Durbin model is chosen to analyze the impact of tourism development on urban carbon emission intensity and green economic efficiency.

Table 3. Spatial econometric model selection.

Statistics		Model (1)	Model (2)	Model (3)	Model (4)
		GEE	CI	GEE	CI
LM Spatial Lag		450.385 ***	37.215 ***	426.884 ***	21.279 ***
Robust LM Spatial Lag		84.494 ***	84.324 ***	79.038 ***	55.872 ***
LM Spatial Error		1692.191 ***	1814.403 ***	1612.687 ***	1702.513 ***
Robust LM Spatial Error		1326.301 ***	1861.513 ***	1264.841 ***	1737.106 ***
Compare SDM with SAR	LR Inspection	89.18 ***	26.86 ***	75.61 ***	30.43 ***
	Wald Inspection	89.01 ***	26.65 ***	75.99 ***	30.17 ***
Compare SDM with SEM	LR Inspection	65.77 ***	42.60 ***	63.78 ***	44.58 ***
	Wald Inspection	42.35 ***	31.01 ***	39.89 ***	31.67 ***
Hausman Inspection		−2.24	404.17 ***	40.01 ***	403.47 ***
Time fixation effect R-squre		0.2189	0.1865	0.0560	0.0339
Individual fixation effect R-squre		0.2660	0.2849	0.2789	0.2404
Double fixed effect R-squre		0.0290	0.2711	0.0124	0.2009

Note: ****, **, and * represent significance levels of 1%, 5%, and 10%, respectively.

4.2. Spatial Spillover Effects of Tourism Development on Urban Carbon Emission Intensity and Green Economic Efficiency

The spatial spillover effect can be decomposed into the direct effect, indirect effect and total effect. Among them, the direct effect includes the direct effect of tourism development on the explained variables and the feedback effect caused by tourism development affecting the explained variables in adjacent areas. The indirect effect represents the spatial spillover effect of tourism development, including the influence of tourism development in neighboring areas on the explained variables in the region and the influence of tourism development in neighboring areas on their own explained variables, which in turn has an impact on the explained variables in the region. The total effect, on the other hand, is the sum of the direct and indirect effects, reflecting the average effect of tourism development on the explanatory variables.

As seen from Table 4, the total effect coefficient (lnTC) of the impact of tourism development on urban green economic efficiency is significantly positive, indicating that the development of tourism is conducive to enhancing urban green economic efficiency. On the one hand, tourism is a resource-dependent industry, and good ecological and environmental conditions are the basis of its development, so tourism investment involves financial support for the ecological and environmental restoration of tourism destinations. On the other hand, tourism development produces change in the industrial structure of destinations, forcing enterprises to conduct energy restructuring, and especially has a crowding-out effect on industries, but due to market demand, tourism development can have a significant impact on the service sector. However, tourism development can optimize the service and manufacturing industries due to market demand and technology spillover, reduce their pollution emissions, and thus improve the efficiency of the destination's green economy. The direct effect of tourism development is not significant. Tourism itself has low pollution emissions, and tourism development does not act directly on green economic efficiency but indirectly enhances urban green economic efficiency by forcing local industrial structure optimization and other forms through high correlation with other industries. For example, for every 1% increase in the tourism development level, the green economic efficiency of neighboring areas is indirectly enhanced by 0.442%, i.e., the promotion effect of tourism development on urban green economic efficiency is mainly manifested as an indirect effect, i.e., spillover effect [60]. Therefore, hypothesis H1 is verified.

Table 4. Benchmark regression results on the impact of tourism development on urban carbon emissions and green economic efficiency.

Variable	GEE			CI		
	Direct Effects	**Indirect Effects**	**Total Effect**	**Direct Effects**	**Indirect Effects**	**Total Effect**
lnTC	0.010	0.442 ***	0.452 ***	0.028	−1.336 ***	−1.307 ***
lnis	−0.034 ***	−0.055	−0.09	−0.022	−2.95 ***	−2.972 ***
lneco	0.026 *	−0.193	−0.167	−0.068	1.235 **	1.168 **
lnfdi	−0.003	−0.231 **	−0.234 **	−0.004 ***	−0.021	−0.024
lnino	0.113 ***	0.011	0.125 ***	−0.006	0.092	0.086
lnenv	0.005	0.062	0.057	0.045 ***	−1.026 **	−0.981 **
R-squared		0.253			0.1548	
Log-likelihood		885.249			−446.1593	
Observations		3640			3640	
City FE		YES			YES	
Year FE		NO			NO	

Note: ****, **, and * represent significance levels of 1%, 5%, and 10%, respectively.

As seen from Table 4, the total effect coefficient of the impact of tourism development on the intensity of urban carbon emissions is significantly negative, indicating that tourism development can mitigate urban carbon emissions and achieve urban emission reduction. On the one hand, because tourism itself is a low-consumption and low-carbon industry, its development is based on the ecological environment. On the other hand, the tourism development model is constantly updated, and green development has been a basic requirement, especially the development of the digital economy in recent years, which provides the basis for the creation of a digital tourism model and, to a large extent, relieves the pressure of urban carbon emission reduction. As tourism development affects the green economic efficiency of cities, the urban carbon reduction effect of tourism development also shows a significant spillover effect, with each 1% increase in the tourism development level indirectly reducing the carbon emission intensity of neighboring areas by 1.336%. The development of tourism is one of the main paths for carbon reduction in cities [32]. Therefore, hypothesis H3 is verified.

From the control variables, the effect of industrial structure on green economic efficiency mainly works as a direct effect, and the effect on carbon emission intensity mainly works as an indirect effect. Tourism development leads to an increase in factor costs, which makes the maximum use of energy structure through reasonable resource allocation and reduces the redundancy of resource inputs and pollution emissions such as carbon dioxide. The results of the effect of economic level on green economy efficiency are not significant, indicating that economic development and green economy development are not equivalent. Meanwhile, the effect of the economic level on carbon emission intensity shows a significant positive direction, which integrally indicates that most of China's cities are still trying to eliminate the severe development model, and are still sacrificing resources and the environment for the improvement of the economic level, which does not correspond to the development of the green economy. To a certain extent, the enhancement of the city's reputation and the brand effect, in addition to the management experience provided by foreign investment, improved production processes, technological innovation and improvement of the business environment caused by the growth of tourism development promote the city's carbon emission reduction and green economy development [6]. The direct and indirect effects of technological innovation on the efficiency of the green economy are significantly positive, and innovation has been an important variable that has helped green economic development to reach a turning point. Under the stimulation of the policy of cultural tourism integration, "tourism +" continues to push out new ideas and become richer in industries, but "tourism + technology" still has serious deficiencies that inhibit economic growth and green development, but the development of the technology level is not enough to significantly reduce carbon emission intensity in most cities at present [63]. The spillover effect of environmental regulation variables is significant [64]; environmental regulation is necessary due to pollution externalities; in the short term, it is inhibitory to economic development, and the direct effect is not obvious. Tourism development causes the agglomeration of the tourism industry, which not only promotes the improvement of the carbon emission efficiency of tourism but also promotes the expansion of the service industry and the development of manufacturing services by forcing the optimization of industrial structure, bringing the "innovation compensation" effect and ultimately achieving a Porter "win-win" [64]. Through the interaction and correlation with foreign investment and environmental regulations, tourism development drives technological innovation and industrial structure optimization, reduces the intensity of urban carbon emissions, and enhances the efficiency of the green economy.

4.3. Heterogeneous Effects of Tourism Development on Urban Carbon Intensity and Green Economic Efficiency in Different Regions

Considering the possible spatial heterogeneity of the impact of tourism development on urban green economic efficiency and carbon emission intensity, this paper divides regions and urban agglomerations, and the criteria for making these divisions are shown in relevant documents (For regional division standards, see the National Bureau of Statistics' "*Methods for the Division of East, West, Central and Northeast Regions*") and the literature [65]. The region is divided into four parts, eastern, central, western and northeastern, and cities are categorized into two groups, urban agglomeration and non-urban agglomeration, to test the regional heterogeneity of tourism development on urban green economic efficiency and carbon emission intensity. Tables 5 and 6 show the spatial heterogeneity impact of tourism development on green economy efficiency and carbon emission intensity, respectively.

Table 5. Regional heterogeneity of the impact of tourism development on urban green economic efficiency.

Variable	Eastern Region	Central Region	Western Region	Northeast Region	Urban Agglomeration	Non-Urban Agglomerations
lnTC	0.043 * (1.800)	0.015 (0.560)	0.047 * (1.700)	0.113 *** (3.54)	0.024 (1.380)	−0.005 (−0.240)
W×lnTC	0.182 * (1.760)	0.330 *** (2.580)	0.184 ** (2.010)	0.245 *** (2.78)	0.183 ** (2.060)	0.413 *** (4.640)
R-squared	0.308	0.256	0.179	0.2210	0.292	0.204
log-likelihood	456.936	446.489	50.027	113.5438	574.890	311.798
Observations	1118	1040	1040	442	2158	1482
City FE	YES	YES	YES	YES	YES	YES
Year FE	NO	NO	NO	NO	NO	NO

Note: ****, **, and * represent significance levels of 1%, 5%, and 10%, respectively.

Table 6. Regional heterogeneity of the impact of tourism development on urban carbon emission intensity.

Variable	Eastern Region	Central Region	Western Region	Northeast Region	Urban Agglomeration	Non-Urban Agglomerations
lnTC	0.047 (0.57)	0.104 (0.98)	0.149 ** (2.03)	−0.224 ** (−1.95)	0.014 (0.24)	0.018 (0.29)
W × lnTC	0.121 (0.39)	−0.849 *** (−2.79)	0.095 (0.44)	0.263 (0.91)	−0.034 (−0.17)	−0.030 (−0.15)
R-squared	0.0281	0.0739	0.0225	0.0778	0.0144	0.0535
log-likelihood	−1007.2842	−986.6959	−945.4118	−455.7187	−1983.2718	−1442.5192
Observations	1118	1040	1040	442	2158	1482
City FE	YES	YES	YES	YES	YES	YES
Year FE	NO	NO	NO	NO	NO	NO

Note: ****, **, and * represent significance levels of 1%, 5%, and 10%, respectively.

In the eastern region, tourism development makes a significant contribution to local and neighboring green economic efficiency, and the effect is inferior compared with the western and northeastern regions, which may be due to the weakening of the marginal effect of tourism development on urban green economic efficiency as the economic level increases, and the economic level implies to a certain extent that technological innovation and industrial structure optimization are also higher in the eastern region due to the developed tourism industry. Green economic development tends to be flat growth. The effect of tourism development in the eastern region on urban carbon emission intensity is not significant, probably because the industrial structure of cities in the eastern region is no longer dominated by the severe secondary industry development model, and tourism development no longer has the extrusion effect on the high carbon emission secondary industry. The local effect of tourism development in the central region is not significant, and the effect of tourism development in the central region on the green economic efficiency and carbon emission intensity of neighboring cities is significant. On the one hand, the level of tourism development in the central region is low, and the promotion effect is not obvious. On the other hand, in the context of integrated development of the central city cluster, many factors can interact with one another, which leads to the obvious spillover effect of tourism development. Tourism development in the western region can not only improve the efficiency of the local green economy but also promote the development of the green economy in neighboring areas, and the urban carbon emission reduction effect of tourism development is mainly local. That is, tourism development in the western region still has an optimization effect on the industrial structure of the local cities but the "resource curse", and the "Dutch disease effect", which inhibit green economic efficiency and carbon emission reduction in cities, are also evident [48]. The effect of tourism development in Northeast China on the green economic efficiency of cities is significant, but the spillover effect on carbon emission reduction is not obvious, which indicates that the development of tourism in Northeast China, as a heavy industrial base, helps to alleviate local pollution emissions and is beneficial to the development of the urban green economy. From the city cluster heterogeneity, the local effect of tourism development on green economic efficiency both inside and outside the city cluster is not significant, and the spatial spillover effect on neighboring cities is significant. There is no significant effect on urban carbon emission reduction, partly because the development of city clusters is not synergistic, and city clusters do not bring due opportunities to specific cities. In summary, there is no "siphon effect" in urban agglomerations, with significant differences between the eastern, central, western and northeastern regions, and again, the spatial spillover effect of tourism development on urban green economic efficiency is more significant than the direct effect.

4.4. Nonlinear Effects of Tourism Development on Carbon Emission Intensity and Urban Green Economic Efficiency

To investigate whether there is a similar phenomenon of tourism development on urban green economic efficiency and carbon emission intensity [64], this paper introduces a quadratic term of the logarithm of tourism development level based on the spatial Durbin econometric model to test the nonlinear characteristic relationship between tourism development and carbon emission intensity and urban green economic efficiency. The results are shown in Table 7. The positive coefficient of the quadratic term of tourism development on urban green economic efficiency indicates that there is a positive U-shaped nonlinear characteristic relationship between tourism development and urban green economic efficiency, and similarly, there is an inverse U-shaped nonlinear characteristic relationship between tourism development and urban carbon emission intensity [32]. Similarly, there is an inverse U-shaped nonlinear relationship between tourism development and urban carbon emission intensity. Furthermore, to accurately measure the degree of nonlinear effects, a semiparametric spatial lag model is introduced, and the nonlinear relationship between tourism development and urban green economic efficiency and carbon emission intensity can be visually observed by drawing the partial derivatives of G(lnTC) in the model, as shown in Figures 3 and 4, where the horizontal coordinates indicate the level of tourism development and the vertical coordinates indicate the marginal effects.

Table 7. Nonlinear effects of tourism development on urban carbon intensity and green economic efficiency.

Variable	GEE		CI	
	x	W × x	x	W × x
lnTC	0.124 ***	0.424 ***	−0.086 **	−0.458 **
lnTC2	0.011 ***	0.034 **	−0.011 ***	−0.016 *
lnis	−0.031 ***	0.116	−0.012	−0.805 ***
lneco	0.023 *	−0.064	−0.071 ***	0.349 ***
lnfdi	−0.003	−0.07 **	−0.004	−0.023
lnino	−0.12 *	0.065 ***	0.003	0.015
lnenv	0.006	0.054	0.047 ***	−0.372 ***
rho	0.544 ***		0.728 ***	
R-squared	0.2748		0.1386	
Log-likelihood	903.8068		−438.8832	
Observations	3640		3640	
City FE	YES		YES	
Year FE	NO		NO	

Note: ****, **, and * represent significance levels of 1%, 5%, and 10%, respectively.

From Figure 3, it can be seen that tourism development has a positive "U" shaped nonlinear effect on urban green economic efficiency, but the curve is always above 0, which means that the marginal effect of tourism development on urban green economic efficiency is always positive, i.e., it always shows a facilitating effect. The curve in Figure 3 shows that although the tourism industry has always had a positive effect on the green economic efficiency of the city, there are roughly three stages. The first stage is lnTC between −8 and −7, with decreasing marginal effects, which indicates that the development of tourism does produce certain pollution in the initial stage or will attract a large number of manufacturing industries to gather in the tourist destination, thus causing some suppression of the marginal effect of tourism development on green economic efficiency. This causes a certain suppression of lnTC from −7 to −2. The second stage is between lnTC from −7 to −2, and the positive marginal effect of tourism development in this stage grows slowly and represents the exploration stage of the green tourism development model. The third stage is the stage after lnTC-2, which fully demonstrates the positive effect of tourism development on urban green economic efficiency. Therefore, hypothesis H2 is verified.

Figure 4 shows that tourism development has an "M" shape on urban carbon intensity, which is different from the inverted "U" shape obtained in the previous paper, and the "U" shape may only be part of the "M" shape. In contrast, the marginal effect of tourism development on urban carbon intensity is overwhelmingly below 0, indicating that tourism development is beneficial for cities to reduce carbon emissions. Similarly, it can be seen that the mitigation effect of tourism development on urban carbon emission intensity is relatively stable until lnTC is −3, with a brief rise in the curve between −3 and −2. The marginal effect ushers in a rapid decline after −2. As with the marginal effect of green economic efficiency, tourism development, after a certain point, has a positive impact. Therefore, hypothesis H4 is verified.

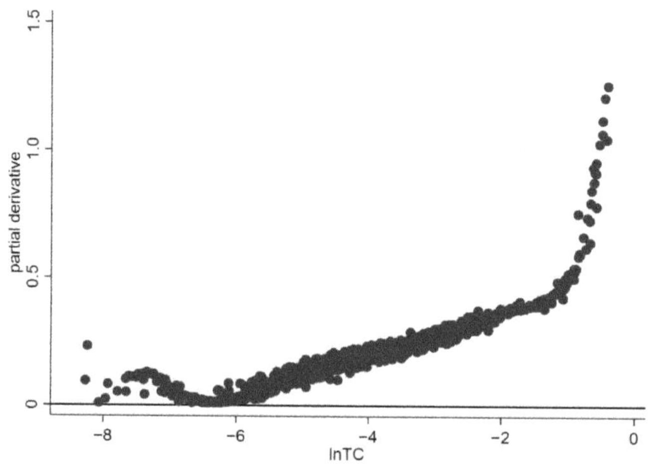

Figure 3. Partial derivative of tourism development on urban green economic efficiency.

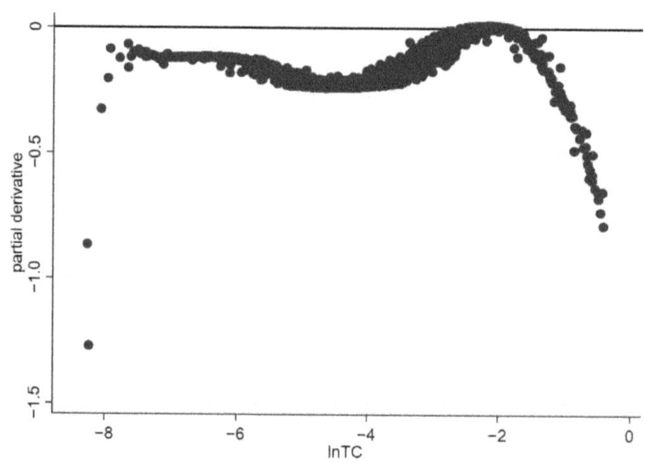

Figure 4. Partial derivative of tourism development on urban carbon emission intensity.

4.5. Mediating Effect of Urban Carbon Emission Intensity on the Role of Tourism Development on Green Economic Efficiency

To examine whether urban carbon emission intensity has mediating utility in the effect of tourism development on urban green economic efficiency, the mediating utility model is used, and a second test is conducted by bootstrapping to ensure the robustness of the test results. Tourism development has a significant positive impact on urban green economic efficiency, which is consistent with the previous results, while urban carbon emission intensity has a significant negative impact on urban green economic efficiency, and urban carbon emissions have an inhibitory effect on the improvement of green economic efficiency, which also confirms the robustness of the mediated utility model. The test result of model (7) shows that urban carbon emission intensity has a significant mediating effect on the influence of tourism development on urban green economic efficiency (Table 8); meanwhile, the result of the bootstrap test shows that the upper and lower bounds of BC do not contain 0 between them, and the test is passed, which proves the mediating utility of urban carbon emission intensity. Therefore, hypothesis H5 is verified.

Table 8. Results of the mediating effect test of urban carbon emission intensity.

Variables	Model (5)	Model (6)	Model (7)
TC	0.603 *** (−14.36)		0.510 *** (−11.69)
CI		−0.723 *** (−18.75)	−0.128 *** (−7.13)
Constant	0.0874 (−1.31)	−1.799 *** (−29.45)	−0.143 (−1.94)
Bootstrap Inspection (Direct effects)		0.51045561 (BC: 0.4023555, 0.6608595)	
Bootstrap Inspection (Indirect effects)		0.09251758 (BC: 0.0558356, 0.1304556)	
Control	YES	YES	YES
Individual fixation effect	YES	YES	YES
Time fixation effect	NO	NO	NO
N	3640	3640	3640
R2	0.6294	0.1578	0.6345
F	1031.22	114.65	903.31

Note: ****, **, and * represent significance levels of 1%, 5%, and 10%, respectively.

4.6. Exogenous Shock Impacts of Low-Carbon City Pilot Policies

The parallel trend test allows the use of multiperiod DID for policy assessment analysis. In this paper, the low-carbon city pilots enacted in China in 2010 and afterward are used as dummy variables, which are low-carbon city pilots recorded as 1 and 0 otherwise. First, the traditional multiperiod DID is used for estimation, as shown in Table 9. The traditional DID results show that with or without control variables, the exogenous shock of the low-carbon city pilot policy has a significant positive effect on city green economic efficiency and carbon emission intensity, and under the spatial DID model, the low-carbon pilot city policy shock on both city green economic efficiency and carbon emission intensity shows a significant spillover effect. As shown in Figures 5 and 6, the kernel density distribution of the estimated coefficients by 1000 randomly generated treatment groups shows that the red curves are all above the horizontal line of 0.1, and the blue curves have p values of mostly approximately 0. The values of the true coefficients are shown as red dashed lines, which are significantly different from the red curves on the left, and the placebo test indicates that the estimation results are not biased. The low-carbon city pilot reflects the realistic path of tourism development to adjust the industrial structure, and tourism development promotes green economic efficiency and reduces the carbon emission intensity of the city in the construction of the low-carbon pilot city, so the exogenous test in this study further verifies the robustness of the effect of tourism development on urban green economic efficiency and urban carbon emission intensity.

Table 9. Exogenous impacts of low-carbon city pilot policies.

Variables	DID				Spatial DID	
	GEE	GEE	CI	CI	GEE	CI
LC	0.245 *** (11.17)	0.176 *** (8.58)	−0.477 *** (−12.80)	−0.078 *** (−2.71)	0.049 *** (2.72)	0.001 (0.02)
W×LC	/	/	/	/	0.906 *** (8.08)	−0.574 *** (−3.88)
Control	NO	YES	NO	YES	YES	YES
R-squared	0.476	0.575	0.7280	0.8506	0.234	0.1688
Observations	3640	3640	3640	3640	3640	3640
City FE	YES	YES	YES	YES	YES	YES
Year FE	NO	NO	NO	NO	NO	NO

Note: ****, **, and * represent significance levels of 1%, 5%, and 10%, respectively.

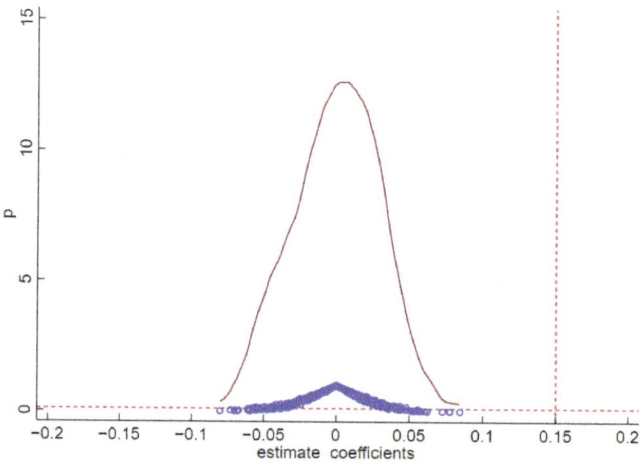

Figure 5. Placebo test of the impact of low-carbon city pilot policies on urban green economic efficiency.

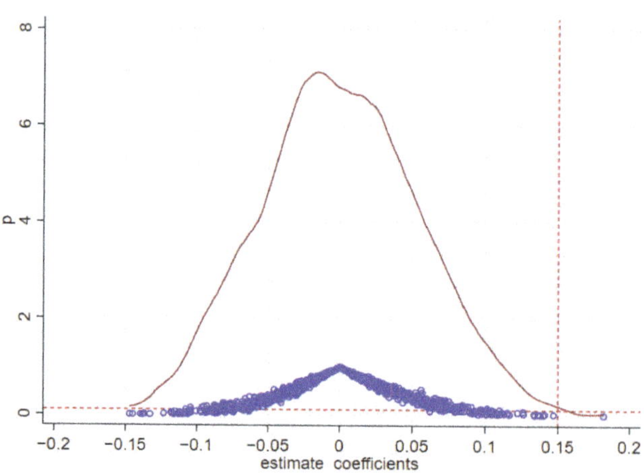

Figure 6. Placebo test of the impact of low-carbon city pilot policies on urban carbon emission intensity.

5. Conclusions and Discussion

5.1. Conclusions

In this paper, based on the mechanism of the influence of tourism development on urban green economic efficiency and carbon emission intensity, we measured the levels of tourism development and urban green economic efficiency and carbon emission intensity of 280 prefecture-level cities from 2007 to 2019 using entropy TOPSIS and superefficiency EBM and tested the spatial spillover effect of tourism development on urban green economic efficiency and carbon emission intensity using the spatial Durbin model. The spatial heterogeneity and nonlinear characteristics of the spillover effects from cities are further analyzed. Finally, the mediating role of urban carbon emission intensity in the impact of tourism development on urban green economic efficiency is examined.

Tourism development has a significant enhancing effect on urban green economic efficiency and carbon emission intensity mitigation, and it also decomposes the role effect. It is found that the green economic effect and carbon emission reduction effect of tourism development are mainly manifested as spillover effects. From the perspective of regional heterogeneity, it is found that the green economic effect and carbon emission reduction effect of tourism development are much less effective in the eastern region than in the central, western and northeastern regions. The results show that the green economic effect and carbon emission reduction effect of tourism development are not affected by urban agglomeration [60], which is basically consistent with Tong Yun's conclusion.

The green economic effect and carbon emission reduction effect of tourism development are nonlinear, with a positive "U" shape and an "M" shape, respectively. It is found that although tourism development has a certain degree of negative impact on green economic efficiency and carbon emission intensity at the early stage, the overall impact is positive. At the same time, the green economic effect and carbon emission reduction effect of tourism development significantly increase after the level of tourism development reaches 0.135 or above, and by calculation, only 13% of the cities have reached this level in 2019.

The results of the intermediary effect show that carbon emission intensity has a significant intermediary effect in the influence of tourism development on the green economic efficiency of cities. Tourism development can achieve a green economic effect through its carbon emission reduction effect. Moreover, the low-carbon city policy not only verifies the positive effect of the policy on carbon emission reduction and the green economic development of cities but also proves that the green economic effect and carbon emission reduction effect of tourism development are robust.

5.2. Discussion

This study complements and enriches the impact and spillover effects of tourism development on urban green economic efficiency and carbon emission intensity. It also examines the spatial heterogeneity of urban clusters and different regions and demonstrates the nonlinear characteristic relationship between tourism development and urban green economic efficiency and carbon emission intensity strength.

The possible marginal contributions of this study are as follows: this study is supported by the data of 280 prefecture-level cities in China from 2010 to 2019 and uses the "low-carbon city pilot" as an exogenous shock to test the spatial spillover effects of tourism development on urban green economic efficiency and carbon emission intensity. At the same time, this study measures the development of the tourism industry in terms of both tourism scale and tourism economy, and the results are more representative. This study explores the nonlinear characteristic relationship between tourism development on urban green economic efficiency and carbon emission intensity, further improving and enriching the study of the spillover effect of tourism development on urban green economic efficiency. Moreover, this paper confirms the mediating role of carbon emission intensity in tourism development for urban green economic efficiency spillover.

Meanwhile, the following policy insights are obtained from this paper. First, the spillover effect of tourism development on urban green economic efficiency and carbon emission intensity has significant spatial heterogeneity, and different regions should develop differentiated strategies according to their development conditions [66]. The marginal spillover effect of tourism development in the eastern region is weakened, and tourism development should be shifted to high-end and low-carbon sectors, strengthening the linkage and integration with other industries, encouraging technological innovation, and providing technical support for resource-saving development models. Second, there is regional heterogeneity in the spillover effects of tourism development on urban green economic efficiency and carbon emission intensity, which highlights the importance of mutual coordination and cooperation within strategic alliances for regional tourism cooperation [67]. On the premise of breaking down administrative barriers, we actively promote the rational matching of resource elements between regions. Third, to promote the green development of cities, it is necessary to focus on both the green economic effects of tourism development and to explore the mode of green development of tourism itself. It is important to focus on the negative impact of tourism development on the environment, and to focus on the sustainable development of tourism at a reasonable pace [60].

In the future, the green economy effect and carbon emission reduction effect of tourism development will be more prominent, but the intermediary role of carbon emission intensity in the impact of tourism development on green economy efficiency may not be clear. On the one hand, global enthusiasm and efforts in carbon reduction will be maintained, and the green transformation of industries is a trend. Meanwhile, the tourism industry, contributing to the implementation of carbon emission reduction in several regions and countries, has outstanding green economy attributes. On the other hand, in many more backward developing regions, tourism is a rough economic pattern due to the lack of experience in tourism development and late development history. Although these countries or regions attract a large number of tourists by virtue of their unique tourism resources, they cause more damage to the ecological environment than before development. Likewise, the issue of carbon emissions is not taken into account in the process of tourism development, resulting in a potentially unsatisfactory relationship between tourism development and green economic effects and carbon reduction effects. In addition, as the COVID-19 epidemic continues to impact the tourism industry, new forms of tourism are emerging. Many real tourism activities are shifted to virtual tourism activities, in which case the negative environmental externalities of the tourism industry itself may be weakened, as well as the EKC effect, positive environmental externality effect and indirect effect of tourism development.

Compared with previous studies, this paper refines the traditional subject of "tourism, carbon dioxide and economic growth" [6] to "tourism, carbon emissions and green economic efficiency", and further confirms Tong's conclusion on the green economic effect of tourism [60]. On the basis of this paper, we further find that the green economy effect of tourism has a positive "U" nonlinear characteristic, which is one of the important points of innovation in this paper. This paper also confirms that tourism development has a significant carbon reduction effect [6,32], and also finds that there is an "M" type nonlinear relationship between tourism development and carbon emission intensity, which is different from the "inverted U" type relationship obtained by Reza et al. [32]. To be more precise, the same "inverted U" type was found in this paper, but after the accurate measurement of the nonlinear effect by the semi-parametric spatial lag model, it was found that the "inverted U" type is only a vague form of the "M". The "inverted U" shape is only a fuzzy form of the "M" shape. In order to further confirm whether the carbon emission reduction of tourism affects the green economy effect of tourism, this paper innovatively uses carbon emission intensity as a mediating variable and concludes that tourism development affects green economy efficiency through carbon emission reduction. Although the article has conducted a detailed study on the relationship between tourism development, carbon emission intensity and green economic efficiency, there are also the following shortcomings. First, the article takes 280 prefecture-level cities in China as examples, but the tourism development of these cities is uneven, which may cause some interference in the results. Subsequent research can further discuss the formation mechanism among the three factors. Second, the impact mechanism is not thoroughly explored in this paper, which leaves room for further discussion of the formation mechanisms between the three. Instead, this paper only discusses the relationship between tourism development, carbon emission intensity and green economy efficiency in detail.

Author Contributions: Conceptualization, X.H. and H.X.; methodology, J.S. and X.H.; validation, X.H.; formal analysis, J.S. and C.C.; investigation, J.S. and X.H.; resources, J.S. and C.C.; writing—original draft preparation, J.S. and X.H.; writing—review and editing, X.H. and J.S.; supervision, Q.H.; funding acquisition, X.H. All authors have read and agreed to the published version of the manuscript.

Funding: This research was funded by the National Social Science Foundation, grant number 21FGLB070, and Natural Science Foundation of Hunan Province, grant number 2022JJ30387.

Institutional Review Board Statement: Ethical review and approval were waived for this study. The national law "Regulations for Ethical Review of Biomedical Research Involving Humans" does not require the full review process for data collection from adults who have adequate decision-making capacity to agree to participate.

Informed Consent Statement: Informed consent was obtained from all subjects involved in the study.

Conflicts of Interest: The authors declare no conflict of interest.

References

1. Raihan, A.; Tuspekova, A. The nexus between economic growth, energy use, urbanization, tourism, and carbon dioxide emissions: New insights from Singapore. *Sustain. Anal. Model.* **2022**, *2*, 100009. [CrossRef]
2. Feng, H.; Xie, R.; Fang, J. Urban Agglomeration Economies and Industrial Energy Efficiency. *Energy* **2018**, *162*, 45–59.
3. Li, M.; Patiño-Echeverri, D.; Zhang, J.J. Policies to promote energy efficiency and air emissions reductions in China's electric power generation sector during the 11th and 12th five-year plan periods: Achievements, remaining challenges, and opportunities. *Energy Policy* **2019**, *125*, 429–444. [CrossRef]
4. Huang, J.B.; Luo, Y.M.; Feng, C. An overview of carbon dioxide emissions from China's ferrous metal industry: 1991–2030. *Resour. Policy* **2019**, *62*, 541–549. [CrossRef]
5. Tao, X.; Wang, P.; Zhu, B.J. Provincial green economic efficiency of China: A non-separable input–output SBM approach. *Appl. Energy* **2016**, *171*, 58–66. [CrossRef]
6. Lee, J.W.; Brahmasrene, T.J. Investigating the influence of tourism on economic growth and carbon emissions: Evidence from panel analysis of the European Union. *Tour. Manag.* **2013**, *38*, 69–76. [CrossRef]
7. Walker, N.L.; Williams, A.P.; Styles, D. Key performance indicators to explain energy & economic efficiency across water utilities, and identifying suitable proxies. *J. Environ. Manag.* **2020**, *269*, 110810.
8. Amigues, J.-P.; Moreaux, M. Competing land uses and fossil fuel, and optimal energy conversion rates during the transition toward a green economy under a pollution stock constraint. *J. Environ. Econ. Manag.* **2019**, *97*, 92–115. [CrossRef]
9. Chen, S.; Golley, J. 'Green'productivity growth in China's industrial economy. *Energy Econ.* **2014**, *44*, 89–98. [CrossRef]
10. Qian, Z.; Liu, X.C. Regional differences in China's green economic efficiency and their determinants. *Resour. Environ.* **2013**, *23*, 104–109.
11. Sueyoshi, T.; Goto, M.J. DEA radial measurement for environmental assessment: A comparative study between Japanese chemical and pharmaceutical firms. *Appl. Energy* **2014**, *115*, 502–513. [CrossRef]
12. Cui, Q.; Li, Y.J. An empirical study on the influencing factors of transportation carbon efficiency: Evidences from fifteen countries. *Appl. Energy* **2015**, *141*, 209–217. [CrossRef]
13. Wang, Q.; Zhao, Z.; Zhou, P.; Zhou, D.J. Energy efficiency and production technology heterogeneity in China: A meta-frontier DEA approach. *Econ. Model.* **2013**, *35*, 283–289. [CrossRef]
14. Liu, Y.; Dong, F.J. How technological innovation impacts urban green economy efficiency in emerging economies: A case study of 278 Chinese cities. *Resour. Conserv. Recycl.* **2021**, *169*, 105534. [CrossRef]
15. Honma, S.; Hu, J.-L. Industry-level total-factor energy efficiency in developed countries: A Japan-centered analysis. *Appl. Energy* **2014**, *119*, 67–78. [CrossRef]
16. Mousavi-Avval, S.H.; Rafiee, S.; Jafari, A.; Mohammadi, A.J. Optimization of energy consumption for soybean production using Data Envelopment Analysis (DEA) approach. *Appl. Energy* **2011**, *88*, 3765–3772. [CrossRef]
17. Song, M.-L.; Zhang, L.-L.; Liu, W.; Fisher, R.J. Bootstrap-DEA analysis of BRICS'energy efficiency based on small sample data. *Appl. Energy* **2013**, *112*, 1049–1055. [CrossRef]
18. Hu, J.L. Ecological total-factor energy efficiency of regions in China. *Energy Policy* **2012**, *46*, 9.

19. Gómez-Calvet, R.; Conesa, D.; Gómez-Calvet, A.R.; Tortosa-Ausina, E.J. Energy efficiency in the European Union: What can be learned from the joint application of directional distance functions and slacks-based measures? *Appl. Energy* **2014**, *132*, 137–154. [CrossRef]
20. Pjz, A.; Lez, A.; Hyla, B.; Yang, Z.C.; Hyh, A.; Xyw, D.J. Green economic efficiency and its influencing factors in China from 2008 to 2017: Based on the super-SBM model with undesirable outputs and spatial Dubin model. *Sci. Total Environ.* **2020**, *741*, 1.
21. Yuan, H.; Feng, Y.; Lee, C.C.; Cen, Y. How does manufacturing agglomeration affect green economic efficiency? *Energy Econ.* **2020**, *92*, 104944. [CrossRef]
22. Li, D.; Shangguan, Z.; Huang, M.; Zhang, X.; Tang, L.J. Impacts of urban development on regional green development efficiency—A case of the Yangtze River Delta in China. *Energies* **2022**, *15*, 4785. [CrossRef]
23. Wang, Z.; Feng, C.J. A performance evaluation of the energy, environmental, and economic efficiency and productivity in China: An application of global data envelopment analysis. *Appl. Energy* **2015**, *147*, 617–626. [CrossRef]
24. Haisheng, C.; Dingqing, N.; Shuiping, Z.; Ying, Y.; Manhong, S. Does the National Credit Demonstration Policy Affect Urban Green Economy Efficiency? Evidence from the Yangtze River Delta Region of China %J International Journal of Environmental Research and Public Health. *Int. J. Environ. Res. Public Health* **2022**, *19*, 1.
25. Wu, D.; Wang, Y.; Qian, W. Efficiency evaluation and dynamic evolution of China's regional green economy: A method based on the Super-PEBM model and DEA window analysis. *J. Clean. Prod.* **2020**, *264*, 121630. [CrossRef]
26. Tone, K.; Tsutsui, M. An epsilon-based measure of efficiency in DEA—A third pole of technical efficiency. *Eur. J. Oper. Res.* **2010**, *207*, 10. [CrossRef]
27. Shuguang, L.; Lei, S.; Yue, H. Measurement and Convergence Test of Green Economic Efficiency of the Yangtze River Economic Belt Under Different Spatial Network Correlation. *Front. Environ. Sci.* **2022**. [CrossRef]
28. NaiGang, C.; Lin, Z.; XiaoTong, G. Spatio-temporal evolution and driving mechanism of green economic efficiency at county level in the Yellow River Delta, China. *Ying Yong Sheng Tai Xue Bao = J. Appl. Ecol.* **2021**, *32*, 12.
29. Manzoor, F.; Wei, L.; Asif, M.; Haq, M.Z.; Rehman, H. The contribution of sustainable tourism to economic growth and employment in Pakistan. *Int. J. Environ. Res. Public Health* **2019**, *16*, 3785. [CrossRef] [PubMed]
30. Peeters, P.; Dubois, G.J. Tourism travel under climate change mitigation constraints. *J. Transp. Geogr.* **2010**, *18*, 447–457. [CrossRef]
31. Satrovic, E.; Muslija, A.J. The empirical evidence on tourism-urbanization-CO_2 emissions nexus. *Adv. Hosp. Tour. Res.* **2019**, *7*, 85–105. [CrossRef]
32. Sherafatian-Jahromi, R.; Othman, M.S.; Law, S.H.; Ismail, N.W. Development; Sustainability, Tourism and CO_2 emissions nexus in Southeast Asia: New evidence from panel estimation. *Environ. Dev. Sustain.* **2017**, *19*, 1407–1423. [CrossRef]
33. Jianping, C.; Jiaquan, D.; Keji, L.; Qiao, Y.; Zhijian, Z. A study on the decoupling state of tourism growth from carbon emissions and its drivers: A new decoupling analytical framework. *J. Tour.* **2022**, *37*, 13–24.
34. Toubes, D.R.; Araújo-Vila, N.J. A Review Research on Tourism in the Green Economy. *Economies* **2022**, *10*, 137. [CrossRef]
35. Wang, S.; Zhou, C.; Li, G.; Feng, K.J. CO_2, economic growth, and energy consumption in China's provinces: Investigating the spatiotemporal and econometric characteristics of China's CO_2 emissions. *Ecol. Indic.* **2016**, *69*, 184–195. [CrossRef]
36. Ben Jebli, M.; Ben Youssef, S.; Apergis, N.J. The dynamic linkage between renewable energy, tourism, CO_2 emissions, economic growth, foreign direct investment, and trade. *Lat. Am. Econ. Rev.* **2019**, *28*, 2. [CrossRef]
37. Litavcová, E.; Chovancová, J.J. Economic Development, CO_2 Emissions and Energy Use Nexus-Evidence from the Danube Region Countries. *Energies* **2021**, *14*, 3165. [CrossRef]
38. Dogan, E.; Aslan, A.J. Exploring the relationship among CO_2 emissions, real GDP, energy consumption and tourism in the EU and candidate countries: Evidence from panel models robust to heterogeneity and cross-sectional dependence. *Renew. Sustain. Energy Rev.* **2017**, *77*, 239–245. [CrossRef]
39. Khan, A.; Chenggang, Y.; Hussain, J.; Bano, S.; Nawaz, A.J. Natural resources, tourism development, and energy-growth-CO_2 emission nexus: A simultaneity modeling analysis of BRI countries. *Resour. Policy* **2020**, *68*, 101751. [CrossRef]
40. Agyeman, F.O.; Zhiqiang, M.; Li, M.; Sampene, A.K.; Dapaah, M.F.; Kedjanyi, E.A.; Buabeng, P.; Li, Y.; Hakro, S.; Heydari, M.J. Probing the Effect of Governance of Tourism Development, Economic Growth, and Foreign Direct Investment on Carbon Dioxide Emissions in Africa: The African Experience. *Energies* **2022**, *15*, 4530. [CrossRef]
41. Yıldırım, S.; Yıldırım, D.Ç.; Aydın, K.; Erdoğan, F.J. Regime-dependent effect of tourism on carbon emissions in the Mediterranean countries. *Environ. Sci. Pollut. Res.* **2021**, *28*, 54766–54780. [CrossRef] [PubMed]
42. Zhang, J.; Zhang, Y.J. Tourism, economic growth, energy consumption, and CO_2 emissions in China. *Tour. Econ.* **2021**, *27*, 1060–1080. [CrossRef]
43. Mikhno, I.; Koval, V.; Shvets, G.; Garmatiuk, O.; Tamošiūnienė, R.J. Green economy in sustainable development and improvement of resource efficiency. *Cent. Eur. Bus. Rev.* **2021**, *10*, 99–113. [CrossRef]
44. Brel, O.; Zaytseva, A.; Kaizer, P. Contribution of tourism to diversification and development of a green economy. *E3S Web Conf.* **2020**, *217*, 9. [CrossRef]
45. Marsiglio, S.J. Economic growth and environment: Tourism as a trigger for green growth. *Tour. Econ.* **2015**, *21*, 183–204. [CrossRef]
46. Goffi, G.; Cucculelli, M.; Masiero, L. Fostering tourism destination competitiveness in developing countries: The role of sustainability. *J. Clean. Prod.* **2019**, *209*, 101–115. [CrossRef]
47. Gossling, S.; Hall, C.M. Swedish Tourism and Climate Change Mitigation: An Emerging Conflict? *Scand. J. Hosp. Tour.* **2008**, *8*, 141–158. [CrossRef]

48. Corden, W.M.; Neary, J.P. Booming sector and de-industrialisation in a small open economy. *Econ. J.* **1982**, *92*, 825–848. [CrossRef]
49. Gunduz, L.; Hatemi, J.A. Is the tourism-led growth hypothesis valid for Turkey? *Appl. Econ. Lett.* **2005**, *12*, 499–504. [CrossRef]
50. Davies, T.; Cahill, S. *Environmental Implications of the Tourism Industry*; Ageconsearch: Saint Paul, MN, USA, 2000; p. 14.
51. Han, L.; Li, L.J. Engineering, Sustainable development of tourism under the background of low-carbon and green economy. *Adv. Mater. Sci. Eng.* **2021**, *2021*, 8587024. [CrossRef]
52. Liu, G.; Shi, P.; Hai, F.; Zhang, Y.; Li, X.J. Study on measurement of green productivity of tourism in the Yangtze River Economic Zone, China. *Sustainability* **2018**, *10*, 2786. [CrossRef]
53. Dzikuć, M.; Piwowar, A.J. Economic Aspects of Low Carbon Development. In Multidisciplinary Digital Publishing Institute. *Energies* **2022**, *15*, 5228. [CrossRef]
54. Jian, L.; Sohail, M.T.; Ullah, S.; Majeed, M.T. Examining the role of non-economic factors in energy consumption and CO_2 emissions in China: Policy options for the green economy. *Environ. Sci. Pollut. Res. Int.* **2021**, *28*, 67667–67676. [CrossRef]
55. Pan, S.-Y.; Gao, M.; Kim, H.; Shah, K.J.; Pei, S.-L.; Chiang, P.-C. Advances and challenges in sustainable tourism toward a green economy. *Sci. Total Environ.* **2018**, *635*, 452–469. [CrossRef] [PubMed]
56. Weixiang, X.; Jianping, Z.; Chengjun, L. The spatial effects of the development of the digital economy on urban carbon emissions. *Geogr. Stud.* **2022**, *41*, 111–129.
57. Yuhua, Z.; Jun, C. Heterogeneity impact of high-speed rail opening on tourism development of station cities: A study based on double difference method. *Tour. Sci.* **2018**, *32*, 79–92.
58. Sun, Y.-Y. Decomposition of tourism greenhouse gas emissions: Revealing the dynamics between tourism economic growth, technological efficiency, and carbon emissions. *Tour. Manag.* **2016**, *55*, 326–336. [CrossRef]
59. Shuai, S.; Fan, Z.J. Modeling the role of environmental regulations in regional green economy efficiency of China: Empirical evidence from super efficiency DEA-Tobit model. *J. Environ. Manag.* **2020**, *261*, 110227. [CrossRef] [PubMed]
60. Yun, T.; Haimeng, L.; Yong, M.; Jun, L.; Rui, Z. The impact of China's tourism economy on urban green development and spatial spillover effects. *Geogr. Stud.* **2021**, *76*, 2504–2521.
61. Fan, J.; Zhou, L.J. Three-dimensional intergovernmental competition and urban sprawl: Evidence from Chinese prefectural-level cities. *Land Use Policy* **2019**, *87*, 104035. [CrossRef]
62. Yujun, L.; Wenda, W.; Rucai, Y. Monte Carlo simulation of Hausman's test of the validity of statistics. *Math. Stat. Manag.* **2014**, *33*, 830–841.
63. Xixi, T.; Qing, X.; Fei, C. Research on the Impact of Tourism Development and Technological Innovation on Economic Growth: Based on the analysis of provincial spatial panel data. *East China Econ. Manag.* **2020**, *34*, 48–55.
64. Mingkang, W.; Yanping, L.; Tao, L. The differential impact of tourism industry agglomeration on environmental pollution: 287 prefecture-level cities. *Reform* **2019**, *02*, 102–114.
65. Fan, Z.; Yuemin, N.; Xiyang, L. The competitiveness of China's urban agglomerations and their impact on regional disparities. *Geogr. Stud.* **2019**, *38*, 1664–1677.
66. Hongbing, Z.; Jing, Z.; Lei, Z.; Shenglang, J. Low carbon transition and sustainable development path of tourism industry. In *IOP Conference Series: Earth and Environmental Science*; IOP Publishing: Bristol, UK, 2017; Volume 64.
67. Lei, Z.; Fang, F.; Xiangming, W. Tourism Development, Spatial Spillover and Economic Growth: Empirical Evidence from China. *J. Tour.* **2014**, *29*, 16–30.

energies

Article

The Green Innovation Effect of Environmental Regulation: A Quasi–Natural Experiment from China

Jiaxin Li [1], Shaoguo Zhan [2,*], Teng Huang [3] and Debo Nie [4]

1 Graduate Institute for Taiwan Studies, Xiamen University, Xiamen 361005, China
2 School of Public Policy and Management, Tsinghua University, Beijing 100084, China
3 College of Economics and Management, Northwest A&F University, Yangling, Xianyang 712100, China
4 NewHuadu Business School, Minjiang University, Fuzhou 350108, China
* Correspondence: zhansg@mail.tsinghua.edu.cn

Abstract: The "Two Control Zones" (TCZ) policy is the first air pollution regulation policy in China. We aim to examine the impact of the TCZ policy on green technological progress applying a difference–in–differences (DID) approach, using a city–level panel data set from 1990 to 2016. We show that the TCZ policy effectively increases the number of green patents of the cities in the two control zones. In particular, the TCZ policy has a significantly positive effect on the quantity and structure of human capital, including the number of inventors of patents and green patents, and the percentage of population with a higher education level. Moreover, the effects are heterogeneous, that is, the TCZ policy has a greater impact on the number of green patents in the control zones, where there are better R&D bases and more foreign investments.

Keywords: environmental regulation; green innovation; TCZ policy; human capital

Citation: Li, J.; Zhan, S.; Huang, T.; Nie, D. The Green Innovation Effect of Environmental Regulation: A Quasi–Natural Experiment from China. *Energies* **2022**, *15*, 7746. https://doi.org/10.3390/en15207746

Academic Editors: Junpeng Zhu and Xinlong Xu

Received: 20 September 2022
Accepted: 17 October 2022
Published: 19 October 2022

1. Introduction

In recent years, global economic development began facing severe challenges, so economies are seeking new drivers for economic growth. Innovation is considered an important means to break through the bottlenecks and shape new advantages in economic development. In the meantime, economic development resulted in serious environmental pollution problems all over the world. To balance environmental protection and sustainable economic development, the ability to innovate green technologies is seen as a potential solution. Green technology innovation refers to an economic behavior that emphasizes environmental performance improvement and can effectively balance economic development and ecological protection issues. Hence, with the increase in the challenges of resources and the environment, it is essential to promote the development of a green economy by promoting green technologies [1–3]. The sustainable development can thus be achieved through the innovation and progress of green technology [4]. Green technologies raised public attention in both academics and industry.

In the 2000s, Porter and Van Der Linde [5] proposed that environmental regulation can reduce the pollution caused by enterprises and incentivize enterprises to innovate to make up for the cost of pollution. Since then, a number of studies explored the relationship between environmental regulation and innovation from the perspectives of different industries, regions, and countries [6]. The Porter hypothesis is examined in developed economies [7,8], but it remains unclear whether it applies to emerging economies. Lanjouw and Mody [9] argued that environmental regulation in emerging economies cannot enhance domestic investment in pollution control technologies or green patents. Instead, it may increase the probability of importing green technology from advanced economies and strengthen foreign patents. It is, therefore, still worthwhile to investigate whether the environmental regulation has a significant impact on green innovation and provide a rigorous theoretical analysis and causal identification framework to test the impacts.

As an emerging economy, China's rapid economic development brought numerous forms of material national wealth. However, it brought a series of environmental problems, such as resource shortage, environmental pollution, and ecological deterioration, which became a public concern. To address these issues, the Chinese government took several environmental regulatory measures. However, we still have little knowledge about whether these regulatory measures have a positive effect on regional technological innovation, especially green innovation capacity. To fill in the gap, this study aims to examine and provide a robust estimation on the impacts of environmental regulation on green technology innovation within the Chinese context. Specifically, we focus on the "Two Control Zones" (TCZ) policy carried out in 1998, which was the first regulation policy for air pollution in China.

To control acid rain and sulfur dioxide pollution effectively, the Chinese government approved and implemented the "Two Control Zones" (TCZ) policy in 1998. The two control zones include the acid rain control zone and the sulfur dioxide control zone. In particular, the acid rain control zone is the region where the average pH value of rainfall is less than or equal to 4.5; the sulfur dioxide control zone is the area where the average sulfur dioxide concentration exceeds the national secondary standard of the past three years. The total area of the two control zones accounts for 11.4% of the total area of the national territory, the total population of the two control zones accounts for about 39% of the country's population and the GDP accounts for 67%, indicating a wide coverage of the impacts of the "two control zones" policy. The cities on the list of the two control zones are subject to strict environmental regulation, including restrictions on high–energy consumption, use of heavy–polluting energy sources, and sulfur dioxide emissions.

In our study, we are interested in whether and how the TCZ policy affects green technology innovation. To provide a robust estimation, we apply our analysis to a city–level panel data set from 1997 to 2016 in China, and examine the effects of environmental regulation policies on green innovation, considering the implementation of TCZ policy as a quasi–natural experiment. We apply a DID model, which is considered as the most effective model for policy evaluation. We find that the TCZ policy effectively increases the number of green patents of cities in the two control zones. In particular, the TCZ policy has a significantly positive effect on the quantity and structure of human capital, including the number of inventors of patents and green patents, the ratio of incumbents and newcomers, and the percentage of population with higher education levels. The effects are heterogeneous, that is, the TCZ policy has a greater impact on the number of green patents of cities in the two control zones where there are more R&D bases and more foreign investment.

To our knowledge, our study makes three main contributions to the literature: (1) This study is among the first to investigate the effects of TCZ policy on green technological progress from the perspective of environmental regulation–influenced regions; (2) we apply a DID technique to address the potential endogenous issue arising from omitted variables. It provides reliable and robust empirical evidence for analyzing the impacts of the environmental policy of TCZ on green technological progress; (3) we examine the mechanism of environmental regulation policy impacting the green technology innovation from different perspectives of human capital. It provides a new perspective to explain how environmental regulation policy affects green technology innovation. In addition, the heterogeneous effects of environmental regulation on green technological innovation are examined in terms of R&D base and foreign investment, thus revealing the comprehensive impacts of environmental regulation on green technological innovation.

The remainder of the paper is organized as follows: Section 2 presents a literature review and theoretical analysis, Section 3 presents the data and empirical design, Section 4 reports the empirical results, followed by the discussion in Section 5, and Section 6 concludes.

2. Literature Review and Theoretical Hypothesis

2.1. Literature Review

Many studies focused on the impact of environmental regulation on environmental quality and economic output. For example, studies argue that environmental regulation can effectively constrain pollutant emissions from firms [10–12] and greenhouse gas emissions [13–15], such as environmental protection taxes and emissions trading systems [12,16,17]. The existing studies also tested the pollution haven hypothesis (pollution haven hypothesis) [18–20], which argues that FDI will increase pollutant emissions [21], in which case areas with lax environmental regulation will be more attractive to FDI than strict areas, becoming pollution havens [22].

Then, a part of the study focuses on the microeconomic behavior of firms. It argues that environmental regulation may have some negative effects, such as reducing firm productivity [23], increasing unemployment [24,25], and reducing firms exports [26], among others. However, others also found positive effects, such as favoring industrial structure upgrading [27–29], boosting total factor productivity [30,31], and improving the capacity utilization of firms [12].

In addition, other studies focused on the influence of environmental regulation on corporate innovation, but they remain inconclusive. Based on the Porter hypothesis, reasonable environmental regulation can promote firm innovation [8]. According to the innovation compensation theory, environmental regulation is a triggering factor for technological change, inducing technological innovation that can compensate for environmental regulation payments [32–34]. Environmental regulation can incentivize companies to green upgrade through advanced technologies, such as cleaner production and green manufacturing [35,36]. Zhao and Sun [37] and You et al. [38] confirmed the validity of Porter's hypothesis.

In contrast, some scholars hold a different opinion that environmental regulation hinders corporate innovation, as strict environmental regulation adds unnecessary costs to firms [39–41]. Influenced by environmental governance costs [42], resources for technological innovation will be squeezed [40,43], which leads to a reduction in innovation activities [44]. Overall, there are many heterogeneities in regions and firms hardly following consistent rules of behavior, and different individuals exhibit differentiated technological innovation behavior under environmental regulatory policy constraints [45,46]. Bitat [47] used a panel of German firms to show that traditional regulatory measures cannot trigger innovative behaviors efficiently on a firm level. Moreover, some studies argued that the impact of environmental regulation on technological innovation is indeterminate and shows a non–linear relationship [48,49].

Given the uncertainty of the above findings, this paper suggests that different environmental regulatory measures and regional characteristics may be responsible for such contrasting results [37], and that the implications of the Porter hypothesis require further research. Moreover, although studies concentrated on the effect of environmental policy implementation on technological innovation, only a few studies examined the effect on green technological innovation [48,50–52]. Related studies show that environmental innovation has a positive impact on firms' competitive capability but may have a negative impact on the ecological footprints [53,54]. There is a positive correlation between green entrepreneurship and green innovation [55]. However, the influence of government behavior on enterprise environmental innovation and upgrade remains uncertain [56,57]. At the same time, the specific impact path of environmental policy on green technology innovation is no further distinction. There are potential endogeneity problems in the existing methods of assessing the effectiveness of environmental regulation.

Based on this, this paper explores how to achieve a win–win outcome for both environmental protection and economic development by studying the impact of environmental regulation on green innovation. The study focused on identifying the direct impact of China's TCZ policy on regional green innovation and the specific effect paths. We seek to expand the theoretical framework between environmental regulation and green innovation.

2.2. Theoretical Hypothesis

Due to the market scale effect and production endowment advantage, enterprises are reluctant to conduct green technology innovation activities. Faced with a market failure dilemma, designing and implementing scientifically sound environmental regulation increasingly became an effective means of addressing energy and environmental issues.

Environmental regulation releases a signal that the pollution will be controlled and regulated by the government effectively, indicating that the environmental quality will be improved. According to existing studies, air pollution is harmful to human health and leads to an increasing probability of cardiovascular and respiratory diseases [58,59]. Thus, air pollution leads to population outflow by significantly increasing residents' willingness to migrate internationally [60]. In contrast, there is a positive relationship between environmental quality and residents' health, implying that the environmental quality is better, and the city has a higher level of residential health [61,62]. The more educated or labor–productive groups are, the more sensitive they are to air pollution [63]. Because the population with high education and labor productivity has more knowledge and skills, they have more choices for work. Therefore, they will choose the cities with better urban environmental quality as the place of working and living. Environmental regulation becomes one of the guarantees of city quality, contributing to the inflow of labor and accumulation of human capital for the target cities.

According to the generalized Hicks theory, the incentive of environmental regulation towards the performance of green technology innovation stems from the implicit compliance costs of firms [50]. Under environmental regulation, companies have to improve their production processes, procedures, or equipment to meet the goal of maintaining legal emission standards over time at a lower cost. In such a case, pollution raises the cost of employing a highly qualified workforce, as they will demand higher salaries to participate in a heavily polluted city. Environmental regulation decreases that cost to a degree. Environmental regulatory policy promotes the internalization of environmental management costs and provides incentives for firms to make green innovation decisions. Thus, environmental regulatory policy, as an exterior compulsory driving force, creates a stimulating effect for green innovation and encourages firms to engage in green technological innovation [64].

At the same time, with the inflow of the workforce, especially high–quality human capital, the accumulation of knowledge and absorptive capacity related to environmental innovation can be increased, leading to improved innovation efficiency [65]. Especially in developing countries, access to external technology spillovers is an important channel for firms to acquire technological innovation capabilities. Under environmental regulations, firms will also have to import more high–quality intermediate goods and capital equipment from outside in the short run to meet higher environmental requirements. The technology spillover effects of trade provide firms with more learning opportunities, thus increasing their level of innovation [66].

Therefore, Hypothesis 1 is proposed according to the mentioned analysis: environmental regulation has a positive effect on green innovation performance.

Hypothesis 2 is proposed according to the mentioned analysis: environmental regulation has a positive impact on green technology innovations by attracting human capital inflow.

3. Data and Empirical Design

3.1. Data

To evaluate the impact of TCZ policy on green innovation performance, the number of green patents of the city is used to measure the development of green technology innovation. Green patent data are from the Chinese invention patent database, and the identification of green patents is based on the International Patent Classification (IPC) system code of the "IPC Green Inventory" published by the World Intellectual Property Organization (WIPO) on 16 September 2010. We can merge the Chinese invention patent database with IPC code to identify whether the patent is green or not. Green inventory patents are those related to non–fossil fuel–based methods of propulsion, such as electric or hydrogen cars and related technologies (e.g., batteries). After classifying whether each patent is a green patent, we build a year–city level green patent database based on the city and year information of the patent.

The cities in two control zones are identified by the state document named "The Official Reply of the State Council Concerning Acid Rain Control Areas and Sulfur Dioxide Pollution Control Areas". The document specifies 175 cities and regions in the two control zones, including 158 prefecture–level cities, 13 regions, and 4 municipalities directly under the central government.

The city–level data comes from China City Statistic Yearbook (CCSY) from 1990 to 2016. The control variables include total population, annual gross regional product, investment in fixed assets, foreign investment utilized, number of students in higher education institutions, number of teachers in higher education institutions, the proportion of employment in the secondary industry, employment at the end of the year, and number of new contracts signed in the current year.

Table 1 summarizes the statistic (observations, mean value, and stand deviation) of the main characteristics we used in this paper. The logarithm of the number of green patents each city applies for is 1.109 on average per year. The average annual total population is 5.652 ten thousand persons. The average annual gross regional output value is CNY 14.895 ten thousand. On average, the investment in fixed assets and foreign investment utilized in each city is CNY 13.809 ten thousand and USD 8.225 ten thousand per year, respectively. The logarithm of the number of students and teachers in higher education institutions is 8.87 and 6.537 per year on average, respectively. Approximately, the logarithm of employment and proportion of employment in the secondary industry is 3.746 and 3.503 on average in each city. The log number of new contracts each city signs is 3.746 on average per year.

Table 1. Summary statistic.

Variables	Obs	Mean	Std. Dev
Number of green patent applications (unit)	9234	1.109	1.409
Tcz × Post dummy (unit)	9234	0.342	0.474
Total population (ten thousand persons)	7857	5.652	0.832
Annual gross reginal product (CNY ten thousand)	7830	14.895	1.561
Investment in fixed assets (CNY ten thousand)	7830	13.809	2.084
Number of students in higher education institutions (persons)	7776	8.870	2.752
Number of teachers in higher education institutions (persons)	7776	6.537	2.059
Foreign investment utilized (USD ten thousand)	7750	8.225	2.898
Proportion of employment in the secondary industry (%)	9000	3.503	0.488
Employment (ten thousand persons)	8947	3.746	1.036
Number of new contracts signed (unit)	7578	3.221	1.589

Notes: Number of green patent applications, number of students in higher education institutions, number of teachers in higher education institutions, the proportion of employment in the secondary industry, and employment number of new contracts signed are measured in logs.

3.2. Empirical Design

To estimate the efficacy of TCZ policy for green technological innovations, a difference-in-differences (DID) model is used. Compared to changes in cities that were never under environmental regulation, we focus on how the number of green patents of cities in two control zones changed when the TCZ policy was enacted.

The DID method is adept at catching pre–existing differences between treated cities and untreated cities, thus eliminating selection bias while controlling for confounding variables that are likely to impact both sets of cities. The estimated equation is as follows:

$$G_{i,t} = \beta Tcz_i \times Post_t + \eta X_{i,t} + \gamma_i + \delta_t + \varepsilon_{i,t} \quad (1)$$

where i represents cities, t represents year; $G_{i,t}$ represents green technological innovations, which are measured as the log of the number of green patents applied by city i at year t; Tcz_i is 1 if city i is in two control zones and equals 0 if city i is not; $Post_t$ equals 1 for all years after 1998 (TCZ policy period) and otherwise equals 0. $Tcz_i \times Post_t$ is the interaction between the Tcz_i and $Post_t$, which captures the average difference change in the number of green patents of treated cities compared to untreated cities. Therefore, we focus on the coefficient measuring the DID effect and we posit β is positive, which means the TCZ policy will increase the number of green patents effectively. X_{it} represents city–level control variables, including total population (Pop), annual gross regional product (GDP), investment in fixed assets (Fixedinvest), foreign investment utilized (FDI), the logarithm of number of students in higher education institutions (Students), the logarithm of number of teachers in higher education institutions (Teachers), the logarithm of the proportion of employment in the secondary industry (Second), the logarithm of employment at the end of the year (Employment) and the logarithm of the number of new contracts signed in the current year (Contracts). The control variables represent the number of labors, development of economy and education, as well as degree of investment and openness. γ_i is a vector of city dummies and δ_t is a vector of year dummies to control city–fixed effects and year–fixed effects, respectively.

4. Empirical Results

4.1. The Impact of TCZ Policy on the Green Innovation Performance

Table 2 shows the DID results on the green innovation performance for TCZ policy corresponding to Equation (1). In column (1), we include no additional control variables, city–fixed effects, or year–fixed effects. The coefficient of the interaction term (Tcz × Post) in column (1) is significantly positive. In column (2) we include control variables, suggesting that TCZ policy increases the number of green patent applications by 71.8% on average and the p–value is less than 0.01. In column (3) we include control variables and control for city–fixed effects, while in column (4) we control for city–fixed effects, and year–fixed effects. Both of the results of column (3) and (4) remain highly significant (at the 1 percent level), and column (4) indicates there is a 53.4% increase in green patent applications of the cities in two control zones compared to the other cities. The finding is consistent with the hypothesis that more green patents were applied by target cities after the TCZ policy was enacted.

Table 2. Baseline Estimate.

	Green Patent Applications			
	(1)	**(2)**	**(3)**	**(4)**
Tcz × Post	0.760 ***	0.718 ***	0.596 ***	0.534 ***
	(0.079)	(0.098)	(0.089)	(0.084)
Pop		−0.451 ***	−0.745 ***	−0.316 ***
		(0.052)	(0.068)	(0.078)
GDP		0.559 ***	0.597 ***	0.219 **
		(0.070)	(0.084)	(0.094)
Fixedinvest		0.183 ***	0.228 ***	−0.069
		(0.040)	(0.047)	(0.048)
Students		−0.013	−0.007	−0.044 *
		(0.027)	(0.032)	(0.023)
Teachers		0.135 ***	0.005	−0.037
		(0.039)	(0.050)	(0.035)
FDI		−0.016	−0.019	−0.018
		(0.018)	(0.021)	(0.018)
Second		−0.053	0.065	0.098
		(0.071)	(0.086)	(0.080)

Table 2. *Cont.*

	Green Patent Applications			
	(1)	**(2)**	**(3)**	**(4)**
Employment		0.254 ***	0.188 ***	0.217 ***
		(0.051)	(0.058)	(0.051)
Contracts		0.063 ***	−0.039	0.078 ***
		(0.021)	(0.027)	(0.029)
Constant	0.153 ***	−8.399 ***	−7.178 ***	−0.054
	(0.018)	(0.420)	(0.598)	(0.930)
City FE	No	No	Yes	Yes
Year FE	No	No	No	Yes
Observations	9234	7464	7463	7463
R^2	0.245	0.699	0.775	0.814

Note: For each regression, the log volume of green patent applications is used as an outcome variable. Controls include total population at the end of the year (Pop), annual gross regional product (GDP), investment in fixed assets (Fixedinvest), foreign investment utilized (FDI), the logarithm of number of students in higher education institutions (Students), the logarithm of number of teachers in higher education institutions (Teachers), the logarithm of proportion of employment in the secondary industry (Second), the logarithm of employment at the end of the year (Employment), and the logarithm of number of new contracts signed in the current year (Contracts). Standard errors in parentheses are clustered at the city–year level. * $p < 0.10$, ** $p < 0.05$, *** $p < 0.01$.

4.2. Parallel Trend Analysis

Parallel growth in treated and control groups is the key identifying assumption of using the DID method. Thereby, we assume that there is the same rate of change in the amounts of green patents applied by cities out of the two control zones as cities in the two control zones, except for the implementation of the TCZ policy. Figure 1 plots the difference in the volume of green patent applications of cities that entered two control zones relative to those cities that did not, using an 8–year window before and after TCZ policy. Figure 1 displays no significant differences in pre–trend, implying that the difference in green patent applications the years before TCZ policy is normalized to 0, and the parallel trends assumption holds. After the year of TCZ policy, the estimated coefficients of TCZ–year interaction terms are significantly positive, suggesting an increase in green patent applications in the treated cities relative to the control group.

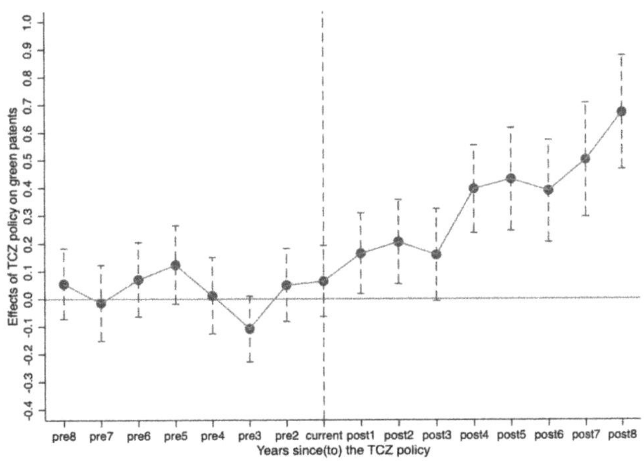

Figure 1. Treatment–year interaction coefficient for city–level green patent applications. Notes: Figure presents coefficient and 95% confidence intervals on Tcz × year interactions from the regression of *Tcz* × year interaction terms, including city–fixed effects and year–fixed effects. TCZ policy was implemented in 1998 (current), the year before TCZ policy was excluded (pre1). Standard errors in parentheses are clustered at the city–year level.

4.3. Validity Checks

4.3.1. Propensity Score Matching DID (PSM–DID)

DID estimation is most appropriate when the experiment is random. Considering that the assignment of the treated group by TCZ policy in our study may be not random, we should first use the Propensity Score Matching (PSM) approach to find and construct some comparable cities as the untreated group, and then evaluate the average impact of the TCZ policy on green patent applications using the DID model to examine whether our basic empirical results remain robust. PSM uses a logistic regression of the outcome variable that equals 1 if the city is in two control zones and equals 0 if it is not, and the independent variables include characteristics before treatment that would influence the "propensity" of cities in TCZ. Cities are matched to kernel values based on their propensity scores.

Firstly, we examine the results of treated and untreated cities before and after matching using the PSM approach. Figure 2 shows city characteristic bias between treatment and control groups before and after matching, implying that the deviation of all characteristics in both groups dropped to zero significantly after matching. From the perspective of kernel density, Figures 3 and 4 display the kernel density of treatment and control groups before and after matching, respectively. We find that the kernel density of the two groups is much closer. The above results indicate the validity of grouping using the PSM approach.

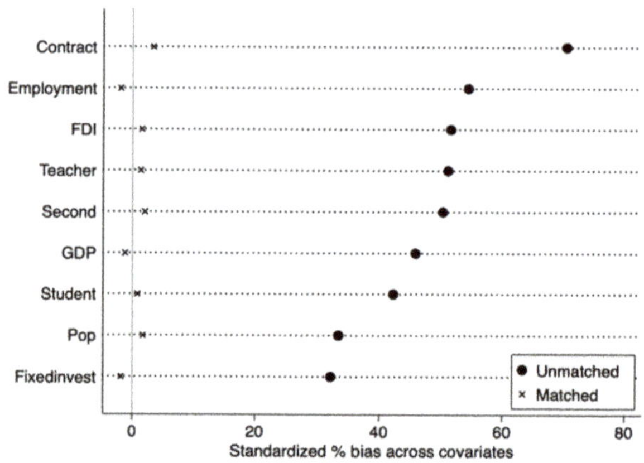

Figure 2. City characteristic bias before and after matching.

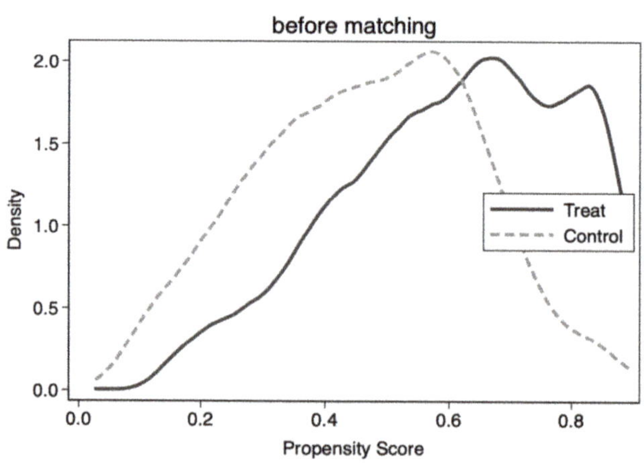

Figure 3. Kernel density of treated and control groups before matching.

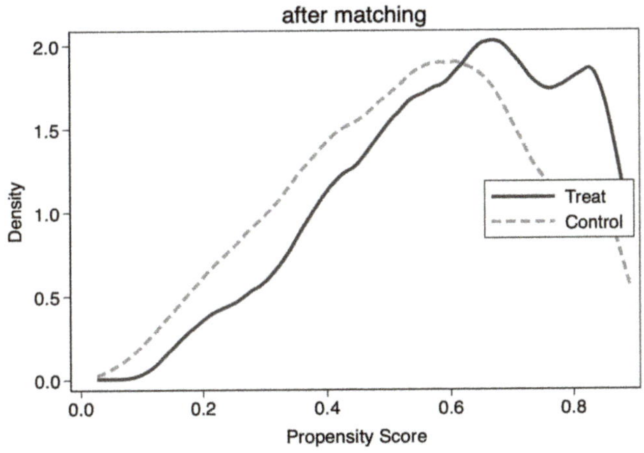

Figure 4. Kernel density of treated and control groups after matching.

Secondly, the PSM–DID results are shown in Table 3. In column (1) the regression includes no fixed effects, while column (2) only includes control city–fixed effects. Both coefficients in column (1) and (2) are significantly positive. The estimated result is controlled for city–fixed effects and year–fixed effects in column (3). The estimate for TCZ policy is significantly positive (at the 1 percent level), implying that targeted cities have 37% more green patent applications when the TCZ policy is enacted. Thus, the interference of unobservable factors in the selection of the treated and untreated groups on the conclusions of this study can be excluded.

Table 3. PSM–DID results.

	Green Patent Applications		
	(1)	(2)	(3)
Tcz × Post	0.564 ***	0.405 ***	0.370 ***
	(0.090)	(0.080)	(0.077)
Control variables	Yes	Yes	Yes
City FE	No	Yes	Yes
Year FE	No	No	Yes
Observations	7052	7051	7051
R^2	0.644	0.736	0.777

Note: For each regression, the log volume of green patent applications is used as outcome variable. Controls include total population at the end of the year (Pop), annual gross regional product (GDP), investment in fixed assets (Fixedinvest), foreign investment utilized (FDI), the logarithm of number of students in higher education institutions (Students), the logarithm of number of teachers in higher education institutions (Teachers), the logarithm of proportion of employment in the secondary industry (Second), the logarithm of employment at the end of the year (Employment), and the logarithm of the number of new contracts signed in the current year (Contracts). Standard errors in parentheses are clustered at the city–year level. *** $p < 0.01$.

4.3.2. Test on the Number of Granted Green Patents

Apart from examining the efficacy of the TCZ policy for the number of green patent applications, the further test is analyzing the policy's effect on the number of green patents granted. We put the log of the number of green patents granted into Equation (1) as the outcome variable instead of the number of green patent applications. The results in Table 4 show that all of the coefficients are statistically significant at the 1 percent level. In column (4), we include control variables, city–fixed effects, and year–fixed effects. The key interaction term's coefficient is 0.565 and statistically significant at the 1% level, indicating that the number of green patents granted increases by 56.5% in two control zones, which examines the robustness of the conclusions of this study.

Table 4. The effect of TCZ policy on green patents granted.

	Green Patent Granted			
	(1)	(2)	(3)	(4)
Tcz × Post	0.675 ***	0.675 ***	0.592 ***	0.565 ***
	(0.069)	(0.069)	(0.073)	(0.073)
Control variables	No	No	Yes	Yes
City FE	Yes	Yes	Yes	Yes
Year FE	No	Yes	No	Yes
Observations	9234	9234	7463	7463
R^2	0.555	0.703	0.616	0.735

Note: The dependent variable in each regression is the log of the number of green patents granted. Controls include total population at the end of the year (Pop), annual gross regional product (GDP), investment in fixed assets (Fixedinvest), foreign investment utilized (FDI), the logarithm of number of students in higher education institutions (Students), the logarithm of number of teachers in higher education institutions (Teachers), the logarithm of the proportion of employment in the secondary industry (Second), the logarithm of employment at the end of the year (Employment), and the logarithm of the number of new contracts signed in the current year (Contracts). Standard errors in parentheses are clustered at the city–year level. *** $p < 0.01$.

5. Further Discussion

5.1. Heterogenous Effect of TCZ Policy

There is large heterogeneity contained by the average TCZ policy effect on green technological innovations. We further conduct our analysis to examine how environmental regulation impact differs for the R&D base and foreign investment utilized. R&D base is measured by cumulative amounts of total patents granted over the past five years, and foreign investment utilized is measured by FDI of the city in that year. We put the two new interaction terms (TCZ × post × R&D and TCZ × post × FDI) into Equation (1), respectively. Table 5 column (1) reports the result of the heterogeneous effect on the R&D base. The coefficient of TCZ × post × R&D is significantly positive, implying that the better the R&D base cities have, the larger the number of green patents they can apply for. Compared to the city with a relatively weaker R&D base, the city with a strong R&D base usually puts more emphasis on innovation activities and accumulates more experience in developing green technology innovation, indicating that it has more ability and recourses to conduct the development of green patents when the environmental regulation regime is enacted. As the results show in column (2), the coefficient of TCZ × post × FDI is positive and statistically significant at 1% level with the value of 0.167. The cities in two control zones with more FDI have better green innovation performance. Foreign firms usually have to face more strict environmental regulations in their home country, resulting in larger amounts of green technologies in the firms. Those target cities are likely to get more technology spillover from multinationals by FDI after the TCZ policy.

Table 5. Heterogenous effect of TCZ policy.

	Green Patent Applications	
	(1)	(2)
TCZ × post × R&D	0.061 **	
	(0.031)	
TCZ × post × FDI		0.167 ***
		(0.032)
Control variables	Yes	Yes
City FE	Yes	Yes
Year FE	Yes	Yes
Observations	7463	7463
R^2	0.871	0.828

Note: For each regression, the log volume of green patent applications is used as an outcome variable. Controls include total population at the end of the year (Pop), annual gross regional product (GDP), investment in fixed assets (Fixedinvest), foreign investment utilized (FDI), the logarithm of number of students in higher education institutions (Students), the logarithm of number of teachers in higher education institutions (Teachers), the logarithm of the proportion of employment in the secondary industry (Second), the logarithm of employment at the end of the year (Employment), and the logarithm of the number of new contracts signed in the current year (Contracts). Standard errors in parentheses are clustered at the city–year level. ** $p < 0.05$, *** $p < 0.01$.

Therefore, the effect of environmental regulation on green innovation is significantly affected by the city's R&D base and FDI. For example, Beijing and Shanghai, as the pilot cities of TCZ policy, both have a high number of accumulated patents and a high level of FDI, and they are also the two cities with the largest number of green innovation patents in China.

5.2. Mechanism

We further examine the channels for cities in two control zones to increase green patents for environmental regulations. Theoretically, human capital is a crucial factor for technology innovation, indicating that the higher quality workforce a city has, the larger the number of green patents the city has. Environmental regulation has a positive impact on pollution reduction and urban quality, which attracts more talents to come to target cities. Thus, human capital is a significant mechanism in TCZ effect.

Table 6 reports the estimated results of the city and year–fixed effects models using the log number of patent inventors (Inventors), the log number of green patent inventors (Ginventors), and the percentage of population with college and higher education (Unipop) as the dependent variables according to Equation (1). We include control variables and control city–fixed effects and year–fixed effects. The coefficients are found to be positive and statistically significant at the 1% level. Column (1) and (2) lists the results of environmental regulation policy impact on patent inventors and green patent inventors. The estimates show that the number of patent inventors increases by 52.3% and the number of green patent inventors increases by 78.4% in treated cities compared to untreated cities by TCZ policy. As the result is shown in column 4, the TCZ policy increases the percentage of the population with high education by 0.9%.

Table 6. Mechanism of human capital.

	Inventors	Ginventors	Unipop
	(1)	(2)	(3)
TCZ × Post	0.523 ***	0.784 ***	0.009 ***.
	(0.094)	(0.114)	(0.002)
Control variables	Yes	Yes	Yes
City FE	Yes	Yes	Yes
Year FE	Yes	Yes	Yes
Observations	7463	7463	747
R^2	0.911	0.782	0.906

Note: The dependent variable in each regression is the log number of patent inventors (Inventors), the log number of green patent inventors (Ginventors), and the percentage of the population with college and higher education (Unipop). Controls include total population at the end of the year (Pop), annual gross regional product (GDP), investment in fixed assets (Fixedinvest), foreign investment utilized (FDI), the logarithm of number of students in higher education institutions (Students), the logarithm of number of teachers in higher education institutions (Teachers), the logarithm of the proportion of employment in the secondary industry (Second), the logarithm of employment at the end of the year (Employment), and the logarithm of the number of new contracts signed in the current year (Contracts). Standard errors in parentheses are clustered at the city–year level. *** $p < 0.01$.

6. Conclusions

To cope with air pollution, the "Two Control Zones (TCZ)" policy was issued and enacted by China's government in 1998. As the first air pollution regulation in China, the impact of the TCZ policy influences the development of following environmental regulations.

The results in our study use a difference in difference model that explores the effect of environmental regulation on green technology innovations and the role of human capital in it. We find evidence consistent with the hypothesis that the TCZ policy significantly increases the number of green patents of cities in two control zones. The result is also robust through the method of PSM–DID and changing the dependent variable. Most importantly, our study also points out the crucial role human capital plays in the mechanism. TCZ policy, as the signal of regulating air pollution and improving urban quality, has a positive effect on the quantity and structure of human capital, leading to providing a talent pool for green technology innovation to reduce pollution.

To exploit the heterogeneity covered under the average treatment effect, the finding shows that TCZ policy is different in the R&D basis and foreign investment utilized. TCZ policy tends to improve more amounts of green patents in the cities with a stronger R&D base or with more FDI. The R&D base provides innovative talents for green technology innovation and FDI provides technology spillover and R&D funding for green technology innovation. The cities with those two characteristics have more ability to undertake the development and application for green patents to cope with TCZ policy.

The findings of this paper provide new insight into the Porter hypothesis, offering some valuable policy recommendations for developing economies. In the context of globalization, developing countries, as a link in the downstream production chain, are highly susceptible to becoming pollution havens. Policymakers of emerging economies draw environmental regulations to control pollution, while promoting the development of green technology innovation by attracting more high–quality human capital. In addition, based on our study, the government should pay more attention to strengthening its R&D base and attracting more FDI, as both of these conditions will enhance the positive impact of environmental regulation on green innovation performance.

There are also some limitations. First, the research sample of this paper is the city–level data in China. We can only control for regional and year effects. In the future, when the green patent data at the corporate level becomes available, we can study it from the perspective of micro firms. Second, green technological innovation can also be subdivided in terms of production processes, such as green process innovation and green product innovation. However, the data refinement is limited, and this paper only uses the number of green patent applications granted to measure the overall green technology innovation of cities. With the increasing availability of data, a comparative analysis of the variability in the impact of environmental regulations on the green production processes of cities will also be worthy of further research.

Author Contributions: Conceptualization, J.L., S.Z. and T.H.; Data curation, J.L. and T.H.; Formal analysis, J.L. and S.Z.; Investigation, S.Z.; Methodology, J.L. and S.Z.; Project administration, J.L.; Supervision, T.H. and D.N.; Validation, T.H. and D.N.; Visualization, T.H.; Writing—review & editing, J.L. and S.Z. All authors have read and agreed to the published version of the manuscript.

Funding: This research received no external funding.

Data Availability Statement: The data presented in this study are available on request from the corresponding author.

Conflicts of Interest: The authors declare no conflict of interest.

References

1. Khorasanizadeh, H.; Honarpour, A.; Park, M.S.-A.; Parkkinen, J.; Parthiban, R. Adoption factors of cleaner production technology in a developing country: Energy efficient lighting in Malaysia. *J. Clean. Prod.* **2016**, *131*, 97–106. [CrossRef]
2. Verdolini, E.; Bosetti, V. Environmental policy and the international diffusion of cleaner energy technologies. *Environ. Resour. Econ.* **2017**, *66*, 497–536. [CrossRef]
3. Raghutla, C.; Chittedi, K.R. Financial development, energy consumption, technology, urbanization, economic output and carbon emissions nexus in BRICS countries: An empirical analysis. *Manag. Environ. Qual. Int. J.* **2020**, *32*, 290–307. [CrossRef]
4. James, D.E.; Jansen, H.M.; Opschoor, J.B. *Economic Approaches to Environmental Problems: Techniques and Results of Empirical Analysis*; Elsevier: Amsterdam, The Netherlands, 2013.
5. Porter, M.E.; van der Linde, C. Towards a New Conception of the Environment–Competitiveness Relationship. *J. Econ. Persp.* **1995**, *9*, 97–118. [CrossRef]
6. Wei, Y.; Qiang, C. 20 Years of Porter Hypothesis—A literature review on the relationship among environmental regulation, innovation and competitiveness. *Sci. Res. Manag.* **2015**, *36*, 65.
7. Kesidou, E.; Demirel, P. On the drivers of eco–innovations: Empirical evidence from the UK. *Res. Policy* **2012**, *41*, 862–870. [CrossRef]

8. Brunnermeier, S.B.; Cohen, M.A. Determinants of environmental innovation in US manufacturing industries. *J. Environ. Econ. Manag.* **2003**, *45*, 278–293. [CrossRef]
9. Lanjouw, J.O.; Mody, A. Innovation and the international diffusion of environmentally responsive technology. *Res. Policy* **1996**, *25*, 549–571. [CrossRef]
10. He, P.; Zhang, B. Environmental tax, polluting plants' strategies and effectiveness: Evidence from China. *J. Policy Anal. Manag.* **2018**, *37*, 493–520. [CrossRef]
11. Fullerton, D.; Karney, D.H. Multiple pollutants, co–benefits, and suboptimal environmental policies. *J. Environ. Econ. Manag.* **2018**, *87*, 52–71. [CrossRef]
12. Du, W.; Li, M. Assessing the impact of environmental regulation on pollution abatement and collaborative emissions reduction: Micro–evidence from Chinese industrial enterprises. *Environ. Impact Assess. Rev.* **2020**, *82*, 106382. [CrossRef]
13. Hashmi, R.; Alam, K. Dynamic relationship among environmental regulation, innovation, CO2 emissions, population, and economic growth in OECD countries: A panel investigation. *J. Clean. Prod.* **2019**, *231*, 1100–1109. [CrossRef]
14. Pei, Y.; Zhu, Y.; Liu, S.; Wang, X.; Cao, J. Environmental regulation and carbon emission: The mediation effect of technical efficiency. *J. Clean. Prod.* **2019**, *236*, 117599. [CrossRef]
15. Neves, S.A.; Marques, A.C.; Patrício, M. Determinants of CO_2 emissions in European Union countries: Does environmental regulation reduce environmental pollution? *Econ. Anal. Policy* **2020**, *68*, 114–125. [CrossRef]
16. Gao, J.; Woodward, A.; Vardoulakis, S.; Kovats, S.; Wilkinson, P.; Li, L.; Xu, L.; Li, J.; Yang, J.; Cao, L.; et al. Haze, public health and mitigation measures in China: A review of the current evidence for further policy response. *Sci. Total Environ.* **2017**, *578*, 148–157. [CrossRef]
17. Wang, Y.; Zuo, Y.; Li, W.; Kang, Y.; Chen, W.; Zhao, M.; Chen, H. Does environmental regulation affect CO_2 emissions? Analysis based on threshold effect model. *Clean Technol. Environ. Policy* **2019**, *21*, 565–577. [CrossRef]
18. Cai, X.; Lu, Y.; Wu, M.; Yu, L. Does environmental regulation drive away inbound foreign direct investment? Evidence from a quasi–natural experiment in China. *J. Dev. Econ.* **2016**, *123*, 73–85. [CrossRef]
19. Zhu, H.; Duan, L.; Guo, Y.; Yu, K. The effects of FDI, economic growth and energy consumption on carbon emissions in ASEAN–5: Evidence from panel quantile regression. *Econ. Model.* **2016**, *58*, 237–248. [CrossRef]
20. Kearsley, A.; Riddel, M. A further inquiry into the pollution haven hypothesis and the environmental Kuznets curve. *Ecol. Econ.* **2010**, *69*, 905–919. [CrossRef]
21. Baek, J. A new look at the FDI–income–energy–environment nexus: Dynamic panel data analysis of ASEAN. *Energy Policy* **2016**, *91*, 22–27. [CrossRef]
22. Demena, B.A.; Afesorgbor, S.K. The effect of FDI on environmental emissions: Evidence from a meta–analysis. *Energy Policy* **2020**, *138*, 111192. [CrossRef]
23. Hancevic, P.I. Environmental regulation and productivity: The case of electricity generation under the CAAA–1990. *Energy Econ.* **2016**, *60*, 131–143. [CrossRef]
24. Hafstead, M.A.; Williams, R.C., III. Unemployment and environmental regulation in general equilibrium. *J. Public Econ.* **2018**, *160*, 50–65. [CrossRef]
25. Zheng, J.; He, J.; Shao, X.; Liu, W. The employment effects of environmental regulation: Evidence from eleventh five–year plan in China. *J. Environ. Manag.* **2022**, *316*, 115197. [CrossRef]
26. Shi, X.; Xu, Z. Environmental regulation and firm exports: Evidence from the eleventh Five–Year Plan in China. *J. Environ. Econ. Manag.* **2018**, *89*, 187–200. [CrossRef]
27. Oueslati, W. Environmental tax reform: Short–term versus long–term macroeconomic effects. *J. Macroecon.* **2014**, *40*, 190–201. [CrossRef]
28. Chen, Y.J.; Li, P.; Lu, Y. Career concerns and multitasking local bureaucrats: Evidence of a target–based performance evaluation system in China. *J. Dev. Econ.* **2018**, *133*, 84–101. [CrossRef]
29. Liu, S.; Liu, C.; Yang, M. The effects of national environmental information disclosure program on the upgradation of regional industrial structure: Evidence from 286 prefecture–level cities in China. *Struct. Chang. Econ. Dyn.* **2021**, *58*, 552–561. [CrossRef]
30. Yuan, B.; Zhang, K. Can environmental regulation promote industrial innovation and productivity? Based on the strong and weak Porter hypothesis. *Chin. J. Popul. Resour. Environ.* **2017**, *15*, 322–336. [CrossRef]
31. Li, J.; Zhao, M.; Yang, Y. Environmental regulation and firms' performance: A quasi–natural experiment from China. *Chin. J. Popul. Resour. Environ.* **2019**, *17*, 278–294. [CrossRef]
32. Ramanathan, R.; Black, A.; Nath, P.; Muyldermans, L. Impact of environmental regulations on innovation and performance in the UK industrial sector. *Manag. Decis.* **2010**, *48*, 1493–1513. [CrossRef]
33. Kneller, R.; Manderson, E. Environmental regulations and innovation activity in UK manufacturing industries. *Resour. Energy Econ.* **2012**, *34*, 211–235. [CrossRef]
34. Ambec, S.; Barla, P. Can environmental regulations be good for business? An assessment of the Porter hypothesis. *Energy Stud. Rev.* **2006**, *14*. [CrossRef]
35. Shao, S.; Hu, Z.; Cao, J.; Yang, L.; Guan, D. Environmental regulation and enterprise innovation: A review. *Bus. Strategy Environ.* **2020**, *29*, 1465–1478. [CrossRef]

36. Zhang, W.; Luo, Q.; Liu, S. Is government regulation a push for corporate environmental performance? Evidence from China. *Econ. Anal. Policy* **2022**, *74*, 105–121. [CrossRef]
37. Zhao, X.; Sun, B. The influence of Chinese environmental regulation on corporation innovation and competitiveness. *J. Clean. Prod.* **2016**, *112*, 1528–1536. [CrossRef]
38. You, D.; Zhang, Y.; Yuan, B. Environmental regulation and firm eco–innovation: Evidence of moderating effects of fiscal decentralization and political competition from listed Chinese industrial companies. *J. Clean. Prod.* **2019**, *207*, 1072–1083. [CrossRef]
39. Feng, C.; Shi, B.; Kang, R. Does environmental policy reduce enterprise innovation?—Evidence from China. *Sustainability* **2017**, *9*, 872. [CrossRef]
40. Yuan, B.; Xiang, Q. Environmental regulation, industrial innovation and green development of Chinese manufacturing: Based on an extended CDM model. *J. Clean. Prod.* **2018**, *176*, 895–908. [CrossRef]
41. Palmer, K.; Oates, W.E.; Portney, P.R. Tightening environmental standards: The benefit–cost or the no–cost paradigm? *J. Econ. Perspect.* **1995**, *9*, 119–132. [CrossRef]
42. Ling Guo, L.; Qu, Y.; Tseng, M.-L. The interaction effects of environmental regulation and technological innovation on regional green growth performance. *J. Clean. Prod.* **2017**, *162*, 894–902.
43. Lin, J.; Xu, C. The impact of environmental regulation on total factor energy efficiency: A cross–region analysis in China. *Energies* **2017**, *10*, 1578. [CrossRef]
44. Andersson, R.; Quigley, J.M.; Wilhelmsson, M. Agglomeration and the spatial distribution of creativity. *Pap. Reg. Sci.* **2005**, *84*, 445–464. [CrossRef]
45. Johnstone, N.; Managi, S.; Rodríguez, M.C.; Haščič, I.; Fujii, H.; Souchier, M. Environmental policy design, innovation and efficiency gains in electricity generation. *Energy Econ.* **2017**, *63*, 106–115. [CrossRef]
46. Luo, Y.; Salman, M.; Lu, Z. Heterogeneous impacts of environmental regulations and foreign direct investment on green innovation across different regions in China. *Sci. Total Environ.* **2021**, *759*, 143744. [CrossRef]
47. Bitat, A. Environmental regulation and eco–innovation: The Porter hypothesis refined. *Eurasian Bus. Rev.* **2018**, *8*, 299–321. [CrossRef]
48. Fan, F.; Lian, H.; Liu, X.; Wang, X. Can environmental regulation promote urban green innovation Efficiency? An empirical study based on Chinese cities. *J. Clean. Prod.* **2021**, *287*, 125060. [CrossRef]
49. Ouyang, X.; Li, Q.; Du, K. How does environmental regulation promote technological innovations in the industrial sector? Evidence from Chinese provincial panel data. *Energy Policy* **2020**, *139*, 111310. [CrossRef]
50. Cai, X.; Zhu, B.; Zhang, H.; Li, L.; Xie, M. Can direct environmental regulation promote green technology innovation in heavily polluting industries? Evidence from Chinese listed companies. *Sci. Total Environ.* **2020**, *746*, 140810. [CrossRef]
51. Liu, M.; Li, Y. Environmental Regulation and Green Innovation: Evidence from China's Carbon Emissions Trading Policy. *Financ. Res. Lett.* **2022**, *48*, 103051. [CrossRef]
52. Wu, W.; Liu, Y.; Wu, C.-H.; Tsai, S.-B. An empirical study on government direct environmental regulation and heterogeneous innovation investment. *J. Clean. Prod.* **2020**, *254*, 120079. [CrossRef]
53. Skordoulis, M.; Ntanos, S.; Kyriakopoulos, G.L.; Arabatzis, G.; Galatsidas, S.; Chalikias, M. Environmental innovation, open innovation dynamics and competitive advantage of medium and large–sized firms. *J. Open Innov. Technol. Mark. Complex.* **2020**, *6*, 195. [CrossRef]
54. Liu, C.; Ni, C.; Sharma, P.; Jain, V.; Chawla, C.; Shabbir, M.S.; Tabash, M.I. Does green environmental innovation really matter for carbon–free economy? Nexus among green technological innovation, green international trade, and green power generation. *Environ. Sci. Pollut. Res.* **2022**, *29*, 67504–67512. [CrossRef]
55. Skordoulis, M.; Kyriakopoulos, G.; Ntanos, S.; Galatsidas, S.; Arabatzis, G.; Chalikias, M.; Kalantonis, P. The Mediating Role of Firm Strategy in the Relationship between Green Entrepreneurship, Green Innovation, and Competitive Advantage: The Case of Medium and Large–Sized Firms in Greece. *Sustainability* **2022**, *14*, 3286. [CrossRef]
56. Ren, S.; Sun, H.; Zhang, T. Do environmental subsidies spur environmental innovation? Empirical evidence from Chinese listed firms. *Technol. Forecast. Soc. Chang.* **2021**, *173*, 121123. [CrossRef]
57. Wen, H.; Lee, C.-C.; Zhou, F. Green credit policy, credit allocation efficiency and upgrade of energy–intensive enterprises. *Energy Econ.* **2021**, *94*, 105099. [CrossRef]
58. Hanna, R.; Oliva, P. The effect of pollution on labor supply: Evidence from a natural experiment in Mexico City. *J. Public Econ.* **2015**, *122*, 68–79. [CrossRef]
59. Holub, F.; Hospido, L.; Wagner, U.J. *Urban Air Pollution and Sick Leaves: Evidence from Social Security Data*; Banco de España: Madrid, Spain, 2020.
60. Qin, Y.; Zhu, H. Run away? Air pollution and emigration interests in China. *J. Popul. Econ.* **2018**, *31*, 235–266. [CrossRef]
61. Bamai, Y.A.; Miyashita, C.; Araki, A.; Nakajima, T.; Sasaki, S.; Kishi, R. Effects of prenatal di (2–ethylhexyl) phthalate exposure on childhood allergies and infectious diseases: The Hokkaido Study on Environment and Children's Health. *Sci. Total Environ.* **2018**, *618*, 1408–1415. [CrossRef]
62. Jarvis, I.; Koehoorn, M.; Gergel, S.E.; van den Bosch, M. Different types of urban natural environments influence various dimensions of self–reported health. *Environ. Res.* **2020**, *186*, 109614. [CrossRef]

63. He, J.; Liu, H.; Salvo, A. Severe air pollution and labor productivity: Evidence from industrial towns in China. *Am. Econ. J. Appl. Econ.* **2019**, *11*, 173–201. [CrossRef]
64. Acemoglu, D.; Aghion, P.; Bursztyn, L.; Hemous, D. The environment and directed technical change. *Am. Econ. Rev.* **2012**, *102*, 131–166. [CrossRef] [PubMed]
65. Jiang, Z.; Wang, Z.; Li, Z. The effect of mandatory environmental regulation on innovation performance: Evidence from China. *J. Clean. Prod.* **2018**, *203*, 482–491. [CrossRef]
66. Shang, L.; Tan, D.; Feng, S.; Zhou, W. Environmental regulation, import trade, and green technology innovation. *Environ. Sci. Pollut. Res.* **2022**, *29*, 12864–12874. [CrossRef]

energies

MDPI

Article

Revisiting the Environmental Kuznets Curve Hypothesis in South Asian Countries: The Role of Energy Consumption and Trade Openness

Bartosz Jóźwik [1,*], Phouphet Kyophilavong [2], Aruna Kumar Dash [3] and Antonina Viktoria Gavryshkiv [1]

1 Department of International Economics, The John Paul II Catholic University of Lublin, 20-950 Lublin, Poland
2 Faculty of Economics and Business Management, National University of Laos, Vientiane 7322, Laos
3 Department of Economics, IBS Hyderabad, IFHE University, Dontanapally, Telangana, Hyderabad 501203, India
* Correspondence: bjozwik@kul.pl

Abstract: South Asian countries have seen remarkable economic growth and development in the past few decades. This has been driven by financial sector reforms, industrialization, and expansion of foreign trade. The present study is designed to identify the long- and short-run relationships among environmental degradation, economic growth, energy consumption, and trade openness in the South Asian region. Our research contributes to the literature by employing a new approach (the NARDL method). We examine annual data for four South Asian countries between 1971 and 2014. We found that there was a long-run equilibrium relationship between environmental degradation, economic growth, energy consumption, and trade openness. The results confirmed the inverted U-shaped EKC hypothesis only for India and Pakistan. However, the long-term coefficients related to energy consumption were statistically significant only in Pakistan. The most interesting finding was that only in Sri Lanka did the long-run coefficients associated with trade openness shocks significantly impact carbon dioxide emissions. These impacts were based on the scale effect. Our study has some policy implications. Foremost, the governments of South Asian countries should promote and subsidize green energy use by increasing R&D spending on renewable energy.

Keywords: environmental Kuznets curve; South Asian countries; trade openness; energy consumption; economic growth; carbon dioxide emissions; NARDL model; ARDL model

Citation: Jóźwik, B.; Kyophilavong, P.; Dash, A.K.; Gavryshkiv, A.V. Revisiting the Environmental Kuznets Curve Hypothesis in South Asian Countries: The Role of Energy Consumption and Trade Openness. *Energies* **2022**, *15*, 8709. https://doi.org/10.3390/en15228709

Academic Editors: Junpeng Zhu and Xinlong Xu

Received: 29 September 2022
Accepted: 15 November 2022
Published: 19 November 2022

1. Introduction

In the past few decades, South Asian countries have seen remarkable economic growth and development. This is attributable to financial sector reforms, industrialization, and expansion of foreign trade. The region's gross domestic product increased more than 17-fold—from 190.7 to 3241.9 billion US dollars from 1960 to 2020, with an average annual growth rate of 4.92%. It is noteworthy that this region's growth rate has been higher compared to the world's. Between 1961 and 1979, the world's growth rate was ahead of the growth rate of South Asian countries several times. However, this relationship changed from 1980 onwards, due to financial sector reform. In the early 1980s, financial sector reform, particularly banking sector reform, was initiated by South Asian countries to increase their competitiveness. As a result, policies have been adopted to restructure public sector banks and allow private sector banks to promote competition in the banking sector, and efforts have been taken to liberalize the financial sector [1]. Between 1980 and 2020, the South Asian region's growth rate was above the world's growth rate. Notably, only in 1984, 2000, and 2020 was the world's growth rate ahead of that of the South Asian region.

Although this region's growth performance is impressive, South Asia is globally perceived as an underprivileged region, where more than 50% of the world's poor live. To eliminate poverty and unemployment, the South Asian region supports fast economic growth. This, however, is done without assessing the vulnerabilities of the environment [2]. The curtailment of energy consumption is not an easy task as it slows down the economic growth and development of a country [3].

Noteworthy, one of the necessary components of the economic growth of a country is trade openness and expansion in foreign trade, which enhances economic activities and energy demand [4]. Trade openness enables many underdeveloped or developing countries to import the latest technology from developed nations, which in turn helps them to produce more output while lowering energy intensity. Trade openness may simultaneously determine income and environmental quality. In South Asian countries, the volume of trade has shown an increasing trend since the early 1980s, which might be due to the financial sector reform in 1980. The merchandise trade, which is the combination of exports and imports, worth 6.6 billion US dollars in 1960, increased to 39.9 billion US dollars in 1980 and reached 1083 billion US dollars in 2018. At the same time, the total amount of CO_2 emissions increased from 0.26 to 1.53 metric tons per capita from 2006 to 2018. In 2018, South Asian countries exported 41% of manufactured products [5]. Trade openness leads to deterioration of the environmental quality due to large-scale production of merchandise goods, which causes higher energy consumption and CO_2 emissions.

Foreign trade expansion and industrialization results in a growing demand for energy consumption. For example, the total fossil fuel energy consumption in the South Asian region amounted to 33.87% in 1971 and more than doubled in 2014 to 71.52%. The consensus believes that the consumption of fossil fuels (coal, natural gas and oil) led to a rapid increase in CO_2 emissions, disrupting environmentally sustainable growth in South Asia. India, Bangladesh, Pakistan, and Sri Lanka consume more fossil fuel compared with other countries of that region. South Asia's percentage share of the world's fossil fuel consumption increased from 40 to 88% from 1971 to 2014.

Many studies have been conducted on the environmental Kuznets curve hypothesis in South Asian countries; for example, the recently published studies by Sadiq et al. [6], Ali et al. [7], Mehmood et al. [8], and Tan et al. [9]. However, the empirical results for those countries are mixed. Most researchers have used the conventional cointegration approaches. In this study, we make several contributions to the current literature. First, we used a method that does not ignore the asymmetry effect. Second, we consider the roles of energy consumption and asymmetric shocks in trade openness in the environmental Kuznets curve. Third, we describe how government programs could influence environmental quality, especially in India and Pakistan, where the long-run coefficient for squared GDP per capita is negative and significant. These coefficients indicate that we should expect increased environmental quality.

Our aim is to identify long-run and short-run relationships between environmental degradation, economic growth, energy consumption, and trade openness in South Asian countries. We examine annual data for four South Asian countries (India, Bangladesh, Sri Lanka, and Pakistan) for the period between 1971 and 2014. The selection of the time period and sample was determined by data availability. All annual time series data come from the World Bank collection of development indicators.

The remainder of the paper is structured as follows: Section 2 reviews literature on environmental degradation in South Asian countries and the linkages between energy consumption, trade openness, and carbon dioxide emissions; Section 3 describes the data and methodology; Section 4 presents the empirical results. In this section, we present both linear and non-linear ARDL models. Section 5 includes conclusions and highlights policy implications.

2. Literature Review

2.1. Environmental Degradation in South Asian Countries

The South Asian region faces large-scale environmental issues compounded by the overlapping factors of growing industrialization, urbanization, population growth, and increasing international trade [10]. In recent years, countries in South Asia have seen growing urbanization and industrialization, which has led to rising rates of greenhouse gas emissions and increasing levels of environmental degradation [11,12].

This region has enjoyed some successes in reducing poverty. This was possible thanks to rapid industrialization and the implementation of liberal economic reforms. India and Bangladesh have been overly involved in expanding heavy industries due to their partial adoption of the development model. This has led to an increased industrial output and acceleration of environmental deterioration. According to Mehmood and Tariq [13], globalization led to an increase in CO_2 emissions in South Asian nations. This trend was observed from 1972 to 2013. It does not mean that rising production is always positively connected with environmental degradation indices; instead, environmental degradation depends on the use of contemporary technology and regulations adopted to protect the environment.

The International Energy Agency (IEA) predicted that during the next few decades, the demand for energy in the South Asian region would increase at a rate more than twice as fast as the average growth rate for the entire world. The rise in economic activity results in higher energy demand, contributing to the economy's expansion and growth. Rahman and Velayutham [14] examined the effect of consumption of renewable and non-renewable energy, and the effect of fixed capital formation on economic growth for a panel of five South Asian countries over the period of 1990–2014. Their findings indicated that these factors positively contributed to economic growth. In this scenario, increased economic activity may hasten the exhaustion of natural resources and lead to environmental deterioration in the absence of sufficient regulations. Greater consumption of resources results in a rise in carbon dioxide emissions and a decline in environmental quality, negatively influencing human health [8,15].

Increasing population growth, widespread poverty, lack of public awareness of environmental issues, failure to properly and robustly implement environmental laws and regulations, and failure to monitor environmental conditions—all these factors contribute to the deterioration of the environment. The vast majority of the unemployed in South Asia are low-skilled workers earning daily wages in the informal sector. One could argue that widespread poverty is the most significant contributor to the deterioration of the environment in this region. People who live below the poverty line are highly reliant on the services provided by ecosystem services, for their livelihoods [16]. They focus on satisfying their immediate needs rather than achieving future security regarding resources. People are driven to desperate measures by lack of financial resources. As a result, they are cutting down forests for fuel, encroaching on marginal lands, and overgrazing grasslands with livestock. A lack of laws and regulations in this area may be linked to the deterioration of the local environment.

As in many other parts of the world, environmental degradation is becoming so severe that it undermines economic growth in South Asia. According to the World Bank [17], South Asian countries should take immediate action to reduce their carbon emissions. If this is not done, the impact will become even more severe. Growth in the economy, which can be encouraged through liberalization and industrialization policies, brings gains from a short-term perspective. In the long run, however, it increases the vulnerability of South Asian countries to environmental deterioration and the risks that are associated with it. The article presents some newly released research results on the relationship between economic expansion and environmental degradation in that region. These results are shown in Table 1.

Table 1. Literature review on environmental degradation in South Asian countries. Recently published papers.

Author(s)	Country/Period	Variables	Results
Khan et al. [18]	5 South Asian countries (1972–2017)	CO_2 emissions, economic growth, non-renewable energy consumption, KOF index of globalization	The results support the inverted U-shaped EKC hypothesis. Research identified the causality between GDP growth and carbon emissions and found bidirectional causality between economic growth and energy use.
Tan et al. [9]	South and Southeast Asian countries (2013–2019)	biodiversity loss, economic growth, agricultural land, corruption	The results strongly support an inverted U-shaped relationship between income and biodiversity loss. Control of corruption and biodiversity loss are negatively associated, while agricultural land has a significant and positive effect on biodiversity loss.
Sadiq et al. [6]	5 South Asian countries (1972–2019)	CO_2 emissions, economic growth, non-renewable energy consumption, KOF index of globalization	The results support the inverted U-shaped EKC hypothesis. Economic growth Granger causes CO_2 emanations. Heavy dependence on fossil energy consumption is not environmentally friendly for sustainable development in this region.
Mehmood et al. [8]	4 South Asian countries (1972–2019)	CO_2 emissions, economic growth, renewable energy consumption, tourism	The results support the inverted U-shaped EKC hypothesis in Pakistan and India. The findings show mixed results regarding the impact of tourism on CO_2 emissions.
Sharma et al. [19]	4 South Asian countries (1990–2016)	carbon intensity, economic growth, renewable energy consumption, stock market capitalization, technological innovations, trade	Stock market development, per capita income, and trade expansion increased carbon intensity in South Asian countries.
Murshed et al. [20]	South Asian countries (1995–2015)	CO_2 emissions, ecological footprints, economic growth, renewable energy consumption	The results confirmed the validity of the EKC hypothesis. The use of renewable energy is associated with environmental betterment in all five South Asian countries. The results imply that economic growth is both the short-run cause and long-run solution to the environmental adversities in South Asian countries.
Fong et al. [21]	9 South-east Asian countries (1993–2012)	SO_2 emissions, NO_x, PM2.5 concentration, economic growth, renewable energy consumption, primary energy intensity, urban population, services sector, foreign direct investment	The results support the inverted U-shaped EKC hypothesis for all pollutants. Spatial spillovers are not found for NOx emissions but are supported for SO_2 and PM2.5 emissions. Most countries are still on the upward sloping portion of the curve.
Ullah and Awan [22]	Developing Asian countries (1973–2010/2016)	CO_2 emissions, SO_2 emission, PM2.5 concentration, economic growth, income inequality, foreign direct investment, trade openness, population density, urban population	The results support the inverted U-shaped EKC hypothesis. Moreover, the findings reveal that income inequality is positively related to CO_2 and SO_2 emissions and PM2.5 concentrations.

2.2. Energy Consumption and Environmental Degradation

Economic prosperity and growth have always constituted part of the policy agenda of every country. They are of utmost importance for South Asian countries, where 40 percent of the world's poor live. Excessive population leads to excessive human activity and excessive consumption of energy, which results in CO_2 emissions. Nowadays, South Asian countries can achieve improved economic growth, but at the cost of environmental degradation caused by increased consumption of natural resources [23]. Consequently, they are confronted with the dual problem of generating higher economic growth while at the same time containing the progression of environmental damage. This is particularly apparent in

South Asian countries, which are desperate to achieve higher economic growth to alleviate poverty and improve the standard of living. By comparison, developed countries face fewer challenges. Energy consumption is one of the most significant variables for rapid economic expansion, industrialization, and urbanization. This consumption in South Asian countries comes from non-renewable energy sources, in particular oil, coal, and natural gas, which in turn drive carbon dioxide emissions.

The connection between rising energy use and worsening environmental conditions is significant from the point of view of economic policy. Much research has been carried out to investigate this nexus, taking into account a variety of energy sources utilized in South Asian nations. For instance, Rahman [24] found that the use of energy had a negative long-term impact on the quality of the environment in a group of 11 Asian countries over the period 1960–2014. Similar findings were published by Dong et al. [25]. They found that using natural gas had a considerable negative influence on CO_2 emissions for a panel of 14 Asia-Pacific nations between 1970–2016. In their study, Munir and Riaz [26] reviewed the data of three South Asian countries (Pakistan, India, and Bangladesh) from 1985 to 2017, and concluded that an increase in the use of gas, coal, and electricity led to a rise in CO_2 emissions. Mujtaba et al. [27] demonstrate that a positive shock in oil prices is associated with an increase in energy consumption, which in turn has a positive and significant influence on CO_2 emissions in India. Additionally, the research findings regarding the amount of foreign direct investment brought into this country lend credence to the pollution haven hypothesis.

It is interesting to note that the consumption of renewable energy has a positive impact on environmental quality in this region. This is something that should be taken into consideration. A significant portion of the existing body of knowledge focuses on this problem. Recent research by Anwar et al. [28] shows that the use of renewable energy sources resulted in lower carbon dioxide emissions in 15 Asian economies from 1990 to 2014. Additionally, Murshed et al. [20] found that increasing the levels of renewable energy consumption and renewable electricity outputs reduced the ecological and carbon footprints of five South Asian economies (Bangladesh, India, Pakistan, Sri Lanka, and Nepal) during the period 1995–2015. Similar results for different regions and countries were published by Shahbaz et al. [29], Ma et al. [30], Ulucak and Yucel [31], and Erdogan et al. [32].

The continued expansion of economic activity in South Asian countries along with a growing population will boost energy consumption in the following decades. It is anticipated that by 2040 the demand for energy in developing countries, including South Asian countries, will be 33 percent higher than it is today. However, the current economic growth patterns in this region, particularly in India, are environmentally unsustainable due to the country's reliance on fossil fuel-based energy consumption and imported crude oil, which significantly degrade the environment. This is especially true in India [29].

2.3. *Trade Openness and Environment Degradation*

The advent of globalization has made it possible for numerous nations to engage in cross-border international transactions. Since the opening of the economy in the early 1990s, a growing body of literature has investigated the impact of trade openness on the environment. This nexus is essential for policy-makers because it will assist them in achieving their goals of simultaneously accelerating economic growth and improving environmental quality. Though the nexus is significant from the point of view of policy-makers, the environmental implications of trade openness have not received much attention in South Asian nations.

The impact of trade openness on pollution is described by means of the scale effect, composition effect, and technology effect. The scale effect is connected with adverse environmental consequences. It is believed that trade openness causes pollutant emissions due to increased economic activity. Trade increases production volume and energy use, which in turn causes an increase in CO_2 emissions and a decline in environmental quality.

The composition effect is based on the principles of factor endowment or the Hecksher Ohlin theory. Therefore, the economy should focus on the industries with a competitive advantage. Countries with abundant labor supply should specialize in and export labor-intensive products. Similarly, countries where capital is abundant should specialize in and export capital-intensive products. The South Asian nations, where there is a plentiful supply of labor, should specialize in and export goods that need much labor. Theoretically, labor-intensive industries should not cause an increase in pollution. However, several study findings indicate that trade openness has a detrimental impact on the environment of lower-income countries, where "dusty industries" are exported by industrialized nations, according to the factor endowment theory and pollution haven hypothesis [33]. The policy-makers feel that the developing and less developed nations pursue rapid economic growth to raise their living standards and combat poverty. Consequently, they are relaxing environmental rules and regulations to attract more foreign direct investors, who take advantage of lax legislation and harm the host country's environment [34].

The technological effects of trade openness bring modern eco-friendly technology that will reduce pollution. The question arises among policy-makers, researchers, and practitioners under what circumstances trade openness benefits the environment. Trade is environmentally beneficial as long as the technological effects outweigh the scale and composition effects. The empirical findings on the trade-environment nexus in developing nations are unresolved in this regard and need to be empirically tested.

There are two camps of opinion on the impact of international trade on CO_2 emissions. On the one hand, it is contended that trade openness enables each nation to gain access to the world market, and in this way increase its market share. Access to global markets encourages less developed and emerging nations to import more environmentally friendly, energy-efficient, and modern technologies to replace the outdated ones. As a result, pollution levels decrease [35,36]. The studies that support a positive association between trade openness and environmental quality are Ahmed et al. [37], Antweiler et al. [38], Copeland and Taylor [39], Cherniwchan [40], Dogan and Turkekul [41], Frankel and Rose [42], Kanjilal and Ghosh [43], and Shahzad et al. [44].

On the other hand, it is claimed that trade openness adversely affects the quality of the environment as it leads to large-scale export production and the establishment of "manufacturing hubs", and what follows higher energy use and higher CO_2 emissions. Foreign trade increases foreign direct investments in the industrial and logistics sectors, which are energy-based activities that lead to an increase in emissions, according to Hakimi and Hamdi [45], and Lopez [46]. Schmalensee et al. [47], and Copeland and Taylor [48] argue that growing global trade is to blame for the depletion of natural resources, which in turn results in higher CO_2 emissions and worse environmental conditions. Other researchers who share the view that trade openness increases pollution include: Al-Mulali and Sheau-Ting [49], Jun et al. [34], Jalil and Feridun [50], Kellenberg [51], Managi and Kumar [52], Shahbaz et al. [53]. Because of this, we decided to verify the hypothesis that trade openness significantly impacts carbon dioxide emissions in South Asian countries.

It is interesting to note that the literature suggests conflicting and mixed impacts between trade openness and CO_2 emissions, demonstrating the inconsistency of the findings. One of the reasons why these results are inconsistent is connected with different levels of economic development in the countries under study. Le et al. [54], for instance, assert that trade openness increases pollution in middle- and low-income nations while reducing it in high-income countries. Similarly, Baek et al. [55] found that trade openness negatively affected the environment's quality in less developed nations due to failure to enforce laws and regulations. On the other hand, strict environmental regulations in industrialized nations drove multinational corporation investment overseas.

Another reason for the inconsistent results is the quality of the economic policy. This is exemplified in the studies undertaken by Grossman and Krueger [56], who argue that the environmental impact of international trade depends on economic policies; Copeland [57], who highlights that trade openness improves environmental quality in the presence of good governance; and Chang [58], who found that trade openness reduced CO_2 emissions in countries with low levels of corruption while increasing CO_2 emissions in countries where corruption was high.

Contrary to the findings above, some authors argue that there is no association or insignificant effect between trade openness and environmental pollution: Farhani et al. [59], Jalil and Mahmud [60], Jayanthakumaran et al. [61], and Sharma [62]. Our study will fill the literature gaps.

3. Data and Methods

The paper examines annual data for four South Asian countries: India, Bangladesh, Sri Lanka, and Pakistan. The period for the analysis (1971–2014), was selected based on data availability. The annual time series data came from the World Bank collection of development indicators, and include the following variables: C—carbon dioxide (CO_2) emissions per capita (in metric tons); Y—GDP per capita (in constant 2010 US$); E—energy consumption per capita (kg of oil equivalent), and T—trade openness (% of GDP). Carbon dioxide emissions are defined as emissions that result from cement manufacturing and fossil fuel combustion. They also include CO_2 emissions produced during the consumption of gas fuels and gas flaring, and liquid and solid fuels. Energy consumption refers to primary energy use; i.e., before it is transformed to other end-use fuels. It is equal to domestic production plus imports and stock changes, minus exports and fuels used in international transport (World Bank Development Indicators). Trade openness is defined as the sum of imports and exports of services and goods measured as a share of GDP.

Table 2 shows a data description. According to the skewness and kurtosis measures, we found that the series of some countries showed evidence of asymmetry, fat tails, and high peaks for all variables. These results indicated that the non-linear ARDL approach is suitable for our analysis. Additionally, we performed the test for parameter instability by Andrews [63] and the Brock, Dechert, and Scheinkman (BDS) test [64] to check the data. The test for parameter instability confirmed the instability for all variables in all countries (Table A1). The BDS test confirmed the failure of the assumption of iid residuals (linear model) for some variables in some countries (Table A2). These results also show that applying the non-linear ARDL approach is appropriate for this study.

Table 2. Descriptive statistics of the variables.

	lnCO_2	lnE	lnY	lnY^2	lnT	lnCO_2	lnE	lnY	lnY^2	lnT
	Bangladesh					India				
Mean	−0.828	2.121	2.679	7.194	1.927	−0.150	2.571	2.827	8.030	9.103
Median	−0.867	2.097	2.637	6.956	1.840	−0.156	2.561	2.780	7.730	9.115
Maximum	−0.384	2.360	2.978	8.870	2.625	0.217	2.804	3.215	10.336	9.203
Minimum	−1.278	1.938	2.508	6.292	1.522	−0.441	2.427	2.582	6.664	8.895
Std. Dev.	0.242	0.119	0.133	0.727	0.367	0.190	0.111	0.196	1.126	0.084
Skewness	0.3826	0.1665	0.0315 *	0.0215 *	0.0426 **	0.5562 *	0.1600	0.1731 **	0.1175*	0.0196 *
Kurtosis	0.0683	0.1445	0.4937 *	0.6683 *	0.0906 **	0.0358 *	0.1663	0.0180 **	0.0479*	0.7892 *
Jarque-Bera	2.211	2.879	4.848 *	5.364 *	5.199 *	2.110	2.866	3.615	3.827	5.171 *
Probability	0.331	0.237	0.089	0.068	0.074	0.348	0.239	0.164	0.148	0.075
	Sri Lanka					Pakistan				
Mean	−0.433	2.562	3.147	9.944	2.793	−0.250	2.592	2.858	8.183	2.409
Median	−0.518	2.509	3.116	9.708	2.76	−0.216	2.619	2.894	8.376	2.396
Maximum	−0.072	2.741	3.545	12.565	3.213	−0.060	2.699	3.023	9.138	2.567
Minimum	−0.699	2.458	2.839	8.058	2.390	−0.511	2.455	2.654	7.041	2.269
Std. Dev.	0.212	0.086	0.206	1.313	0.261	0.136	0.078	0.114	0.648	0.077
Skewness	0.3742 ***	0.1176 ***	0.3589 *	0.2415 *	0.8526 ***	0.1211 *	0.1997 ***	0.2594 *	0.3200 *	0.7817
Kurtosis	0.0000 ***	0.0004 ***	0.0374 *	0.0728 *	0.0000 ***	0.0258 *	0.0002 ***	0.0129 *	0.0093 *	0.2074
Jarque-Bera	4.987*	5.033*	2.502	2.701	3.360	3.998	4.467	3.203	3.055	1.155
Probability	0.083	0.081	0.286	0.259	0.186	0.135	0.107	0.202	0.217	0.561

Sources: The authors' estimation. Note: *, ** and *** show the significance at the 10%, 5% and 1% level, respectively.

Our model is based on the EKC hypothesis, which postulates an association between economic growth and environmental degradation. The pattern of economic growth can affect environmental quality in many ways. According to Grossman and Krueger [65], this influence can occur through three channels: scale effect, composition effect, and technique effect. Following the literature (e.g., Soytas et al. [66], Shahbaz et al. [67], Kyophilavong et al. [68], Kisswani et al. [69], Jóźwik et al. [70], and Soylu et al. [71]), we assume that the EKC has an inverted U-shape. This means that at the initial stage of development, countries focus more on economic growth, which results in increasing environmental pollution and decreasing environmental quality. Once their threshold level of income (i.e., beyond some level of per capita income) has been achieved, they become more concerned about the environment by implementing more restrictive environmental laws and regulations and encouraging investment in eco-friendly projects. As a result, the pollution level is reduced and environmental quality increases.

Our aim is to identify the long-run relationship and causality between environmental degradation, economic growth, energy consumption, and trade openness in South Asian countries. This association can be expressed as follows:

$$CO_2 = f\left(E, Y, Y^2, T\right) \tag{1}$$

All data in the model have been transformed into natural logarithms. Thus, the ARDL model (Equation (2)) and NARDL model (Equation (3)) are rewritten as:

$$\ln CO_{2\,t} = \alpha + \beta_1 \ln E_t + \beta_2 \ln Y_t + \beta_3 (\ln Y_t)^2 + \beta_4 \ln T_t + \varepsilon_t \tag{2}$$

$$\ln CO_{2\,t} = \alpha + \beta_1 \ln E_t + \beta_2 \ln Y_t + \beta_3 (\ln Y_t)^2 + \beta_4^+ \ln T_t^+ + \beta_4^- \ln T_t^- + \varepsilon_t, \tag{3}$$

where CO_2 is carbon dioxide emissions in metric tons per capita in year t, E_t is energy consumption in kilogram of oil equivalent per capita, Y_t is real GDP per capita (in constant prices 2010 US$), Y_t^2 is real GDP per capita squared, T_t defines trade openness (% of GDP), T_t^+ and T_t^- represent positive and negative shocks of foreign trade (trade openness), and ε_t is the error term. As was pointed out earlier all the data were collected from the World Bank (World Development Indicators).

The sign of the coefficient β_1, which is associated with energy consumption, is usually positive, indicating that an increase in energy consumption, which leads to higher economic growth, triggers CO_2 emissions. But recent research has suggested that the impact of energy consumption on environmental quality is heavily conditional and dependent on energy sources; for example, Fatima et al. [72], Saidi and Omri [73], Ma et al. [30], and Shahbaz [29]. In our research, it is essential to note that the majority of South Asian countries have traditionally been overwhelmingly dependent on non-renewable fossil fuels to meet their increasing energy demand [20,74].

The signs of coefficients β_2, and β_3 associated with GDP per capita can have positive and negative values. According to the inverted U-shaped EKC hypothesis, the relationship requires that β_2 should be positive and β_3 should be negative [75,76]. If coefficient β_3 is statistically insignificant, there is a monotonic increase in the relationship between CO_2 emissions per capita and real GDP per capita.

In liberalized South Asian countries, the expected sign of coefficient β_4 associated with GDP per capita is positive. According to Copeland and Taylor [39], the environmental effects of trade liberalization can be classified into five categories: scale effects, structural effects, technology effects, direct effects, and regulation effects. Three of them were explained earlier. The expected sign of the coefficient for trade openness is negative if trade openness promotes energy-efficient technology through the import of new technologies, encouraging cleaner domestic products, and imposing stricter environmental regulations [77]. On the other hand, the coefficient is positive if trade openness increases pollution-intensive export and promotes a pollution haven for foreign direct investment [56,67,78].

The ARDL framework of Equation (2) can be written as:

$$\begin{aligned}\Delta \ln CO_{2_t} = {} & \alpha + \beta_0 \ln CO_{2_{t-1}} + \beta_1 \ln E_{t-1} + \beta_2 \ln Y_{t-1} + \beta_3 (\ln Y_{t-1})^2 + \beta_4 \ln T_{t-1} \\ & + \sum_{i=1}^{p} \zeta_0 \Delta \ln CO_{2_{t-i}} + \sum_{i=0}^{r} \zeta_1 \Delta \ln E_{t-i} + \sum_{i=0}^{r1} \zeta_2 \Delta \ln Y_{t-i} \\ & + \sum_{i=0}^{r1} \zeta_3 \Delta (\ln Y_{t-i})^2 + \sum_{i=0}^{r1} \zeta_4 \Delta \ln T_{t-i} + \varepsilon_t \end{aligned} \quad (4)$$

where Δ denotes the operator, r denotes the lag lengths, and ε_t is the error term. The null hypothesis is that there is no relationship (cointegration) between CO_2 emissions and the determinant variables, and the alternative hypothesis states that a long-run relationship (cointegration) between the variables exists.

Additionally, we investigate an asymmetric impact of trade openness on CO_2 emissions. To do this, we apply the NARDL approach, which has been widely used in empirical studies since the mid-1990s, when a substantial body of work considered the joint issues of non-linearity and non-stationarity. Among the recently published studies, we can mention Rahman and Ahmad [79], Qamruzzaman et al. [80], Sheikh et al. [81], and Mujtaba et al. [82]. The main idea of an asymmetric impact is that a positive shock may have a larger absolute effect in the short run while a negative shock has a larger absolute effect in the long run (or vice-versa). The NARDL has several advantages compared to the ARDL model [83]. First, the ARDL approach does not consider the asymmetric relationship between the variables. The positive and negative variations of independent variables have the same effect on the dependent variable. Second, the NARDL approach enables us to test for hidden cointegration, which helps differentiate between linear cointegration, non-linear cointegration, and lack of cointegration. The concept of hidden cointegration (which means that no cointegration is detected when using conventional techniques, but cointegration is found between positive and negative components of the series) was developed by Granger and Yoon [84].

The NARDL framework of Equation (3) can be written as:

$$\begin{aligned}\Delta \ln CO_{2_t} = {} & \alpha + \delta_0 \ln CO_{2_{t-1}} + \delta_1 \ln E_{t-1} + \delta_2 \ln Y_{t-1} + \delta_3 (\ln Y_{t-1})^2 + \delta_4^+ \ln T_{t-1}^+ \\ & + \delta_4^- \ln T_{t-1}^- \\ & + \sum_{i=1}^{p} \zeta_0 \Delta \ln CO_{2_{t-i}} + \sum_{i=0}^{r} \zeta_1 \Delta \ln E_{t-i} + \sum_{i=0}^{r} \zeta_2 \Delta \ln Y_{t-i} \\ & + \sum_{i=0}^{r} \zeta_3 \Delta (\ln Y_{t-i})^2 + \sum_{i=0}^{r} \left(\zeta_4^+ \Delta \ln T_{t-i}^+ + \zeta_4^- \Delta \ln T_{t-i}^- \right) + \varepsilon_t \end{aligned} \quad (5)$$

where T_t^+ and T_t^- represent positive and negative shocks of foreign trade (trade openness). The long-run and short-run changes are represented by coefficients δ_i and ζ_i, respectively.

The short-run NARDL model estimations with an error correction mechanism can be estimated with the following equation:

$$\begin{aligned}\Delta \ln CO_{2_t} = {} & \alpha + \sum_{i=1}^{p} \varphi_0 \Delta \ln CO_{2_{t-i}} + \sum_{i=0}^{p} \varphi_1 \Delta \ln E_{t-1} + \sum_{i=0}^{p} \varphi_2 \Delta \ln Y_{t-1} \\ & + \sum_{i=0}^{p} \varphi_3 \Delta (\ln Y_{t-1})^2 + \sum_{i=0}^{p} (\varphi_4^+ \ln T_{t-1}^+ + \varphi_4^- \ln T_{t-1}^-) + \psi ECM_{t-1} \end{aligned} \quad (6)$$

The long-run symmetry and asymmetry are tested with the standard Wald test. The asymmetric cumulative dynamic multipliers effect on $\ln CO_2$ of a unit change in $\ln T_t^+$ and $\ln T_t^-$ can be obtained as follows:

$$\begin{aligned} m_h^+ &= \sum_{i=0}^{h} \frac{\Delta \ln CO_{2_{t+i}}}{\Delta \ln T_t^+} \\ m_h^- &= \sum_{i=0}^{h} \frac{\Delta \ln CO_{2_{t+i}}}{\Delta \ln T_t^-} \end{aligned} \tag{7}$$

Finally, we applied the asymmetry causality test developed by Hatemi [85]. The causality testing is asymmetric in the sense that positive and negative shocks may have different causal impacts.

4. Results and Discussion

In the first step, we use the Augmented Dickey-Fuller and Phillips-Perron unit root tests to check if all variables are stationary. The null hypothesis of the ADF and Perron tests is that the variable contains a unit root, and the alternative is that the variable is generated by a stationary process. The results of the tests with intercept and trend can be found in Table A3. The null hypothesis can be rejected at the 1% level of significance for all variables at the first difference. This implies that all variables used in this study are integrated on the order of one I(1).

After confirming the ordering of the integration, we apply the ARDL and NARDL approaches to examine long-run relationships (cointegration) and estimate the coefficients. To implement these approaches, the selection of appropriate lag length is necessary. We chose one lag based on the results of Akaike's information criterion and Schwarz's Bayesian information criterion. Tables 3 and 4 provide the results of ARDL and NARDL tests for cointegration. In the NARDL test, the null hypothesis of no cointegration between variables was rejected at the 10% level of significance in Bangladesh, India and Pakistan, and at 5% in Sri Lanka. The estimated F-statistics were larger than the critical upper bounds. The results of the NARDL test were more significant (Table 4). The null hypothesis was rejected at the 1% level of significance in India and Pakistan, and at 5% in Bangladesh. We also rejected the null hypothesis for Sri Lanka, accepting that the F-statistic was slightly smaller than the upper bound at the 10% level of significance. In summary, these results show that all equations are co-integrated.

Table 3. Results of the ARDL test for cointegration. Model: LnCo2 = f(LnE, LnY, LnY^2, LnT^+, LnT^-).

Country	F-Statistic	Result
Bangladesh	25.571 *	Cointegration
India	7.840 *	Cointegration
Sri Lanka	3.071 **	Cointegration
Pakistan	8.038 *	Cointegration
Critical Value for F-Statistic	Lower Bound I(0)	Upper bound I(1)
1%	3.29	4.37
5%	2.56	3.49
10%	2.2	3.09

Sources: The authors' estimation. Note: * and ** show the significance at 10%, 5% and level respectively.

The differences between coefficients estimated by the ARDL and NARDL approach are highlighted in Table 5. The NARDL estimation captures richer insights into the asymmetric effects of trade openness on CO_2 emissions. As specified in Equation (3), trade openness is split into positive and negative shocks in the NARDL model. Table 5 compares the long-run and short-run coefficients.

Table 4. Results of the NARDL test for cointegration. Model: LnCo2 = f(LnE, LnY, LnY^2, LnT^+, LnT^-).

Country	F-Statistic	Result
Bangladesh	3.473 **	Cointegration
India	18247.93 ***	Cointegration
Sri Lanka	2.901 *	Cointegration
Pakistan	4.291 ***	Cointegration
Critical Value for F-Statistic	Lower Bound I(0)	Upper bound I(1)
1%	3.06	4.15
5%	2.39	3.38
10%	2.08	3.00

Sources: The authors' estimation. Note: *, ** and *** show the significance at 10%, 5% and 1% level respectively.

Table 5. Results of ARDL and NARDL tests.

Variables	Bangladesh	India	Sri Lanka	Pakistan	Bangladesh	India [1]	Sri Lanka	Pakistan
ARDL Analysis Results								
	Long-run coefficients				Short-run coefficients [1]			
lnE	0.781	0.555	1.074	0.889 *	1.543 *	0.000	2.040 *	1.248 *
lnY	3.022	7.594 ***	−10.268 **	5.246 *	2.522	−0.012	21.455 *	15.451 *
lnY^2	−0.315	−1.097 ***	1.427 **	−0.813 *	−0.159	0.002	−3.472 *	−2.614 *
lnT	−0.122 ***	−1.755	1.554 *	0.076	−0.041	−0.999 *	0.524 **	0.016
C	−8.107	0.798	10.605	−11.081 *				
ECT_{t-1}					−0.938 *	0.001 *	−0.611 *	−0.707 *
NARDL Analysis Results								
	Long-run coefficients				Short-run coefficients			
ΔlnE	5.550	0.978	0.827	0.945 ***	37.464 ***	0.007	1.976 ***	1.153 ***
ΔlnY	−0.108	0.021	−12.173 **	9.985 ***	−7.349 ***	−0.003	19.006 ***	−2.176
ΔlnY^2	−3.625	−1.180 *	1.663 **	−1.787 ***	0.940 ***	0.004	−3.111 ***	0.429
ΔlnT_NEG	−0.075	−2.209 ***	1.594 **	0.157 ***	−0.106 **	−1.002 ***	−0.171	0.115 *
ΔlnT_POS	0.255	0.272	1.877 **	0.461	0.179 ***	−0.996 ***	1.087 ***	−0.056
C	−7.366	−1.177	18.562 *	−16.708 ***				
ECT_{t-1}					−0.289 ***	0.011 ***	−0.684 ***	−0.555 ***

Sources: The authors' estimation. **Notes:** [1] The lag length for CO_2 in India is 2; thus, additional coefficients were estimated: 0.825 * ($\Delta LNCO2_{t-1}$); −0.001 (ΔLNE_{t-1}); −0.020 ** (ΔLNY_{t-1}); 0.004 ** ($\Delta LNY^2{}_{t-1}$); 0.824 * (ΔLNT_{t-1}). *, ** and *** show the significance at the 10%, 5% and 1% level, respectively.

As outlined above, the signs of the coefficients associated with GDP per capita can have positive and negative values. Based on the inverted U-shaped EKC hypothesis, the relationship requires that β_3 should be negative (and β_2 should be positive). We observe similar coefficients in the ARDL and NARDL models. Similarly, Dong et al. [25], Murshed et al. [20], Khan et al. [18], and Sadiq et al. [6] proved the inverted U-shaped EKC hypothesis in that region. Dong et al. [25] pointed out that the turning points lie at $1181.60 in Pakistan, $1861.49 in India, and $1937.23 in Bangladesh, while the turning years were estimated in 2041, 2039, and 2048, respectively. Other studies indicate that using renewable energy is associated with environmental betterment [20], and sustainable development policies can revisit the conflict between globalization and environmental degradation [6,18].

In the NARDL model, the long-run coefficient for squared GDP per capita is negative and significant in India (−1.180 *) and Pakistan (−1.787 ***), while it is positive in Sri Lanka (1.663 **). The coefficients for India and Pakistan indicate that we should expect increased environmental quality. Notably, the Indian government has taken many initiatives to reduce environmental degradation in recent years. For example, the International Solar Alliance's launch summit was co-chaired by Prime Minister Narendra Modi and French President Emmanuel Macron in March 2018, demonstrating India's leadership in supporting renewable energy (ISA). In January 2019, the Ministry for Environment introduced the National Clean Air Program (NCAP), which gives the states and union government a framework to tackle air pollution. Since 2018, India's 2019 climate change index (CCPI) performance has improved from 14th to 11th place [86]. Pakistan has recently given serious thought to addressing the world's escalating environmental concerns, according to the United National Development Program 2020. Several Acts have been promulgated along with some policies and public sector initiatives currently in effect. For example, clean and green initiatives have been implemented; environmental protection agencies at the federal and provincial levels have been strengthened; environmental laboratories and courts, national environment quality standards, the National Energy Efficiency and Conservation Authority (NEECA), and national environmental quality standards have all been developed [87].

Another potential environmental problem is that the coefficient associated with GDP per capita is relatively high in Pakistan. Let us recall at this point that the coefficients of GDP per capita indicate the scale effect, which is associated with adverse environmental consequences. It is highly probable that high trade openness causes pollutant emissions due to increased economic activity. Our study corroborates the findings by Ullah et al. [88] and Khan et al. [89], who found that trade liberalization (trade openness) led to increased CO_2 emissions in Pakistan. This positive relationship can be explained by scale effects where large-scale manufacturing operations, particularly in fossil-fueled and export-oriented industries, increase emissions of pollutants. This is because in the early stages of the development process, more emphasis is placed on economic growth than on pollution control. At this stage, less developed countries are often "hungry" for rapid economic growth to fight against poverty. The negative sign of β_2 in Sri Lanka should definitely be assessed in a positive way.

The results of the long-run coefficients associated with energy consumption, both in the ARDL and NARDL model, surprised us. Usually, energy consumption significantly impacts the dioxide carbon emissions in such a way that there is a positive long-run relationship between these two (cf. Wang et al. [90], Gierałtowska et al. [91] and Verbič et al. [92]). Energy consumption should likely be associated with other factors. This relationship is visible in developed countries. For example, Wang et al. [90] indicate that energy intensity and foreign direct investment and urbanization strongly impact carbon dioxide emissions. In our research, these long-run coefficients are significant only in Pakistan (ARDL 0.889 * and NARDL 0.945 ***). By contrast, these coefficients are highly significant in the short run in Bangladesh (37.464 ***), Sri Lanka (1.976 ***), and Pakistan (1.153 ***) in the NARDL model.

The most interesting finding was that the long-run coefficients associated with trade openness shocks, both negative and positive, significantly impacted CO_2 emissions only in Sri Lanka (at the significance level of 5%). These research results did not support the hypothesis that trade openness significantly impacts carbon dioxide emissions in South Asian countries. The estimated coefficients of trade openness with positive and negative shocks are 1.887 and 1.594, respectively. Therefore, increasing trade openness by 1% increases carbon dioxide emissions by 1.887%, while reducing trade openness decreases carbon dioxide emissions by 1.594%. These impacts are based on the scale effect. The primary contributors to Sri Lanka's economy are tourism, tea export, textile and garment manufacturing, rice and other agricultural goods, and food products. Gasimli et al. state that domestic investors do not use environmentally friendly technology [93]. Additionally, imported technology in the form of machinery does not have a positive impact on the environment. In the cases of India and Pakistan, trade openness coefficients are significant at 1% only for negative shocks. For example, in India, an increase in trade openness has no significant impact on carbon dioxide emissions, while a reduction by 1% increases carbon dioxide emissions by 2.209%. Otherwise, a recent study by Shahbaz et al. [94] reports that the discussion on the energy-led growth of India necessitates the cross-border movement of resources, which influences the carbon dioxide emissions pattern. As the Indian import portfolio was majorly dependent on crude oil, the import substitution policies have reduced the import of crude oil and other petroleum products and, consequently, the level of carbon dioxide emissions.

The results for short-run trade openness coefficients, for positive and negative shocks, are significant in Bangladesh and India. Moreover, positive and negative shocks perform considerably differently in Bangladesh. For example, a positive shock (0.179 ***) impact is greater than a negative shock (−0.106 **), which demonstrates that positive shocks have more profound effects than negative shocks. This proves the significant impact of trade openness on the environment in the short run. But in 2021, Sharma et al. [19] published a paper describing the importance of importing innovative solutions to reduce environmental degradation in the long run. Domestic enterprises will try to import innovative technological solutions to improve their energy efficiency and reduce their carbon footprint.

Finally, we examine the stability of the model. Table 6 presents the diagnostic tests for serial correlation, heteroscedasticity, normality, and Ramsey. The diagnostic tests of the ARDL model indicate problems with serial correlation in all countries, heteroscedasticity in Sri Lanka, and non-linearity in Bangladesh and India. However, we found no serial correlation, heteroscedasticity problem, or normality problems in the NARDL. This diagnostic test confirmed that the NARDL was more appropriate than the ARDL model.

Table 6. Diagnostic checks of the ARDL and NARDL tests.

Test	Bangladesh	India	Sri Lanka	Pakistan
ARDL Analysis Results				
Serial Correlation	0.000	0.000	0.060	0.006
Heteroscedasticity	0.174	0.172	0.090	0.629
Normality	0.784	0.391	0.819	0.553
Ramsey	0.092	0.013	0.714	0.431
NARDL Analysis Results				
Serial Correlation	0.341	0.112	0.029	0.364
Heteroscedasticity	0.717	0.731	0.107	0.152
Normality	0.664	0.598	0.853	0.612
Ramsey	0.318	0.036	0.601	0.841

Sources: The authors' estimation. Note: They are *p* values.

5. Conclusions and Recommendations

In recent years, environmental pollution has become a global threat. In this study, we attempted to establish the short-run and long-run relationships among environmental degradation, economic growth, energy consumption, and trade openness in South Asian countries. Additionally, we verified the hypothesis that trade openness significantly impacts carbon dioxide emissions in South Asian countries. To do so, we used annual data for four South Asian countries (India, Bangladesh, Sri Lanka, and Pakistan) covering the period between 1971 and 2014. Our selection of countries for the study was based on the availability and uniformity of data in that period. We used the linear ARDL and non- linear ARDL (NARDL) model, which allowed us to analyze the impact of positive and negative shocks in trade openness on CO_2 emissions. Both methods show the long-run equilibrium relationship between environmental degradation, economic growth, energy consumption, and trade openness. The empirical outcome shows that the environmental Kuznets curve holds for India and Pakistan out of the four analyzed countries.

In the NARDL model, the long-run coefficients for squared GDP per capita are statistically significant and negative for India and Pakistan, while for Sri Lanka they are statistically significant and positive. Bangladesh's squared GDP per capita is negative but not statistically significant. According to the environmental Kuznets curve, the coefficients for India and Pakistan indicate that environmental quality is expected to improve as income increases in the long run. The estimated long-run coefficients associated with energy consumption in the ARDL and NARDL models surprised us. They are statistically significant only in Pakistan. This indicates that energy consumption significantly aggravated environmental degradation only in Pakistan. This may be associated with poor institutional quality due to political instability in Pakistan.

The most interesting finding was that the long-run coefficients associated with trade openness shocks, both negative and positive, significantly impact CO_2 emissions only in

Sri Lanka. These impacts are based on the scale effect. On the other hand, the results for short-run trade openness coefficients, for positive and negative shocks, are significant in Bangladesh and India. In Bangladesh, positive shock increases carbon dioxide emissions, while negative shock decreases them. However, positive and negative shocks in India reduce environmental pollution. These research results did not support the hypothesis that trade openness significantly impacts carbon dioxide emissions in South Asian countries.

This study has some policy implications. But first, we assumed that if the environmental Kuznets curve is confirmed over a long period in India and Pakistan, there is a high probability that this relationship will exist for a long period. Then we can propose some recommendations. South Asian countries' governments require adequate policy directions to use clean energy while producing output and generating income. Like other low- and middle-income countries, they have limited environmental regulatory capacity. Due to poverty, low-income populations rely on timber wood for food and heating in the winter, causing significant pollution. The region's reliance on fossil fuel energy consumption is not environmentally friendly for long-term development. The consensus believes that developing renewable energies, including wind, solar, and hydroelectric power plants, will replace the infrastructure powered by fossil fuels.

With increased income in this region, governments should prioritize green growth, which is critical for sustainable development. Such actions have been taken in the past. For example, the Pradhan Mantri Ujjwala Yojana (PMUY) is a flagship scheme of India launched on 1 May 2016, by Hon'ble Prime Minister Shri Narendra Modi. The program aims to make clean cooking fuels such as LPG available to rural and deprived households that would otherwise rely on traditional cooking fuels such as firewood, coal, or cow-dung cakes. From a practical point of view, the Indian government should focus on maintaining an affordable price for LPG cylinders, along with taking more steps toward poverty reductions and keeping inflation at a desirable level, especially nowadays when its rate is high. Otherwise, poor people will revert to traditional food preparation methods, which can cause severe health and environmental problems. Consequently, Ujjwala Yojana policy paralysis may occur, leading to increased carbon dioxide emissions.

To combat environmental pollution, the governments in South Asian countries should promote and subsidize green energy by increasing their R&D spending, among others. The fifth-largest economy in the world, India, should take the lead in reducing pollution in the region. Usually, as income levels rise, so does the demand for a cleaner environment, putting pressure on the government to enact stricter environmental regulations. Governments should focus on developing advanced technology, implementing strict environmental policies, and introducing carbon pricing for polluting industries to contribute to sustainable development. Policy-makers should implement some measures to raise environmental standards without lowering income and output levels. Additionally, the financial sector should support companies and households that use environmentally friendly projects to reduce pollution. These findings should be helpful both to policy-makers when developing environmental and trade policies in the South Asian region, and practitioners. There is also a need for more and more awareness to be created among the students at primary, secondary, and tertiary education levels for effective energy utilization and moving toward green energy. All these efforts may provide desirable outcomes. We assume that the success or failure of any policy depends on people's acceptance or rejection of a policy. Therefore, collective efforts are required to reduce pollution.

Although our study has some limitations, it has the scope for further research. The first limitation refers to the sample size. Based on data availability, we examined annual data only for four South Asian countries from 1971–2014. Second, the analysis uses a limited number of factors determining economic growth and environmental degradation. We recommend that other essential variables, such as institutional quality, financial sector development, and urbanization should be considered to understand the relationship between energy use and CO_2 emissions in South Asian countries. Moreover, this study did not examine the specific effects of renewable and non-renewable energy sources on emissions in South Asian countries. Finally, our research can act as a baseline study for other South Asian countries, as the issues discussed pertain to most developing countries. Therefore, the policy recommendations discussed in the study can be generalized.

Author Contributions: Conceptualization: B.J., A.K.D., P.K. and A.V.G.; methodology, P.K. and B.J.; software, P.K. and B.J.; formal analysis, P.K., B.J. and A.V.G.; data curation, P.K., B.J. and A.V.G.; writing—original draft preparation, B.J., A.K.D. and A.V.G.; writing—review & editing, B.J.; project administration and funding acquisition, B.J. All authors have read and agreed to the published version of the manuscript.

Funding: The APC was funded by the John Paul II Catholic University of Lublin.

Data Availability Statement: Publicly available datasets were analyzed in this study. This data can be found here: https://data.worldbank.org (accessed on 28 September 2022).

Conflicts of Interest: The authors declare no conflict of interest. The funders had no role in the design of the study; collection, analyses, or interpretation of data; writing of the manuscript; or the decision to publish the results.

Appendix A

Table A1. The parameter instability test results.

	Bangladesh	India	Sri Lanka	Pakistan
Sup LR	27.730 ***	39.963 ***	6.387 ***	11.134 ***
Sub Wald	138.650 ***	199.814 ***	31.935 ***	55.67 ***
Exp LR	10.432 ***	16.876 ***	1.836 ***	3.877 ***
Exp Wald	65.891 ***	96.474 ***	13.387 ***	25.204 ***
Mean LR	5.556 ***	17.400 ***	2.656 ***	4.842 ***
Mean Wald	27.781 ***	86.999 ***	13.279 ***	24.211 ***

Sources: The authors' estimation. Note: *** shows the significance at the 1% level.

Table A2. The Brock, Dechert, and Scheinkman (BDS) tests results.

	Dimension	$\ln CO_2$	lnE	lnY	$\ln Y^2$	lnT
Bangladesh	2	1.057	0.509	4.06 ***	3.863 ***	2.592 ***
	3	0.407	−0.436	4.838 ***	4.566 ***	2.336 **
	4	0.379	−0.759	5.552 ***	5.3 ***	2.448 **
	5	0.704	0.036	5.908 ***	5.66 ***	2.536 **
	6	0.593	−0.208	6.192 ***	5.969 ***	2.552 **
India	2	0.021	−0.917	0.218	−0.433	0.122
	3	−0.889	−2.39 **	0.316	−0.345	−1.235
	4	−0.781	−1.679 *	−0.037	−0.78	−0.993
	5	−1.14	−1.333	0.272	−0.39	−0.963
	6	−1.079	−1.044	0.572	−0.089	−0.735
Sri Lanka	2	−2.209 **	1.617	1.968 **	2.319 **	1.196
	3	−2.083 **	0.719	1.112	0.974	1.587
	4	−1.852 *	0.576	−0.154	−0.335	1.84 *
	5	−1.169	0.642	−0.237	−0.669	1.341
	6	−0.564	0.619	−0.606	−1.183	1.335
Pakistan	2	0.351	−1.247	0.366	0.348	−0.202
	3	0.004	−1.497	0.226	0.031	0.072
	4	−0.465	−1.938*	−0.408	−0.695	−0.035
	5	−0.073	−1.605	−0.53	−0.505	0.232
	6	−0.097	−1.118	−0.921	−0.804	−0.495

Sources: The authors' estimation. Note: *, ** and *** show the significance at the 10%, 5% and 1% level, respectively.

Table A3. ADF and PP unit root tests results.

Variable	ADF Test				PP Test			
	At Level		At First Difference		At Level		At First Difference	
	Intercept	With Trend	Intercept	With Trend	Intercept	With Trend	Intercept	With Trend
Bangladesh								
lnCO_2	−1.726 (0)	−13.956 *** (0)	−33.796 *** (0)	−32.765 *** (0)	−1.957 (3)	−9.249 *** (5)	−30.575 *** (1)	−29.935 *** (3)
lnE	1.679 (1)	−1.001 (0)	−8.252 *** (0)	−8.656 *** (0)	2.126 (6)	−1.001 (0)	−8.255 *** (2)	−9.015 *** (3)
lnY	2.894 (0)	−2.132 (0)	−0.934 (2)	−12.936 *** (0)	3.312 (1)	−2.129 (3)	−8.810 *** (4)	−15.548 *** (1)
lnY^2	3.351 (0)	−1.672 (0)	−0.674 (2)	−12.549 *** (0)	3.810 (1)	−1.652 (3)	−8.100 *** (4)	−14.466 *** (1)
lnT	0.049 (0)	−2.648 (0)	0.262 *** (0)	−6.514 *** (0)	−0.063 (1)	−2.990 (8)	−5.556 *** (1)	−6.566 *** (5)
lnT_NEG	−8.490 *** (9)	−10.474 *** (9)	−8.007 *** (9)	−5.809 *** (9)	−10.376 *** (13)	−6.230 *** (7)	−6.291 *** (1)	−7.510 *** (5)
lnT_POS	0.417 (0)	−1.719 (0)	−5.547 *** (0)	−5.509 *** (0)	0.319 (2)	−1.821 (1)	−5.558 *** (2)	−5.523 *** (2)
India								
lnCO_2	1.694 (0)	−1.064 (0)	−7.228 *** (0)	−7.863 *** (0)	1.917 (2)	−1.106 (3)	−7.209 *** (3)	−7.735 *** (3)
lnE	3.795 (0)	−0.169 (0)	−4.793 *** (0)	−6.152 *** (0)	3.669 (2)	−0.345 (3)	−5.023 *** (4)	−6.210 *** (3)
lnY	3.305 (0)	−1.830 (0)	−6.388 *** (0)	−8.280 *** (0)	5.396 (5)	−1.940 (4)	−6.386 *** (4)	−14.602 *** (10)
lnY^2	4.040 (0)	−1.327 (0)	−5.741 *** (0)	−8.158 *** (0)	6.890 (6)	−1.363 (6)	−5.802 *** (4)	−14.638 *** (12)
lnT	2.429 (0)	−0.318 (0)	−3.225 ** (1)	−7.844 *** (0)	2.939 (1)	−0.226 (3)	−6.722 *** (4)	−7.717 *** (3)
lnT_NEG	2.662 (0)	−2.076 (3)	−2.745 * (1)	−3.128 (1)	2.391 (3)	−0.754 (4)	−6.080 *** (4)	−6.784 *** (4)
lnT_POS	−1.908 (0)	−1.586 (0)	−7.137 *** (0)	−7.404 *** (0)	−2.409 (8)	−1.460 (2)	−7.267 *** (4)	−8.838 *** (9)
Sri Lanka								
lnCO_2	−0.070 (0)	−2.243 (0)	−7.258 *** (0)	−7.409 *** (0)	0.099 (3)	−2.178 (2)	−7.258 *** (0)	−7.411 *** (1)
lnE	0.078 (0)	−2.335 (0)	−7.290 *** (0)	−6.521 *** (1)	0.436 (5)	−2.157 (2)	−7.399 *** (2)	−7.840 *** (6)
lnY	3.038 (0)	−0.709 (0)	−5.867 *** (0)	−6.445 *** (0)	3.203 (4)	−0.767 (1)	−5.870 *** (2)	−6.432 *** (2)
lnY^2	3.830 (0)	−0.233 (0)	−5.226 *** (0)	−6.135 *** (0)	4.149 (5)	−0.330 (2)	−5.258 *** (2)	−6.144 *** (2)
lnT	3.830 (0)	−0.233 (0)	−5.226 *** (0)	−6.135 *** (0)	4.149 (5)	−0.330 (2)	−5.258 *** (2)	−6.144 *** (2)
lnT_NEG	−2.304 (0)	−3.464 * (0)	−6.790 *** (0)	−6.872 *** (0)	−2.362 (3)	−3.456 * (3)	−6.826 *** (2)	−6.905 *** (2)
lnT_POS	−0.137 (0)	−2.570 (0)	−6.099 *** (0)	−6.022 *** (0)	−0.115 (3)	−2.719 (1)	−6.119 *** (4)	−6.035 *** (4)
Pakistan								
lnCO_2	−0.690 (0)	−1.808 (0)	−8.538 *** (0)	−8.947 *** (0)	−0.695 (1)	−2.173 (3)	−8.303 *** (2)	−9.035 *** (1)
lnE	−2.111 (0)	0.349 (0)	−5.085 *** (0)	−5.768 *** (0)	−1.986 (2)	0.243 (1)	−5.110 *** (2)	−5.769 *** (1)
lnY	−1.846 (1)	−1.468 (1)	−5.675 *** (0)	−5.969 *** (0)	−1.117 (3)	−1.303 (3)	−5.758 *** (3)	−5.982 *** (2)
lnY^2	−1.588 (1)	−1.581 (1)	−5.565 *** (0)	−5.749 *** (0)	−0.934 (3)	−1.452 (3)	−5.612 *** (2)	−5.754 *** (1)
lnT	−2.052 (0)	−4.724 *** (0)	−6.958 *** (0)	−6.778 *** (0)	−2.293 (3)	−4.782 *** (2)	−7.475 *** (4)	−7.282 *** (4)
lnT_NEG	−0.251 (0)	−3.165 (0)	−7.298 *** (0)	−7.263 *** (0)	−0.004 (6)	−3.165 (0)	−8.482 *** (7)	−9.317 *** (8)
lnT_POS	−0.52 (0)	−2.6 (0)	−6.132 *** (0)	−6.062 *** (0)	−0.518 (2)	−2.741 (1)	−6.128 *** (2)	−6.055 *** (2)

Note: *, ** and *** show the significance at 10%, 5% and 1% level respectively.

References

1. Nasreen, S.; Anwar, S.; Ozturk, I. Financial stability, energy consumption and environmental Quality: Evidence from South Asian economies. *Renew. Sust. Energy Rev.* **2017**, *67*, 1105–1122. [CrossRef]
2. Gill, A.R.; Viswanathan, K.K.; Hassan, S. The Environmental Kuznets Curve (EKC) and the environmental problem of the day. *Renew. Sust. Energy Rev.* **2018**, *81*, 16361642.
3. Ahmad, N.; Du, L.; Tian, X.L.; Wang, J. Chinese growth and dilemmas: Modelling energy consumption, CO_2 emissions and growth in China. *Qual. Quant.* **2019**, *53*, 315–338. [CrossRef]
4. Sadorsky, P. Energy consumption, output and trade in South America. *Energy Econ.* **2012**, *34*, 476–488. [CrossRef]
5. United Nations Department for Economic and Social Affairs. *World Economic Situation and Prospects 2020*; United Nations: New York, NY, USA, 2020; pp. 140–145.
6. Sadiq, M.; Kannaiah, D.; Yahya Khan, G.; Shabbir, M.S.; Bilal, K.; Zamir, A. Does sustainable environmental agenda matter? The role of globalization toward energy consumption, economic growth, and carbon dioxide emissions in South Asian countries. *Environ. Dev. Sustain.* **2022**, 1–20. [CrossRef]
7. Ali, M.; Tariq, M.; Azam Khan, M. Re-examining the Kuznets Curve Hypothesis for South Asian Countries: New Evidences. *J. Asian Afr. Stud.* **2022**, 00219096221097744. [CrossRef]
8. Mehmood, U.; Tariq, S.; Ul Haq, Z.; Azhar, A.; Mariam, A. The role of tourism and renewable energy towards EKC in South Asian countries: Fresh insights from the ARDL approach. *Cogent Soc. Sci.* **2022**, *8*, 2073669. [CrossRef]
9. Tan, Y.L.; Yiew, T.H.; Lau, L.S.; Tan, A.L. Environmental Kuznets curve for biodiversity loss: Evidence from South and Southeast Asian countries. *Environ. Sci. Pollut. Res.* **2022**, *29*, 64004–64021. [CrossRef]
10. European Foundation for South Asian Studies. Environmental Degradation in South Asia and China's Belt and Road Initiative. Available online: https://www.efsas.org/publications/articles-by-efsas/environmental-degradation-in-south-asia-and-china-bri/ (accessed on 19 August 2022).
11. Siddique, H.M.A. Industrialization, energy consumption, and environmental pollution: Evidence from South Asia. *Environ. Sci. Pollut. Res.* **2022**, 1–9. [CrossRef]
12. Siddique, H.M.A. Impact of Industrialization and Openness on Environmental Pollution in South Asia. 2021. Available online: https://ssrn.com/abstract=3980442 (accessed on 28 September 2022).
13. Mehmood, U.; Tariq, S. Globalization and CO_2 emissions nexus: Evidence from the EKC hypothesis in South Asian countries. *Environ. Sci. Pollut. Res.* **2020**, *27*, 37044–37056. [CrossRef]
14. Rahman, M.M.; Velayutham, E. Renewable and non-renewable energy consumption-economic growth nexus: New evidence from South Asia. *Renew. Energy* **2020**, *147*, 399–408. [CrossRef]
15. Mishra, A.K.; Dash, A.K. Connecting the Carbon Ecological Footprint, Economic Globalization, Population Density, Financial Sector Development, and Economic Growth of Five South Asian Countries. *Energy Res. Lett.* **2022**, *3*, 32627. [CrossRef]
16. United Nations Environment Programme and Development Alternatives. *South Asia Environment Outlook 2014*; UNEP, SAARC and DA: Bangkok, Thailand, 2014.
17. World Bank. Building a Climate-Resilient in South Asia. Available online: https://www.worldbank.org/en/news/feature/2018/04/20/building-a-climate-resilient-south-asia (accessed on 23 September 2021).
18. Khan, M.B.; Saleem, H.; Shabbir, M.S.; Huobao, X. The effects of globalization, energy consumption and economic growth on carbon dioxide emissions in South Asian countries. *Energy Environ.* **2022**, *33*, 107–134. [CrossRef]
19. Sharma, R.; Shahbaz, M.; Sinha, A.; Vo, X.V. Examining the temporal impact of stock market development on carbon intensity: Evidence from South Asian countries. *J. Environ. Manag.* **2021**, *297*, 113248. [CrossRef] [PubMed]
20. Murshed, M.; Haseeb, M.; Alam, M.S. The Environmental Kuznets Curve hypothesis for carbon and ecological footprints in South Asia: The role of renewable energy. *GeoJournal* **2022**, *87*, 2345–2372. [CrossRef]
21. Fong, L.S.; Salvo, A.; Taylor, D. Evidence of the environmental Kuznets curve for atmospheric pollutant emissions in Southeast Asia and implications for sustainable development: A spatial econometric approach. *Sustain. Dev.* **2020**, *28*, 1441–1456. [CrossRef]
22. Ullah, S.; Awan, M.S. Environmental Kuznets Curve and Income Inequality: Pooled Mean Group Estimation for Asian Developing Countries. *Forman J. Econ. Stud.* **2019**, *15*, 157–179. [CrossRef]
23. Chakravarty, D.; Mandal, S.K. Is economic growth a cause or cure for environmental degradation? Empirical evidences from selected developing economies. *Environ. Sustain. Indic.* **2020**, *7*, 100045. [CrossRef]
24. Rahman, M.M. Do population density, economic growth, energy use and exports adversely affect environmental quality in Asian populous countries? *Renew. Sust. Energy Rev.* **2017**, *77*, 506–514. [CrossRef]
25. Dong, K.; Sun, R.; Li, H.; Liao, H. Does natural gas consumption mitigate CO_2 emissions: Testing the environmental Kuznets curve hypothesis for 14 Asia-Pacific countries. *Renew. Sust. Energy Rev.* **2018**, *94*, 419–429. [CrossRef]
26. Munir, K.; Riaz, N. Energy consumption and environmental quality in South Asia: Evidence from panel non-linear ARDL. *Environ. Sci. Pollut. Res.* **2019**, *26*, 29307–29315. [CrossRef] [PubMed]
27. Mujtaba, A.; Jena, P.K.; Joshi, D.P.P. Growth and determinants of CO_2 emissions: Evidence from selected Asian emerging economies. *Environ. Sci. Pollut. Res.* **2021**, *28*, 39357–39369. [CrossRef] [PubMed]
28. Anwar, A.; Sinha, A.; Sharif, A.; Siddique, M.; Irshad, S.; Anwar, W.; Malik, S. The nexus between urbanization, renewable energy consumption, financial development, and CO_2 emissions: Evidence from selected Asian countries. *Environ. Dev. Sustain.* **2022**, *24*, 6556–6576. [CrossRef]

29. Shahbaz, M.; Topcu, B.A.; Sarıgül, S.S.; Vo, X.V. The effect of financial development on renewable energy demand: The case of developing countries. *Renew. Energy* **2021**, *178*, 1370–1380. [CrossRef]
30. Ma, X.; Ahmad, N.; Oei, P.-Y. Environmental Kuznets curve in France and Germany: Role of renewable and nonrenewable energy. *Renew. Energy* **2021**, *172*, 88–99. [CrossRef]
31. Ulucak, Z.S.; Yucel, A.G. Can renewable energy be used as an effective tool in the decarbonization of the Mediterranean region: Fresh evidence under cross-sectional dependence. *Environ. Sci. Pollut. Res.* **2021**, *28*, 52082–52092. [CrossRef]
32. Erdogan, S.; Okumus, I.; Guzel, A.E. Revisiting the Environmental Kuznets Curve hypothesis in OECD countries: The role of renewable, non-renewable energy, and oil prices. *Environ. Sci. Pollut. Res.* **2020**, *27*, 23655–23663. [CrossRef]
33. Anouliès, L. Are trade integration and the environment in conflict? The decisive role of countries' strategic interactions. *Int. Econ.* **2016**, *148*, 1–15. [CrossRef]
34. Jun, W.; Mahmood, H.; Zakaria, M. Impact of trade openness on environment in China. *J. Bus. Econ. Manag.* **2020**, *21*, 1185–1202. [CrossRef]
35. Helpman, E. Explaining the structure of foreign trade: Where do we stand? *Rev. World Econ.* **1998**, *134*, 573–589. [CrossRef]
36. Sbia, R.; Shahbaz, M.; Hamdi, H. A contribution of foreign direct investment, clean energy, trade openness, carbon emissions and economic growth to energy demand in UAE. *Econ. Model.* **2014**, *36*, 191–197. [CrossRef]
37. Ahmed, K.; Shahbaz, M.; Kyophilavong, P. Revisiting the emissions-energy-trade nexus: Evidence from the newly industrializing countries. *Environ. Sci. Pollut. Res.* **2016**, *23*, 76767691. [CrossRef] [PubMed]
38. Antweiler, W.; Copeland, B.R.; Taylor, S. Is free trade good for the environment. *Am. Econ. Rev.* **2001**, *91*, 877–908. [CrossRef]
39. Copeland, B.R.; Taylor, M.S. Trade, growth, and the environment. *J. Econ. Lit.* **2004**, *42*, 7–71. [CrossRef]
40. Cherniwchan, J. Trade liberalization and the environment: Evidence from NAFTA and U.S. manufacturing. *J. Int. Econ.* **2017**, *105*, 130–149. [CrossRef]
41. Dogan, E.; Turkekul, B. CO_2 emissions, real output, energy consumption, trade, urbanization and financial development: Testing the EKC hypothesis for the USA. *Environ. Sci. Pollut. Res.* **2016**, *23*, 1203–1213. [CrossRef]
42. Frankel, J.; Rose, A. Is trade good or bad for the environment? Sorting out the causality. *Rev. Econ. Stat.* **2005**, *87*, 85–91. [CrossRef]
43. Kanjilal, K.; Ghosh, S. Environmental Kuznets curve for India: Evidence from tests for cointegration with unknown structural breaks. *Energy Policy* **2013**, *56*, 509–515. [CrossRef]
44. Shahzad, S.J.H.; Kumar, R.R.; Zakaria, M.; Hurr, M. Carbon emission, energy consumption, trade openness and financial development in Pakistan: A revisit. *Renew. Sustain. Energy Rev.* **2017**, *70*, 185–192. [CrossRef]
45. Hakimi, A.; Hamdi, H. Trade liberalization, FDI inflows, environmental quality and economic growth: A comparative analysis between Tunisia and Morocco. *Renew Sustain. Energy Rev.* **2016**, *58*, 1445–1456. [CrossRef]
46. Lopez, R. The environment as a factor of production: The effects of economic growth and trade liberalization. *J. Environ. Econ. Manag.* **1994**, *27*, 163–184. [CrossRef]
47. Schmalensee, R.; Stoker, T.M.; Judson, R.A. World carbon dioxide emissions: 1950–2050. *Rev. Econ. Stat.* **1998**, *80*, 15–27. [CrossRef]
48. Copeland, B.; Taylor, M.S. *International Trade and the Environment: A framework for Analysis*; NBER Working Paper; NBER: Cambridge, MA, USA, 2001.
49. Al-Mulali, U.; Sheau-Ting, L. Econometric analysis of trade, exports, imports, energy consumption and CO_2 emission in six regions. *Renew. Sust. Energy Rev.* **2014**, *33*, 484–498. [CrossRef]
50. Jalil, A.; Feridun, M. The impact of growth, energy and financial development on the environment in China: A cointegration analysis. *Energy Econ.* **2011**, *33*, 284–291. [CrossRef]
51. Kellenberg, D.K. An empirical investigation of the pollution haven effect with strategic environment and trade policy. *J. Int. Econ.* **2009**, *78*, 242–255. [CrossRef]
52. Managi, S.; Kumar, S. Trade-induced technological change: Analyzing economic and environmental outcomes. *Econ. Model.* **2009**, *26*, 721–732. [CrossRef]
53. Shahbaz, M.; Khraief, N.; Uddin, G.S.; Ozturk, I. Environmental Kuznets curve in an open economy: A bounds testing and causality analysis for Tunisia. *Renew. Sust. Energy Rev.* **2014**, *34*, 325–336. [CrossRef]
54. Le, T.H.; Chang, Y.; Park, D. Trade openness and environmental quality: International evidence. *Energy Policy* **2016**, *92*, 45–55. [CrossRef]
55. Baek, J.; Cho, Y.; Koo, W.W. The environmental consequences of globalization: A country specific time-series analysis. *Ecol. Econ.* **2009**, *68*, 2255–2264. [CrossRef]
56. Grossman, G.M.; Krueger, A.B. *Environmental Impacts of a North American Free Trade Agreement*; NBER Working Paper; NBER: Cambridge, MA, USA, 1991.
57. Copeland, B.R. Policy endogeneity and the effects of trade on the environment. *Agric. Resour. Econ. Rev.* **2005**, *34*, 1–15. [CrossRef]
58. Chang, S.-C. The effects of trade liberalization on environmental degradation. *Qual. Quant.* **2015**, *49*, 235–253. [CrossRef]
59. Farhani, S.; Chaibi, A.; Rault, C. CO_2 emissions, output, energy consumption, and trade in Tunisia. *Econ. Model.* **2014**, *38*, 426–434. [CrossRef]
60. Jalil, A.; Mahmud, S.F. Environment Kuznets curve for CO_2 emissions: A cointegration analysis for China. *Energy Policy* **2009**, *37*, 5167–5172. [CrossRef]

61. Jayanthakumaran, K.; Liu, Y. Openness and the environmental Kuznets curve: Evidence from China. *Econ. Model.* **2012**, *29*, 566–576. [CrossRef]
62. Sharma, S.S. Determinants of carbon dioxide emissions: Empirical evidence from 69 countries. *Appl. Energy* **2011**, *88*, 376–382. [CrossRef]
63. Andrews, D.W. Tests for parameter instability and structural change with unknown change point. *Econometrica* **1993**, *61*, 821–856. [CrossRef]
64. Broock, W.A.; Scheinkman, J.A.; Dechert, W.D.; LeBaron, B. A test for independence based on the correlation dimension. *Econom. Rev.* **1996**, *15*, 197–235. [CrossRef]
65. Grossman, G.M.; Krueger, A.B. Economic growth and the environment. *Q. J. Econ.* **1995**, *110*, 353–377. [CrossRef]
66. Soytas, U.; Sari, R.; Ewing, B.T. Energy consumption, income, and carbon emissions in the United States. *Ecol. Econ.* **2007**, *62*, 482–489. [CrossRef]
67. Shahbaz, M.; Lean, H.H.; Shabbir, M.S. Environmental kuznets curve hypothesis in Pakistan: Cointegration and granger causality. *Renew. Sustain. Energy Rev.* **2012**, *16*, 2947–2953. [CrossRef]
68. Kyophilavong, P.; Shahbaz, M.; Anwar, S.; Masood, S. The energy-growth nexus in Thailand: Does trade openness boost up energy consumption? *Renew. Sust. Energy Rev.* **2015**, *46*, 265–274. [CrossRef]
69. Kisswani, K.M.; Harraf, A.; Kisswani, A.M. Revisiting the environmental Kuznets curve hypothesis: Evidence from the ASEAN-5 countries with structural breaks. *Appl. Econ.* **2019**, *51*, 1855–1868. [CrossRef]
70. Jóźwik, B.; Gavryshkiv, A.V.; Kyophilavong, P.; Gruszecki, L.E. Revisiting the Environmental Kuznets Curve Hypothesis: A Case of Central Europe. *Energies* **2021**, *14*, 3415. [CrossRef]
71. Soylu, Ö.B.; Adebayo, T.S.; Kirikkaleli, D. The Imperativeness of Environmental Quality in China Amidst Renewable Energy Consumption and Trade Openness. *Sustainability* **2021**, *13*, 5054. [CrossRef]
72. Fatima, T.; Shahzad, U.; Cui, L. Renewable and nonrenewable energy consumption, trade and CO_2 emissions in high emitter countries: Does the income level matter? *J. Environ. Plan. Manag.* **2020**, *64*, 1227–1251. [CrossRef]
73. Saidi, K.; Omri, A. Reducing CO_2 emissions in OECD countries: Do renewable and nuclear energy matter? *Prog. Nucl. Energy* **2020**, *126*, 103425. [CrossRef]
74. Sharma, S.; Kishan, R.; Doig, A. Low-Carbon Development in South Asia: Leapfrogging to a Green Future, Climate Action Network-South Asia. 2014. Available online: https://in.boell.org/sites/default/files/low-carbon_south_asia_report.pdf (accessed on 30 June 2020).
75. Shahbaz, M.; Sinha, A. Environmental Kuznets Curve for CO_2 Emission: A Literature Survey. *J. Econ. Stud.* **2019**, *46*, 106–168. [CrossRef]
76. Özcan, B.; Öztürk, I. (Eds.) *Environmental Kuznets Curve (EKC): A Manual*; Academic Press: Cambridge, MA, USA, 2019.
77. Kyophilavong, P. *Trade Liberalization, Pollution, and Poverty: Evidence from Lao PDR*; Economy and Environment Program for Southeast Asia: Singapore, 2011.
78. Halicioglu, F. An econometric study of CO_2 emissions, energy consumption, income and foreign trade in Turkey. *Energy Policy* **2009**, *37*, 1156–1164. [CrossRef]
79. Rahman, Z.U.; Ahmad, M. Modeling the relationship between gross capital formation and CO_2 (a)symmetrically in the case of Pakistan: An empirical analysis through NARDL approach. *Environ. Sci. Pollut. Res.* **2019**, *26*, 8111–8124. [CrossRef]
80. Qamruzzaman, M.; Jianguo, W. The asymmetric relationship between financial development, trade openness, foreign capital flows, and renewable energy consumption: Fresh evidence from panel NARDL investigation. *Renew. Energy* **2020**, *159*, 827–842. [CrossRef]
81. Sheikh, U.A.; Tabash, M.I.; Asad, M. Global Financial Crisis in Effecting Asymmetrical Co-integration between Exchange Rate and Stock Indexes of South Asian Region: Application of Panel Data NARDL and ARDL Modelling Approach with Asymmetrical Granger Causility. *Cogent Bus. Manag.* **2020**, *7*, 1843309. [CrossRef]
82. Mujtaba, A.; Jena, P.K. Analyzing asymmetric impact of economic growth, energy use, FDI inflows, and oil prices on CO_2 emissions through NARDL approach. *Environ. Sci. Pollut. Res.* **2021**, *28*, 30873–30886. [CrossRef] [PubMed]
83. Shin, Y.; Yu, B.; Greenwood-Nimmo, M. Modelling Asymmetric Cointegration and Dynamic Multipliers in an ARDL Framework. In *Festschrift in Honor of Peter Schmidt*; Horrace, W.C., Sickles, R.C., Eds.; Springer Science and Business Media: New York, NY, USA, 2014; pp. 281–314.
84. Granger, C.W.; Yoon, G. Hidden cointegration. *U Calif. Econ. Work. Pap.* **2002**. [CrossRef]
85. Hatemi-J, A. Asymmetric causality tests with an application. *Empir. Econ.* **2012**, *43*, 447–456. [CrossRef]
86. Press Information Bureau. Government of India Ministry of Environment, Forest and Climate Change. Available online: https://pib.gov.in/newsite/PrintRelease.aspx?relid=194865 (accessed on 20 July 2020).
87. Environmental Sustainability in Pakistan. Development Advocate Pakistan, Volume 7, Issue 2, (Editors), Maheen Hassan and Umer Akhlaq Malik, United Nations Development Programme, ISBN: 978-969-8736-31-20. Available online: https://www.undp.org/pakistan/publications/environmental-sustainability-pakistan (accessed on 20 July 2020).
88. Ullah, I.; Rehman, A.; Khan, F.U.; Shah, M.H.; Khan, F. Nexus between Trade, CO_2 Emissions, Renewable Energy, and Health Expenditure in Pakistan. *Int. J. Health Plann.* **2019**, *35*, 818–831. [CrossRef] [PubMed]
89. Khan, A.; Safdar, S.; Nadeem, H. Decomposing the effect of trade on environment: A case study of Pakistan. *Environ. Sci. Pollut. Res.* **2022**, 1–18. [CrossRef] [PubMed]

90. Wang, J.; Mamkhezri, J.; Khezri, M.; Karimi, M.S.; Khan, Y.A. Insights from European nations on the spatial impacts of renewable energy sources on CO_2 emissions. *Energy Rep.* **2022**, *8*, 5620–5630. [CrossRef]
91. Gierałtowska, U.; Asyngier, R.; Nakonieczny, J.; Salahodjaev, R. Renewable Energy, Urbanization, and CO_2 Emissions: A Global Test. *Energies* **2022**, *15*, 3390. [CrossRef]
92. Verbič, M.; Satrovic, E.; Muslija, A. Environmental Kuznets curve in Southeastern Europe: The role of urbanization and energy consumption. *Environ. Sci. Pollut. Res.* **2021**, *28*, 57807–57817. [CrossRef]
93. Gasimli, O.; Haq, I.U.; Naradda Gamage, S.K.; Shihadeh, F.; Rajapakshe, P.S.K.; Shafiq, M. Energy, trade, urbanization and environmental degradation nexus in Sri Lanka: Bounds testing approach. *Energies* **2019**, *12*, 1655. [CrossRef]
94. Shahbaz, M.; Sharma, R.; Sinha, A.; Jiao, Z. Analyzing nonlinear impact of economic growth drivers on CO_2 emissions: Designing an SDG framework for India. *Energy Policy* **2021**, *148*, 111965. [CrossRef]

energies

MDPI

Article

Innovation Input, Climate Change, and Energy-Environment-Growth Nexus: Evidence from OECD and Non-OECD Countries

Zhuohang Li [1], Tao Shen [2,*], Yifen Yin [3] and Hsing Hung Chen [4]

1 Faculty of Business, City University of Macau, Macao, China
2 The Institute for Sustainable Development, Macau University of Science and Technology, Macao, China
3 Faculty of Humanities and Social Science, Macao Polytechnic University, Macao, China
4 Faculty of Business, Macau University of Science and Technology, Macao, China
* Correspondence: tshen@must.edu.mo

Abstract: With economic growth and rising incomes, increasing consumption of fossil energy is leading to environmental pollution and climate change, which requires increased innovative inputs to promote the efficiency of renewable energy use. Considering the important impact of innovation input and climate change on renewable energy consumption, greenhouse gas emissions, and green economic growth, this study uses simultaneous equation and sys-GMM model to explore the dynamic nexus of innovation input, climate change, and energy-environment-growth in OECD and non-OECD countries, with panel data covering 2000 to 2019. The empirical results show that renewable energy consumption in non-OECD countries significantly promoted green economic growth, while OECD countries did the opposite. Moreover, renewable energy consumption significantly reduces greenhouse gas emissions caused by climate change, especially for OECD countries. When the level of economic growth exceeds a certain inflection point, greenhouse gas emissions begin to turn from positive to negative, which further verifies the EKC hypothesis. In addition, this study found that innovation input has significantly increased renewable energy consumption, reduced greenhouse gas emissions, and promoted green economic growth in OECD countries. Finally, this study also found that the impact of innovation input in OECD and non-OECD countries on the energy-environment-growth nexus is greater in the short term and more significant in the medium and long term, while the impact of climate change on the energy-environment nexus in OECD and non-OECD countries is more significant in the medium and long term.

Keywords: innovation input; climate change; renewable energy consumption; greenhouse gas emissions; green economic growth; simultaneous equation model

Citation: Li, Z.; Shen, T.; Yin, Y.; Chen, H.H. Innovation Input, Climate Change, and Energy-Environment-Growth Nexus: Evidence from OECD and Non-OECD Countries. *Energies* **2022**, *15*, 8927. https://doi.org/10.3390/en15238927

Academic Editor: David Borge-Diez

Received: 31 October 2022
Accepted: 21 November 2022
Published: 25 November 2022

1. Introduction

Increasing fossil energy consumption aggravates the problems of energy shortage and environmental pollution, resulting in an increase in greenhouse gas emissions [1,2]. While long-term large-scale greenhouse gas emissions are the key reason for extreme weather change [3]. To strengthen the governance of global climate and environment to promote green economic growth, the Paris Climate Agreement clearly puts forward the development of a low-carbon economy [4]. However, the increasing weather events further exacerbate the energy consumption for temperature regulation [5]. Driven by economic growth and increasing income, the energy consumption and greenhouse gas emissions in Organization for Economic Co-operation and Development (OECD) countries tend to be higher than that in non-OECD countries, and then needs to increase innovation input and improve the utilization rate of renewable energy [6]. Therefore, this paper explores the dynamic nexus of innovation input, climate change, and energy-environment-growth in OECD and non-OECD countries, which helps the policymaker to formulate differentiated energy and environmental policies to promote innovation input, increase renewable energy consumption and achieve green economic growth

Most studies have shown that energy consumption in response to temperature change varies greatly among countries in different climate regions [7–9]. Specifically, OECD countries are generally located in high latitudes with huge temperature differences, that is, hot summer and cold winter, while non-OECD countries are on the contrary In addition, according to the Environmental Kuznets Curve (EKC) hypothesis, when the economic income of OECD countries reaches a certain level, they begin to gradually pay attention to the improvement of the ecological environment [10]. With the support of high innovation input, energy efficiency and renewable energy consumption began to improve significantly, and greenhouse gas emissions gradually decreased [11]. In this context, based on the panel data of 35 OECD and 36 non-OECD countries from 2000–2019, this paper further examines the nexus of energy-environment-growth under the differentiated innovation input, which provides theoretical support for the EKC hypothesis.

Compared with the existing literature [12–14], using simultaneous equation and system generalized method of moments (sys-GMM) model is an effective method to explore the dynamic nexus of innovation input, climate change, and energy-environment-growth. As we all know, from the perspective of the production function, the framework of energy-environment-growth includes three important equations: production equation, energy consumption equation, and pollution equation, and each equation provides a reference for further research in this field [15,16]. Moreover, cross-validation shows that the three equations should not be studied separately, which confirms that the simultaneous equations can effectively estimate the dynamic nexus of renewable energy consumption, greenhouse gas emissions, and green economic growth, and help to generate reliable empirical research conclusions [17,18].

This paper is dedicated to exploring the impact of innovative inputs, climate change on renewable energy, consumption of greenhouse gas emissions, and green economic growth. The contributions of this paper are in the following four aspects: First, this paper creatively introduces innovation inputs and climate change into the energy-environment-growth research framework to study their effects on renewable energy consumption, greenhouse gas emissions, and green economic growth. Second, this paper analyzes the differences in the effects caused by the heterogeneity of the sample intervals, and examines the dynamic relationship between innovation inputs, climate change, and energy-environmental growth comprehensively and systematically in the short (2015–2019), medium (2010–2019), and long term (2000–2019), respectively, further confirming the EKC hypothesis. Third, this paper uses frontier simultaneous equations and sys-GMM models to reveal the dynamic relationship among innovation inputs, climate change, and energy-environmental growth, which can better solve the heteroskedasticity, autocorrelation, and endogeneity problems in the model. Fourth, considering the accuracy and comprehensiveness of variable calculation this paper uses principal component analysis to construct the green economic growth index from four dimensions: economic development, resources and environment, globalization, and urban construction (see Table 1). Finally, according to the research results, this paper provides targeted suggestions for the government to develop differentiated energy and environmental policies to promote carbon emission reduction and green economic growth.

Table 1. Indicator system of green economic growth.

Primary Index	Secondary Index	Tertiary Indicators	Symbol	Unit
Green economic growth index	Economic development	Per capita GDP	X1	Dollar
		Final consumption expenditure	X2	Dollar
		Inflation consumer Prices	X3	Dollar
		Taxes on income, profits, and capital gains	X4	
	Resource environment	Per capita energy consumption	X5	kg of oil
		Total natural resources rents in GDP	X6	%
		CO_2 emissions	X7	Kt
		Forest area	X8	Sq.km
	Globalization	Proportion of exports of goods and services in GDP	X9	%
		Proportion of trade in GDP	X10	%
	Urban construction	Agriculture, forestry, and fishing, value added per worker	X11	Dollar
		Population growth	X12	%

2. Literature Review

Energy-environment-growth nexus studies the causality among energy consumption, environmental pollution, and economic growth. Considerable foregoing discussions about this nexus have employed the method of the Granger causality test [19], while the simultaneous equation model is less familiar. To be specific, the granger causality test can only detect whether there is a causal relationship between the concerned variables, but not the relationship sign and sensitivity. However, the simultaneous equation model does not have this limitation. It can not only detect the sign and sensitivity between variables, but also add other essential control variables to avoid missing variables.

In recent decades, the energy-environment-growth nexus has been the subject of a great deal of academic research. There are three branches of research in the literature that deals with the relationships between target variables. The first branch of research focuses on the relationship between economic growth and environmental pollution. Existing literature relies heavily on the Environmental Kuznets Curve (EKC) hypothesis when studying the relationship between the two variables. Stern [20] asserts that the degree of environmental degradation first increased and then decreased with the increase of the GNP per capita. In addition, the degree of environmental degradation is usually measured by air pollution. Some empirical studies verify the EKC hypothesis, such as Naseem et al. studied the relationship between economic development and pollutant gas emissions in OECD and non-OECD countries [21]. And Nasir and Ur-Rehman [22] and Saboori et al. [23] confirmed the existence of the EKC hypothesis by examining the long-term relationship between greenhouse gas emissions (GHGs) and income in Malaysia and Pakistan, respectively.

The second branch investigates the relationship between energy consumption and economic growth. Since the initial study of Kraft [24], the nexus between energy and economy has been the focus of discussion among scholars [25–27]. However, in the existing literature, scholars have several different views on the existence and direction of the causal relationship between these two variables. Soytas and Sari [28] believed that there is no significant causal relationship between energy consumption and economic growth, and supports the neutral hypothesis. Huang et al. [29] pointed out that in middle-income and high-income countries, the economy can affect energy consumption, and supported the conservation hypothesis. In addition, Saidi and Hammami [30] indicated that energy consumption has a significant stimulative effect on economic growth, which supported the feedback hypothesis that there is a two-way causal relationship between the two variables [31,32]. The third branch is related to energy consumption and GHGs. There is a consensus that energy consumption is the main source of GHGs [33–35].

2.1. Climate Change and the Energy-Environment-Growth Nexus

According to the vast majority of literature now available, the energy-environment-growth nexus studies which use the simultaneous equation model have not considered that climate affects energy and the environment in many ways, although extreme temperature changes could distinctly affect energy consumption, and thus GHGs. For example, Considine [36] evaluated the driving factors of GHGs, and the results of the linear logit model indicated the impact of weather changes on GHGs is considerable. That is, the hot summer increases the demand for air conditioning and electricity, which in turn increases energy consumption. While cold winter increases the demand for heating fuel, such as coal, oil, and natural gas. Studies have shown that the consumption of traditional energy sources in both OECD and non-OECD countries will inevitably lead to an increase in greenhouse gases and thus affect economic growth [37].

The environment affects the economy and energy in many ways, and there is heterogeneity in the impact of the environment on economic growth and energy efficiency in OECD and non-OECD countries [38,39]. At the same time, the abnormal temperature will affect the economy in many aspects, causing damage to green economic growth [40,41]. The emergence of extreme temperatures hinders short-term and long-term economic development and affects indicators such as employment and profitability [42,43]. However, the impact of climate change on green economic growth has not been widely studied in the existing literature.

2.2. *Innovation Input and the Energy-Environment-Growth Nexus*

On the role of innovation input in economic growth, a large amount of literature gives almost the same conclusion. In contrast, there are fewer studies on the impact of innovation input on GHGs and energy consumption, especially the impact of innovation input on the energy-environment-growth nexus. Chen and Lei [44] suggested that technological innovation has played an important role in improving energy efficiency and reducing energy consumption. But technological innovation has a greater impact on countries with higher GHGs than on countries with lower GHGs. Zakari et al. studied the factors influencing green innovation in OECD and non-OECD countries respectively [45]. And Khan et al. [26] examined that technological innovation can reduce GHGs and boost economic growth in BRICS countries. The improvement of innovation input is helpful to develop renewable energy and improve energy efficiency, to ensure energy security and achieve green economic growth.

2.2.1. Innovation Input and Economic Growth

A large number of existing literature believed that innovation input is the pillar of economic growth, a key factor to promote green economic growth, and even the power and source of human social development [46,47]. From the perspective of neoclassical economics, Thompson [48] elaborated his point of view: with the development of an innovation economy, social capital will grow internally with the increase of monopolistic competitors' profits and production. In other words, innovation input and economic growth can promote each other and develop together. Adak [49] cites structural changes in Turkey's economy over the past 35 years as evidence of the impact of technological progress and innovation input on economic growth. In this model, innovation input has become a key endogenous variable of the total production function, and innovation input has brought high productivity and rapid positive growth to the economy.

2.2.2. Innovation Input and Environment Pollution

With regard to innovation input and environment pollution, most scholars believe that the impact of innovation input on GHGs is linear and one-way [50,51]. In a detailed analysis of G20 countries, Erdoğan et al. [52] argue that innovation input in different sectors will have different impacts on GHGs. Increased innovation input in the industrial sector leads to reduced GHGs, while increased innovation input in the construction sector leads to the opposite result. At the same time, a few studies believe that there is a non-linear two-way relationship between them. For instance, Carrión-Flores and Innes [53] pointed out that there is a two-way causal relationship between innovation input and environmental pollution. Innovation input was an important driving force of environmental pollution, and strict pollution control targets will promote the improvement of innovation input.

2.2.3. Innovation Input and Energy Consumption

Sun et al. [54] examined the relationship between innovation input and energy consumption, and testified that innovation input has a positive impact on improving energy efficiency and reducing energy intensity. Wurlod and Noailly [55] analyzed the impact of innovation input on the energy intensity of 14 industrial sectors in 17 OECD countries, and found that innovation input contributed to the decline of energy intensity in most industrial sectors. In conclusion, these studies compelling indicate that climate change and innovation input play a very important role within the energy-environment-growth nexus. Therefore, in the following study on the energy-environment-growth nexus, this paper introduced the two variables of climate change and innovation input.

In general, the existing literature mostly studies the correlation between fossil energy consumption, carbon emissions, and economic growth. It is found that when the economic income level is low, fossil energy consumption and carbon emissions are more, while when the economic income level is high, it is the opposite. The difference is that this paper creatively introduces innovation input and climate change into the framework of renewable energy consumption, carbon emissions, and green economic growth further analyzes the dynamic nexus of innovation input, climate change and energy-environment-growth, and then verifies the EKC hypothesis.

3. Methodology

3.1. Production Function

Many countries are interested in energy-environment-growth nexus, which has gradually become a worldwide problem [56]. In this context, this paper draws on the research results of other scholars and regards energy consumption as a production factor within the nexus [57]. Moreover, innovation input is rarely included in the nexus, which reflects a country's science and technology level. In summary, the augmented Cobb-Douglas production function is as follows:

$$geg = Ak^{\alpha 1}e^{\alpha 2} \tag{1}$$

where *geg* is the green economic growth, which is calculated by the PCA method and consists of twelve indicators as shown in Table 1 [58,59]. *A* is the total factor productivity, *k* is the capital per capita, and *e* is the proportion of renewable energy consumption.

After logging, *i* denotes the country and *t* denotes the time period as follows:

$$geg_{it} = \alpha_{it} + \alpha_1 k_{it} + \alpha_2 e_{it} \tag{2}$$

Assuming that green economic growth depends on innovation input it becomes:

$$a_{it} = \alpha_0 + \alpha_4 ino_{it} + \varepsilon_{1,it} \tag{3}$$

Combining Equations (3) and (4), we can get:

$$geg_{it} = \alpha_0 + \alpha_1 k_{it} + \alpha_2 e_{it} + \alpha_3 ino_{it} + \varepsilon_{1,it} \tag{4}$$

Because capital, renewable energy consumption, and innovation are conducive to green economic growth, they are expected to have a positive impact on green economic growth, indicating that α_1, α_2, and α_3 should be positive. Since pollution is not significant in the estimation and obeys the standard production function theory, we do not introduce pollution as an explanatory variable.

3.2. Energy Consumption Function

Referring to previous literature on the energy-environment-growth nexus [60,61], innovation input and climate change are included in the energy consumption function as follows:

$$e_{it} = \beta_0 + \beta_1 geg_{it} + \beta_2 ind_{it} + \beta_3 ino_{it} + \beta_4 stemp_{it} + \beta_5 wtemp_{it} + \varepsilon_{2,it} \quad (5)$$

where *e* is the proportion of renewable energy consumption, *geg* is green economic growth, *ind* is industrialization, *ino* is innovation input, *stemp* is the average temperature of three months in summer and wtemp is the average temperature of three months in winter, ε_2 is the error term.

3.3. Pollution Function

To update the pollution function [62], innovation input and climate change have been included in the pollution function as follows:

$$pol_{it} = \gamma_0 + \gamma_1 geg_{it} + \gamma_2 geg_{it}^2 + \gamma_3 e_{it} + \gamma_4 ino + \gamma_5 stemp_{it} + \gamma_6 wtemp_{it} + \gamma_7 urb_{it} + \gamma_8 poli + \varepsilon_{3,it} \quad (6)$$

where *pol* denotes greenhouse gas emissions; *geg* denotes green economic growth; geg^2 denotes geg squared; *e* denotes the proportion of renewable energy consumption; *urb* denotes urbanization; *poli* denotes climate policy measured by whether the Kyoto Protocol is signed before 2016 or participation in the Paris Agreement after 2016. If the sample has participated in the above two agreements, we will record it as 1, on the contrary, we will record it as 0. Besides, stemp and wtemp respectively represent the average temperature of three months in summer and winter; and ε is the error term. All variables are logarithmic except stemp and wtemp.

From Equations (4)–(6), A three-dimensional simultaneous equation framework is used to analyze the energy-environment-growth nexus. In conclusion, the structural equations look as follows:

$$\begin{aligned} geg_{it} &= \alpha_0 + \alpha_1 k_{it} + \alpha_2 e_{it} + \alpha_3 ino_{it} + \varepsilon_{1,it} \\ e_{it} &= \beta_0 + \beta_1 geg_{it} + \beta_2 ind_{it} + \beta_3 ino_{it} + \beta_4 stemp_{it} + \beta_5 wtemp_{it} + \varepsilon_{2,it} \\ pol_{it} &= \gamma_0 + \gamma_1 geg_{it} + \gamma_2 geg_{it}^2 + \gamma_3 e_{it} + \gamma_4 ino + \gamma_5 stemp_{it} + \gamma_6 wtemp_{it} + \gamma_7 urb_{it} + \gamma_8 poli + \varepsilon_{3,it} \end{aligned} \quad (7)$$

3.4. The Estimation Method

As shown in Figure 1, based on the theoretical framework, the system estimation is applied to study the nexus of energy-environment-growth.

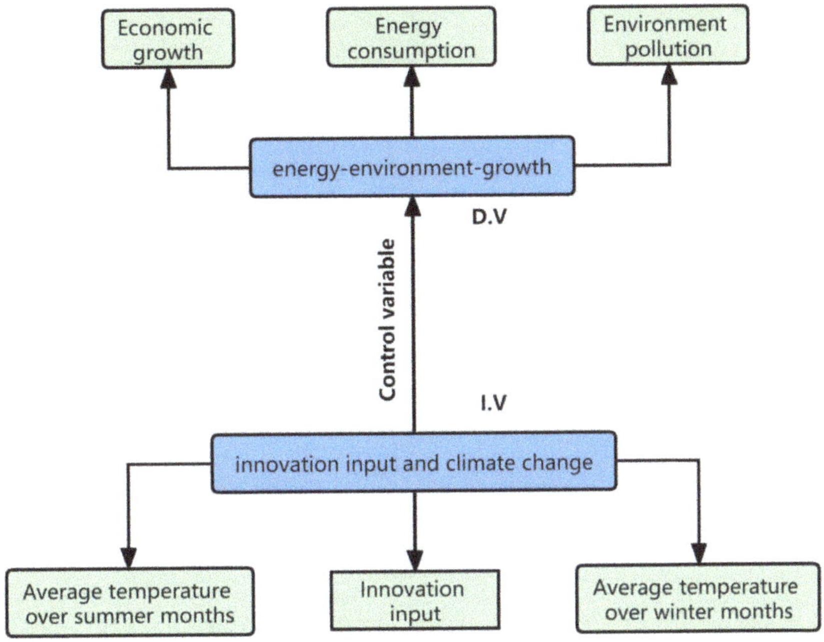

Figure 1. Conceptual framework.

The generalized method of moments (GMM) was first proposed by Hansen and has become one of the most popular measurement methods. Arellano and Bond [63] proposed a first difference GMM (diff-GMM) estimation method. However, Blundell and Bond [64] have found the first-order diff-GMM estimation method is vulnerable to the influence of weak instrumental variables and gets biased estimation results. To overcome the influence of weak instrumental variables, Arellano and Bover [65] and Blundell and Bond [64] proposed another more effective method system GMM (sys-GMM). With the energy-environment-growth nexus, Saidi and Hammami [66] and Sekrafi and Sghaier [67] used diff-GMM in their studies, while Bhattacharya et al. [68] used sys-GMM in the interrelationship of energy-environment-growth. The main advantage of these methods over other methods is that they rely on internal instruments for estimation. However, in the case of a reverse causal relationship, external instruments are the best. However, finding external tools is a difficult task, which varies across units and periods. Fortunately, Farhadi et al. [69] concluded that the internal tools used are different, and sys-GMM is the best choice to control the endogenous nature of explanatory variables.

$$y_{it} = x_{it}\beta + \varphi y_{i,t-1} + c_i + \varepsilon_{it} \tag{8}$$

where t denotes time, and i denotes the cross-section units (countries). It appears that the error terms consist of the fixed individual effects c_i and the idiosyncratic shocks ε_{it}. The properties of fixed individual effects and idiosyncratic shocks are attributed as

$$E(c_i) = E(\varepsilon_{it}) = E(c_i\varepsilon_{it}) = 0 \tag{9}$$

By taking the difference to eliminate the individual effects c_i from Equation (8) resulting in:

$$\Delta y_{it} = (\Delta x)_{it}\beta + \varphi(\Delta y_{i,t-1}) + \Delta\varepsilon_{it} \tag{10}$$

where Δ denotes the first difference operator.

Since Roodman [70] indicated that the diff-GMM and sys-GMM estimator is suitable for data sets with large groups and few periods, the current energy-environment-growth studies do not always follow this rule. However, if groups are too small, the test of cluster robust standard error and sequence correlation becomes inaccurate. Another problem is that the quantity of instruments is quadratic in the periods, which may lead to overfitting the equation because there are too many instruments compared to the sample capacity. To overcome the problem, the quantity of instruments is expected to be less than the groups. To achieve this goal, we can limit the lag of the instruments and collapse the instrument matrix. Table 2 provides the definition and source of the variable. Descriptive statistics are shown in Table 3, which is divided into OECD and non-OECD groups.

Table 2. The definition and source of variables.

Variables	Definition	Source	Calculation by the Author
Dependent variable			
Energy consumption	Renewable energy consumption (% of total)	IEA	
Economic growth	It covers four aspects: Economic development, Resource environment, Globalization, and Urban Construction	Khan et al. (2021) and Zhou et al. (2022)	Calculation by PCA
Environment pollution	Total greenhouse gas emissions (kt of CO2 equivalent)	World Bank's World Development Indicators	
Independent variable			
Innovation input	% Research and development expenditure of total GDP	World Bank's World Development Indicators	
Average temperature over summer months	Average temperatures in June, July, and August for countries with capitals in the Northern Hemisphere, and January, February, and December for countries with capitals in the Southern Hemisphere	World Bank: monthly average temperatures for countries; CIA (2018): latitudes of country capitals	Calculation of average temperatures over summer months based on monthly data
Average temperature over winter months	Average temperatures in January, February, and December for countries with capitals in the Northern Hemisphere, and June, July, and August for countries with capitals in the Southern Hemisphere	World Bank: monthly average temperatures for countries; CIA (2018): latitudes of country capitals	Calculation of average temperatures over winter months based on monthly data
Control variable			
Climate Policy	Signing Kyoto Protocol before 2016 or Joining the Paris Agreement after 2016	Kyoto Protocol and The Paris Agreement	The number 1 represents joining the Paris Agreement or Kyoto Protocol and the number 0 represents no joining
Industrialization	% Value added of industry of total GDP	World Bank's World Development Indicators	Interpolated
Capital	Capital stock at constant 2010 national prices (in mil. 2010USD)	Penn World Table	Divided by population
Urbanization	% Urban population of the total population	Penn World Table	

Table 3. Descriptive statistics.

Variable	Group	Obs	Mean	Std.Dev	Min	Max
pol	OECD	700	11.21	1.852	4.094	13.81
	Non-OECD	720	11.35	1.800	7.534	16.33
geg	OECD	700	1.173	0.575	0.345	3.222
	Non-OECD	720	0.656	0.577	0.211	11.37
k	OECD	700	2.851	1.537	0.860	14.49
	Non-OECD	720	1.288	1.008	−2.232	3.633
e	OECD	700	2.568	0.926	−0.368	4.113
	Non-OECD	720	2.617	1.755	−5.021	4.545
ino	OECD	700	0.277	0.756	−2.040	1.600
	Non-OECD	720	−1.156	1.203	−5.482	0.954
ind	OECD	700	3.249	0.245	2.353	3.856
	Non-OECD	720	3.360	0.287	2.301	4.252
stemp	OECD	700	19.91	3.663	12.86	31.42
	Non-OECD	720	24.62	5.416	8.830	37.01
wtemp	OECD	700	4.326	7.432	−14.19	26.62
	Non-OECD	720	12.60	12.23	−26.85	31.01
urb	OECD	700	4.379	0.593	4.020	8.946
	Non-OECD	720	4.031	0.353	2.901	4.605
poli	OECD	700	0.844	0.363	0	1
	Non-OECD	720	0.850	0.357	0	1

4. Results

4.1. Data Source

For econometric analysis of the proposed models, this paper uses the panel data of 35 OECD and 36 non-OECD countries from 2000 to 2019, which are from the World Bank, Penn World Table, and IEA. In addition, the sample interval is divided into short-term (2015–2019), medium-term (2010–2019), and long-term (2000–2019) for longitudinal comparison. To make the data stable, all variables except temperature are logarithmically transformed.

4.2. The Results of the Production Function

The results of the production functions for the OECD and non-OECD sample groups are shown in Tables 4 and 5. Firstly, in all models, the number of countries is significantly greater than the number of instrumental variables, and the Hansen test presents the instrumental variables are valid at a risk level of 0.05. Furthermore, the results of the Arellano-Bond test indicate that the estimators are consistent. According to the function estimation results, the coefficient of capital per capita is positive in both tables, indicating that capital and wealth are conducive to the development of a green economy in any country and that this effect is more pronounced in non-OECD countries. In contrast, the coefficients on renewable energy consumption are both significant, which is a good indication of the importance of renewable energy for the growth of a green economy. Finally, in agreement with other results in the literature, the coefficients of the innovation input variables are positive and significant, indicating that innovation input has a significant effect on green economic growth in both OECD and non-OECD countries, confirming the importance of innovation input in promoting energy restructuring, increasing the utilization of renewable energy and thus achieving green economic growth. Similar to the effect of the capital per capita variable, the effect of innovation inputs on green economic growth is greater in non-OECD countries due to their lower overall strength than in OECD countries.

Table 4. The production function (OECD).

Variables	Group (1)	Group (2)	Group (3)	Group (4)
	Diff-GMM	Sys-GMM	Diff-GMM	Sys-GMM
L.geg	0.925 ***	0.919 ***	0.934 ***	0.910 ***
	(0.00407)	(0.00218)	(0.00574)	(0.00301)
k	0.0251 ***	0.0124 ***	0.0282 ***	0.0104 ***
	(0.00329)	(0.000664)	(0.00499)	(0.000803)
e	−0.0369 ***	−0.0289 ***	−0.0382 ***	−0.0271 ***
	(0.00255)	(0.00170)	(0.00277)	(0.00203)
ino			0.00769 *	0.0184 ***
			(0.00412)	(0.00166)
Constant		0.146 ***		0.150 ***
		(0.00436)		(0.00571)
Observations	630	665	630	665
Sample	35	35	35	35
AR(1)	0.00221	0.00183	0.00223	0.00189
AR(2)	0.785	0.597	0.808	0.598
Hansen test	0.427	0.586	0.381	0.658

Note: Robust standard errors in parentheses, *** $p < 0.01$, * $p < 0.1$.

Table 5. The production function (non-OECD).

Variables	Group (1)	Group (2)	Group (3)	Group (4)
	Diff-GMM	Sys-GMM	Diff-GMM	Sys-GMM
L.geg	0.130 ***	0.493 ***	0.185 ***	0.489 ***
	(0.00928)	(0.00724)	(0.0102)	(0.00730)
k	0.249 ***	0.202 ***	0.322 ***	0.178 ***
	(0.00274)	(0.00327)	(0.00372)	(0.00462)
e	0.0471 ***	0.0691 ***	0.0839 ***	0.0595 ***
	(0.00548)	(0.00283)	(0.00586)	(0.00529)
ino			0.124 ***	0.0306 ***
			(0.00298)	(0.00391)
Constant		−0.103 ***		−0.00680
		(0.00524)		(0.0129)
Observations	648	684	648	684
Sample	36	36	36	36
AR(1)	0.248	0.230	0.256	0.228
AR(2)	0.310	0.303	0.306	0.306
Hansen test	0.385	0.527	0.358	0.531

Note: Robust standard errors in parentheses, *** $p < 0.01$.

4.3. The Results of Energy Consumption Function

The results of the estimated energy consumption function are shown in Tables 6 and 7, the results show that these estimates are consistent, and the number of instrumental variables is significantly less than that of all model countries. According to the model estimation results, it can be seen that: firstly, the coefficient of the green economic growth variable shows a positive value in the sample group of OECD countries, indicating that green economic growth can promote renewable energy consumption. In contrast, according to the estimation results for the sample of non-OECD countries, the relationship between green economic growth and the renewable energy consumption is the opposite. Secondly, for the effect of industrialization on renewable energy consumption, industrialization is able to promote renewable energy consumption in OECD countries, while in non-OECD countries, the effect of industrialization on renewable energy is negative. This may be due to the fact that most of the OECD countries are more developed economies and therefore have more developed industries and more diverse and sophisticated energy systems than the non-OECD countries. Whereas the non-OECD countries, most of which are developing countries, are at a stage of industrialization where they are using a lot of fossil fuels. Therefore, industrialization in non-OECD countries is negatively correlated with renewable energy consumption.

Table 6. The energy consumption function (OECD).

Variables	Group (1)	Group (2)	Group (3)	Group (4)	Group (5)	Group (6)	Group (7)	Group (8)
	Diff-GMM	Sys-GMM	Diff-GMM	Sys-GMM	Diff-GMM	Sys-GMM	Diff-GMM	Sys-GMM
L.e	0.835 ***	0.947 ***	0.844 ***	0.939 ***	0.767 ***	0.943 ***	0.774 ***	0.947 ***
	(0.0649)	(0.0187)	(0.0534)	(0.0274)	(0.0820)	(0.0316)	(0.0852)	(0.0288)
geg	0.171 ***	0.117 ***	0.0328	0.0581	0.217 ***	0.0977 ***	0.0791	0.0673
	(0.0482)	(0.0371)	(0.0634)	(0.0460)	(0.0499)	(0.0323)	(0.0691)	(0.0438)
ind	−0.131	0.0472	0.122	0.143 ***	−0.223 *	-2.92×10^{-5}	−0.0136	0.135 **
	(0.0982)	(0.0434)	(0.141)	(0.0523)	(0.118)	(0.0420)	(0.173)	(0.0529)
ino			0.213 ***	0.126 ***			0.186 ***	0.108 **
			(0.0664)	(0.0488)			(0.0696)	(0.0500)
stemp					−0.00681	−0.00893 **	−0.00251	−0.00668 *
					(0.00440)	(0.00444)	(0.00445)	(0.00401)
wtemp					0.00502 *	-9.13×10^{-5}	0.00399	0.000999
					(0.00288)	(0.00292)	(0.00296)	(0.00237)
Constant		−0.138		−0.391 **		0.225		−0.265
		(0.208)		(0.186)		(0.172)		(0.195)
Observations	630	665	630	665	630	665	630	665
Sample	35	35	35	35	35	35	35	35
AR(1)	0.00326	0.00193	0.00108	0.00109	0.00320	0.00189	0.00170	0.00111
AR(2)	0.865	0.816	0.959	0.843	0.915	0.840	0.976	0.850
Hansen test	0.481	0.734	0.696	0.621	0.495	0.676	0.545	0.525

Note: Robust standard errors in parentheses, *** $p < 0.01$, ** $p < 0.05$, * $p < 0.1$.

Table 7. The energy consumption function (non-OECD).

Variables	Group (1)	Group (2)	Group (3)	Group (4)	Group (5)	Group (6)	Group (7)	Group (8)
	Diff-GMM	Sys-GMM	Diff-GMM	Sys-GMM	Diff-GMM	Sys-GMM	Diff-GMM	Sys-GMM
L.e	0.672 ***	0.871 ***	0.403 ***	0.873 ***	0.726 ***	0.824 ***	0.409 ***	0.815 ***
	(0.0493)	(0.0318)	(0.0728)	(0.0332)	(0.0570)	(0.0309)	(0.103)	(0.0339)
geg	−0.137 **	−0.0291	−0.0146	−0.0326	−0.101	−0.0954	0.0125	−0.0488
	(0.0665)	(0.0462)	(0.0720)	(0.0499)	(0.0633)	(0.0592)	(0.0884)	(0.0581)
ind	0.0184	−0.259 ***	0.161	−0.246 ***	−0.0459	−0.297 ***	0.272	−0.279 ***
	(0.0845)	(0.0494)	(0.130)	(0.0544)	(0.0882)	(0.0533)	(0.201)	(0.0551)
ino			−0.108 ***	0.00606			−0.146 ***	−0.00251
			(0.0307)	(0.0223)			(0.0509)	(0.0266)
stemp					−0.0185 ***	−0.0337 ***	0.00104	−0.0373 ***
					(0.00691)	(0.00695)	(0.0123)	(0.00831)
wtemp					0.0142 ***	0.00416	0.0264 ***	0.00360
					(0.00432)	(0.00558)	(0.00594)	(0.00652)
Constant		1.231 ***		1.195 ***		2.328 ***		2.352 ***
		(0.244)		(0.251)		(0.286)		(0.287)
Observations	648	684	648	684	648	684	648	684
Sample	36	36	36	36	36	36	36	36
AR(1)	0.273	0.234	0.371	0.234	0.270	0.232	0.345	0.231
AR(2)	0.164	0.149	0.226	0.145	0.192	0.154	0.953	0.160
Hansen test	0.192	0.212	0.349	0.176	0.133	0.106	0.0967	0.0597

Note: Robust standard errors in parentheses, *** $p < 0.01$, ** $p < 0.05$.

Compared with similar studies, it is confirmed that innovation input is positive to the renewable energy consumption of OECD countries, and the innovation input is negatively related to the renewable energy consumption of non-OECD countries, which may be due to the fact that compared to developed OECD countries that develop renewable energy technologies, innovation input in non-OECD countries is not reflected in the application of renewable energy. This may be due to the fact that innovation input in non-OECD countries is not reflected in the use of renewable energy compared to R&D in developed OECD countries. Finally, in the results of the estimation of the effect of climate variables on the consumption of renewable energy, the estimates for the OECD and non-OECD country samples are largely consistent, with renewable energy consumption being negatively correlated with summer temperatures and positively correlated with winter temperatures.

4.4. The Results of the Pollution Function

Tables 8 and 9 show the results of the pollution function, these instrumental variables appear to be effective, because the number of instrumental variables is less than that of these countries. According to the model estimation results, first, the estimated results of green economic growth and its squared term coefficient are basically the same in both OECD and non-OECD country samples, both show positive primary squared term coefficient and negative squared term coefficient, indicating an inverted U-shaped relationship between green economic growth and greenhouse gas emissions, which also confirms that renewable energy consumption plays an important role in reducing greenhouse gas emissions. Second, renewable energy consumption is significantly and negatively correlated with greenhouse gas emissions, a result that is undoubtedly consistent with the objective rule.

Table 8. The pollution function (OECD).

Variables	Group (1)	Group (2)	Group (3)	Group (4)	Group (5)	Group (6)	Group (7)	Group (8)	Group (9)	Group (10)
	Diff-GMM	Sys-GMM	Diff-GMM	Sys-GMM	Diff-GMM	Sys-GMM	Diff-GMM	Sys-GMM	Diff-GMM	Sys-GMM
L.pol	0.486 ***	0.932 ***	0.396 ***	0.929 ***	0.434 ***	0.939 ***	0.337 ***	0.933 ***	0.339 ***	0.949 ***
	(0.0190)	(0.0123)	(0.0231)	(0.0212)	(0.0278)	(0.00804)	(0.0382)	(0.00834)	(0.0407)	(0.0121)
geg	0.246 ***	0.0218	0.666 ***	0.186 ***	0.263 ***	0.142 ***	0.577 ***	0.202 ***	0.523 ***	0.170 ***
	(0.0293)	(0.0390)	(0.0777)	(0.0488)	(0.0310)	(0.0283)	(0.0789)	(0.0462)	(0.0971)	(0.0522)
geg2	−0.0591 ***	−0.0283 **	−0.173 ***	−0.0754 ***	−0.0640 ***	−0.0582 ***	−0.146 ***	−0.0753 ***	−0.126 ***	−0.0727 ***
	(0.00847)	(0.0131)	(0.0239)	(0.0165)	(0.00909)	(0.0111)	(0.0277)	(0.0159)	(0.0356)	(0.0138)
e	−0.0750 ***	−0.0211 ***	−0.0844 ***	−0.00393	−0.0864 ***	−0.0114 ***	−0.0873 ***	−0.0115 ***	−0.0772 ***	−0.000620
	(0.00550)	(0.00226)	(0.00746)	(0.00409)	(0.00914)	(0.00208)	(0.0107)	(0.00270)	(0.0105)	(0.00468)
ino			−0.0715 ***	−0.0288 ***			−0.0462 **	−0.00687	−0.0503 **	−0.00740
			(0.0205)	(0.00779)			(0.0208)	(0.00558)	(0.0223)	(0.00756)
stemp					−0.00797 ***	−0.00295 ***	−0.00763 ***	−0.00273 ***	−0.00844 ***	−0.00160
					(0.000931)	(0.000890)	(0.00110)	(0.000725)	(0.00118)	(0.00104)
wtemp					0.000670	0.00482 ***	0.000523	0.00544 ***	0.000498	0.00389 ***
					(0.000535)	(0.000776)	(0.000819)	(0.000740)	(0.000924)	(0.000853)
urb									0.0255 **	0.0282 ***
									(0.0112)	(0.00645)
poli										0.0105 **
										(0.00415)
Constant		0.847 ***		0.720 ***		0.684 ***		0.700 ***		0.374 ***
		(0.139)		(0.230)		(0.0934)		(0.0829)		(0.124)
Observations	630	665	630	665	630	665	630	665	630	665
Sample	35	35	35	35	35	35	35	35	35	35
AR(1)	0.000294	0.000137	0.000385	0.000123	0.000526	0.000150	0.00126	0.000158	0.00188	0.000112
AR(2)	0.597	0.474	0.927	0.522	0.639	0.281	0.939	0.261	0.976	0.392
Hansen test	0.977	0.995	0.985	0.989	0.980	0.998	0.993	0.998	0.993	0.997

Note: Robust standard errors in parentheses, *** $p < 0.01$, ** $p < 0.05$.

Table 9. The pollution function (non-OECD).

Variables	Group (1)	Group (2)	Group (3)	Group (4)	Group (5)	Group (6)	Group (7)	Group (8)	Group (9)	Group (10)
	Diff-GMM	Sys-GMM	Diff-GMM	Sys-GMM	Diff-GMM	Sys-GMM	Diff-GMM	Sys-GMM	Diff-GMM	Sys-GMM
L.pol	0.928 ***	1.009 ***	0.914 ***	0.980 ***	0.923 ***	0.995 ***	0.913 ***	1.002 ***	0.785 ***	0.986 ***
	(0.00832)	(0.00982)	(0.0149)	(0.0109)	(0.00801)	(0.00855)	(0.0226)	(0.00978)	(0.0334)	(0.0155)
geg	0.0553 ***	−0.113 ***	0.0641 ***	−0.0139	0.0687 ***	−0.0865 ***	0.0822 ***	−0.0356 **	0.0755 ***	−0.0398 ***
	(0.0137)	(0.0179)	(0.0205)	(0.0149)	(0.0155)	(0.0238)	(0.0270)	(0.0175)	(0.0231)	(0.0129)
geg2	−0.00377 ***	0.00968 ***	−0.00447 **	0.000539	−0.00492 ***	0.00721 ***	−0.00611 ***	0.00294 *	−0.00585 ***	0.00317 ***
	(0.00129)	(0.00176)	(0.00185)	(0.00145)	(0.00147)	(0.00210)	(0.00210)	(0.00158)	(0.00178)	(0.00111)
e	−0.0403 ***	−0.0143 ***	−0.0421 ***	−0.0384 ***	−0.0390 ***	−0.0273 ***	−0.0408 ***	−0.0324 ***	−0.0325 ***	−0.0351 ***
	(0.00229)	(0.00160)	(0.00281)	(0.00360)	(0.00328)	(0.00547)	(0.00238)	(0.00464)	(0.00448)	(0.00574)
ino			0.00475 **	−0.00849 **			0.000567	−0.0100 ***	0.0105	0.00264
			(0.00214)	(0.00368)			(0.00572)	(0.00318)	(0.00685)	(0.00542)
stemp					0.00342 **	−0.00264 *	0.00331	−0.00414 ***	0.000712	−0.00359 **
					(0.00173)	(0.00159)	(0.00220)	(0.000676)	(0.00174)	(0.00143)
wtemp					−0.00223 **	0.00148 ***	−0.000910	0.00217 ***	−0.00149	0.00163 ***
					(0.000971)	(0.000520)	(0.00195)	(0.000287)	(0.00151)	(0.000352)
urb									0.356 ***	−0.0749 ***
									(0.0519)	(0.0281)
poli										0.0121 **
Constant		0.0237		0.342 ***		0.243 ***		0.161		0.655 ***
		(0.103)		(0.129)		(0.0891)		(0.117)		(0.169)
Observations	648	684	648	684	648	684	648	684	648	684
Sample	36	36	36	36	36	36	36	36	36	36
AR(1)	0.000407	0.000177	0.000378	0.000353	0.000253	0.000351	0.000283	0.000522	0.000548	0.000468
AR(2)	0.425	0.482	0.420	0.419	0.341	0.526	0.350	0.551	0.382	0.561
Hansen test	0.971	0.988	0.965	0.997	0.965	0.997	0.967	0.997	0.966	0.997

Note: Robust standard errors in parentheses, *** $p < 0.01$, ** $p < 0.05$, * $p < 0.1$.

In addition, innovation input is also negatively correlated with greenhouse gas emissions, suggesting that an increase in innovation input can significantly improve the utilization efficiency of fossil energy and increase the proportion of renewable energy consumption. This is also an important determinant of the reduction of greenhouse gas emissions, such as the stronger the innovation input, the lower the greenhouse gas emissions. Finally, the temperature variable has a significant impact on greenhouse gas emissions, because climate change inevitably increases the consumption of energy for temperature regulation.

4.5. The Heterogeneity of Sample Interval

To analyze the dynamic relationship between climate change, innovation input, and energy-environment-growth in terms of differences between sample zones, the sample was divided into three phases, as shown in Tables 10 and 11. From a green economy growth perspective, innovation investment has a significant contribution to green economy development in both the OECD and non-OECD country samples, and the intensity of the effect decreases gradually depending on the short, medium, and long term of the period. This also suggests that for a green economy to be sustainable, countries need to invest in innovation in the long term. From the perspective of energy consumption, the contribution of innovative input to renewable energy consumption is also significant, with the intensity of the contribution decreasing in the short, medium, and long term.

Table 10. Differences of sample interval (OECD).

Sample Interval	Variables	Production		Energy Consumption		Pollution	
		Diff-GMM	Sys-GMM	Diff-GMM	Sys-GMM	Diff-GMM	Sys-GMM
2015–2019	ino	1.190 ***	−0.0164	0.402 *	0.447 **	−0.582 **	0.0229
		(0.370)	(0.0182)	(0.244)	(0.208)	(0.582)	(0.0154)
	stemp Wt			0.0108	0.0149	0.0173 *	−0.00528
				(0.0126)	(0.0143)	(0.0103)	(0.00387)
	wtemp			−0.00319	−0.00367	0.000773	−0.00652 **
				(0.00414)	(0.00483)	(0.00647)	(0.00264)
2010–2019	ino	0.486 **	−0.00290	0.121	−0.0371	−0.186 **	−0.00415
		(0.197)	(0.0190)	(0.245)	(0.152)	(0.0759)	(0.00392)
	stemp Wt			−0.00888	−0.00320	0.00403	0.00378 ***
				(0.0155)	(0.0128)	(0.00409)	(0.00102)
	wtemp			0.0120 **	0.0171 ***	−0.00161	−0.0021 ***
				(0.00561)	(0.00566)	(0.00115)	(0.00042)
2000–2019	ino	0.00769 *	0.0184 ***	0.186 ***	0.108 **	−0.0503 **	−0.00740
		(0.00412)	(0.00166)	(0.0696)	(0.0500)	(0.0223)	(0.00756)
	stemp Wt			−0.00251	−0.00668 *	−0.00844 ***	−0.00160
				(0.00445)	(0.00401)	(0.00118)	(0.00104)
	wtemp			0.00399	0.000999	0.000498	0.00389 ***
				(0.00296)	(0.00237)	(0.000924)	(0.000853)

Note: Robust standard errors in parentheses, *** $p < 0.01$, ** $p < 0.05$, * $p < 0.1$.

Table 11. Differences of sample interval (non-OECD).

Sample Interval	Variables	Production		Energy Consumption		Pollution	
		Diff-GMM	Sys-GMM	Diff-GMM	Sys-GMM	Diff-GMM	Sys-GMM
2015–2019	ino	1.014 *	−0.0358	0.0137 *	−0.00337	0.0131	−0.0206 *
		(1.181)	(0.0355)	(0.0786)	(0.0546)	(0.0602)	(0.0117)
	stemp Wt			−0.00278	0.0103	−0.00128	−0.000340
				(0.0137)	(0.00646)	(0.00662)	(0.00331)
	wtemp			−0.0180	−0.00542	-8.57×10^{-5}	0.00107
				(0.0118)	(0.00447)	(0.0100)	(0.00152)
2010–2019	ino	0.563 ***	0.0882 ***	0.0304	0.0838 *	−0.0961 ***	−0.00994 **
		(0.0976)	(0.0200)	(0.126)	(0.0745)	(0.0260)	(0.00431)
	stemp Wt			−0.0270	−0.0318 *	0.00655 *	0.000566
				(0.0260)	(0.0167)	(0.00362)	(0.000621)
	wtemp			0.0504 **	0.0394 ***	−0.00647 ***	2.57×10^{-5}
				(0.0207)	(0.0116)	(0.00189)	(0.000346)
2000–2019	ino	0.124 ***	0.0306 ***	−0.146 ***	−0.00251	0.0105	0.00264
		(0.00298)	(0.00391)	(0.0509)	(0.0266)	(0.00685)	(0.00542)
	stemp Wt			0.00104	−0.0373 ***	0.000712	−0.00359 **
				(0.0123)	(0.00831)	(0.00174)	(0.00143)
	wtemp			0.0264 ***	0.00360	−0.00149	0.00163 ***
				(0.00594)	(0.00652)	(0.00151)	(0.000352)

Note: Robust standard errors in parentheses, *** $p < 0.01$, ** $p < 0.05$, * $p < 0.1$.

In addition, the impact of climate change factors represented by temperature on renewable energy consumption is not significant in the short term and only begins to be significant in the medium to long term, which may make it possible that because climate change is not evident to people in the short term, it is often experienced over a long period of time before significant climate change is manifested, which in turn affects people's energy consumption activities. From the perspective of greenhouse gas emissions, innovation input has a dampening effect on greenhouse gas emissions in OECD countries that diminishes over time. In non-OECD countries, on the other hand, innovation input has a significant dampening effect only in the medium term. The effect of climate variables on GHG emissions in different countries is weaker in the short term than in the medium and long term.

In summary, compared with similar studies, it is confirmed that innovation input and climate change are important variables affecting renewable energy consumption, greenhouse gas emissions, and green economic growth. In contrast to other literature, this paper finds that the effects of industrialization and innovation inputs on energy consumption are significantly different in OECD countries. This should be because OECD countries are mostly developed countries, while non-OECD countries are developing countries, and being an OECD country or not indicates being at different stages of economic development. The different development realities and needs lead to different effects of industrialization and innovation inputs. In addition, the paper finds that the intensity of the impact of innovation inputs on economic growth, energy consumption, and pollution emissions varies over time, which indicates the time lag in the application of technological innovations generated by innovation inputs, and on the other hand the fact that new technology will eventually fall behind over time, which is the reason for the need to constantly innovate inputs and technological innovations.

5. Conclusions and Policy Recommendations

The results of this paper show that: firstly, both renewable energy consumption and innovation inputs have a significant impact on green economic growth, and the impact of innovation inputs is stronger in non-OECD countries. Secondly, green economic development, industrialization, and innovation inputs all boost renewable energy consumption in OECD countries, while the opposite is true for non-OECD countries. Third, while climate change increases energy consumption, renewable energy consumption significantly reduces greenhouse gas emissions in both OECD and non-OECD countries, especially for OECD countries with high renewable energy consumption and high energy efficiency. Fourth, innovation inputs contribute to green economic growth in both OECD and non-OECD countries. Innovative inputs have significantly increased renewable energy consumption and reduced greenhouse gas emissions in OECD countries. Finally, innovation inputs have a large impact on the energy-environment-growth nexus in the short term, while the impact is more significant in the medium to long term. At the same time, the impact of climate change on the energy-environment nexus in OECD and non-OECD countries is more significant in the medium to long term.

Based on the above empirical results, the policy implications are as follows:

(1) Renewable energy consumption promotes green economic growth and vice versa. Therefore, OECD and non-OECD countries should speed up the transformation and upgrading, increase the proportion of new and renewable energy sources, promote the low carbonization of the energy system, fully develop and utilize renewable energy such as solar energy, thermal energy, wind energy, biofuels and nuclear energy, and build an efficient and clean energy consumption system. In addition, OECD countries can also build an industrial chain system for energy storage and then to the application link, realize the coordinated development of the upstream, middle and downstream, and produce high-quality, high-tech, and high-value-added green products, to achieve green economic growth.

(2) Renewable energy consumption reduces greenhouse gas emissions caused by climate change. Therefore, OECD and non-OECD countries should give priority to promoting the development of renewable energy and adjusting the energy structure, gradually increasing the proportion of non-fossil energy consumption, and accelerating the construction of a clean, low-carbon, safe, and efficient energy system. At the same time, non-OECD countries should also strengthen the macro policy guidance and legal

protection functions related to renewable energy development and set sustainable development goals and strategic ideas.

(3) Innovation investment promotes green economy growth in OECD and non-OECD countries. Therefore, OECD and non-OECD countries should pay close attention to the iterative trend of global renewable energy technologies and increase financial support for renewable energy technology R&D, increase support for energy conservation and emission reduction technology R&D through financial allocations, tax exemptions, simplified administrative approvals, and scientific and technological innovation incentives, support the development of high-tech industries, establish and improve the energy conservation and emission reduction technology industrial system, thus achieving continuous innovation and development of technology. In addition, OECD and non-OECD countries should also strengthen the guiding role of financial funds in technology research and development, enrich the construction of research and development mechanisms, open up production, learning, and research channels, and attach importance to the long-term applicability and social effects of renewable technology selection and deployment.

(4) The impact of innovation input on the energy-environment-growth nexus is greater in the short term and more significant in the medium and long term. Therefore, OECD and non-OECD countries should set up a long-term renewable energy development strategy, clarify the long-term goals of renewable energy development and make long-term arrangements for the research and development of key renewable energy technologies such as solar and nuclear energy. At the same time, OECD countries should increase innovation investment, pay attention to renewable energy talent training, research and development, and industrial system construction, establish specialized R&D institutions, support the development of renewable energy scientific research, technology development, and industrial services, and train generations of talents with innovative consciousness, core technologies, and challenging spirit, thus promoting the long-term development of renewable energy technology progress and industrialization.

(5) The impact of climate change on the energy-environment nexus is more significant in the medium and long term. Therefore, OECD and non-OECD countries should formulate long-term sustainable policies to cope with climate change, closely follow the implementation of policy objectives, and give flexibility to dynamic adjustment of policies. OECD countries should pay attention to the binding role of laws, timely study and launch climate law, and "legalize" the medium and long-term emission reduction targets. At the same time, OECD and non-OECD countries should strengthen practical cooperation with countries along the belt and road in the fields of green production capacity cooperation and green financial standards, to build a fair and reasonable global climate governance system with win-win cooperation.

Author Contributions: Conceptualization, Z.L.; Methodology, Z.L.; Formal analysis, T.S. and Y.Y.; Investigation, H.H.C.; Data curation, T.S.; Writing—review & editing, Z.L. and Y.Y.; Project administration, T.S. All authors have read and agreed to the published version of the manuscript.

Funding: This research was funded by [Macau Polytechnic University] grant number [RP/FCHS-01/2022]. And The APC was funded by [Macau Polytechnic University].

Data Availability Statement: The data presented in this study are available on request from the corresponding author. The data are not publicly available due to privacy or ethical restrictions.

Conflicts of Interest: The authors declare no conflict of interest.

References

1. Xu, X.-l.; Chen, H.H. Examining the efficiency of biomass energy: Evidence from the Chinese recycling industry. *Energy Policy* **2018**, *119*, 77–86. [CrossRef]
2. Mauree, D.; Naboni, E.; Coccolo, S.; Perera, A.T.D.; Nik, V.M.; Scartezzini, J.-L. A review of assessment methods for the urban environment and its energy sustainability to guarantee climate adaptation of future cities. *Renew. Sustain. Energy Rev.* **2019**, *112*, 733–746. [CrossRef]
3. Reder, A.; Iturbide, M.; Herrera, S.; Rianna, G.; Mercogliano, P.; Gutiérrez, J.M. Assessing variations of extreme indices inducing weather-hazards on critical infrastructures over Europe—The INTACT framework. *Clim. Change* **2018**, *148*, 123–138. [CrossRef]
4. Creti, A.; Nguyen, D.K. Energy, Climate and Environment: Policies and International Coordination. *Energy Policy* **2020**, *140*, 111226. [CrossRef]
5. Zhang, H. Antinomic policy-making under the fragmented authoritarianism: Regulating China's electricity sector through the energy-climate-environment dimension. *Energy Policy* **2019**, *128*, 162–169. [CrossRef]
6. Ulucak, R.; Danish, O.B. Relationship between energy consumption and environmental sustainability in OECD countries: The role of natural resources rents. *Resour. Policy* **2020**, *69*, 101803. [CrossRef]
7. Sanquist, T.F.; Orr, H.; Shui, B.; Bittner, A.C. Lifestyle factors in U.S. residential electricity consumption. *Energy Policy* **2012**, *42*, 354–364. [CrossRef]
8. Auffhammer, M. Climate Adaptive Response Estimation: Short and long run impacts of climate change on residential electricity and natural gas consumption. *J. Environ. Econ. Manag.* **2022**, *114*, 102669. [CrossRef]
9. Arminen, H.; Menegaki, A.N. Corruption, climate and the energy-environment-growth nexus. *Energy Econ.* **2019**, *80*, 621–634. [CrossRef]
10. Bashir, M.F.; Ma, B.; Bashir, M.A.; Bilal; Shahzad, L. Scientific data-driven evaluation of academic publications on environmental Kuznets curve. *Environ. Sci. Pollut. Res. Int.* **2021**, *28*, 16982–16999. [CrossRef]
11. Chen, Y.; Lee, C.-C. Does technological innovation reduce CO_2 emissions? Cross-country evidence. *J. Clean. Prod.* **2020**, *263*, 121550. [CrossRef]
12. Auffhammer, M.; Mansur, E.T. Measuring climatic impacts on energy consumption: A review of the empirical literature. *Energy Econ.* **2014**, *46*, 522–530. [CrossRef]
13. Omri, A.; Ben Mabrouk, N.; Sassi-Tmar, A. Modeling the causal linkages between nuclear energy, renewable energy and economic growth in developed and developing countries. *Renew. Sustain. Energy Rev.* **2015**, *42*, 1012–1022. [CrossRef]
14. Zhang, M.; Zhang, K.; Hu, W.; Zhu, B.; Wang, P.; Wei, Y.-M. Exploring the climatic impacts on residential electricity consumption in Jiangsu, China. *Energy Policy* **2020**, *140*, 111398. [CrossRef]
15. Omri, A. CO_2 emissions, energy consumption and economic growth nexus in MENA countries: Evidence from simultaneous equations models. *Energy Econ.* **2013**, *40*, 657–664. [CrossRef]
16. Khan, S.; Peng, Z.; Li, Y. Energy consumption, environmental degradation, economic growth and financial development in globe: Dynamic simultaneous equations panel analysis. *Energy Rep.* **2019**, *5*, 1089–1102. [CrossRef]
17. Omri, A.; Kahouli, B. Causal relationships between energy consumption, foreign direct investment and economic growth: Fresh evidence from dynamic simultaneous-equations models. *Energy Policy* **2014**, *67*, 913–922. [CrossRef]
18. Adewuyi, A.O.; Awodumi, O.B. Biomass energy consumption, economic growth and carbon emissions: Fresh evidence from West Africa using a simultaneous equation model. *Energy* **2017**, *119*, 453–471. [CrossRef]
19. Granger, C.W.J. Investigating causal relations by econometric models and cross-spectral methods. *Econom. J. Econom. Soc.* **1969**, *37*, 424–438. [CrossRef]
20. Stern, D.I. The Rise and Fall of the Environmental Kuznets Curve. *World Dev.* **2004**, *32*, 1419–1439. [CrossRef]
21. Naseem, S.; Guang Ji, T. A system-GMM approach to examine the renewable energy consumption, agriculture and economic growth's impact on CO_2 emission in the SAARC region. *GeoJournal* **2020**, *86*, 2021–2033. [CrossRef]
22. Nasir, M.; Ur-Rehman, F. Environmental Kuznets Curve for carbon emissions in Pakistan: An empirical investigation. *Energy Policy* **2011**, *39*, 1857–1864. [CrossRef]
23. Saboori, B.; Sulaiman, J.; Mohd, S. Economic growth and CO_2 emissions in Malaysia: A cointegration analysis of the Environmental Kuznets Curve. *Energy Policy* **2012**, *51*, 184–191. [CrossRef]
24. Kraft, J.; Kraft, A. On the relationship between energy and GNP. *J. Energy Dev.* **1978**, *3*, 401–403.
25. Shahbaz, M.; Destek, M.A.; Okumus, I.; Sinha, A. An empirical note on comparison between resource abundance and resource dependence in resource abundant countries. *Resour. Policy* **2019**, *60*, 47–55. [CrossRef]
26. Khan, A.; Chenggang, Y.; Hussain, J.; Bano, S.; Nawaz, A. Natural resources, tourism development, and energy-growth-CO_2 emission nexus: A simultaneity modeling analysis of BRI countries. *Resour. Policy* **2020**, *68*, 101962. [CrossRef]
27. Wang, R.; Tan, J.; Yao, S. Are natural resources a blessing or a curse for economic development? The importance of energy innovations. *Resour. Policy* **2021**, *72*, 102042. [CrossRef]
28. Soytas, U.; Sari, R. Energy consumption, economic growth, and carbon emissions: Challenges faced by an EU candidate member. *Ecol. Econ.* **2009**, *68*, 1667–1675. [CrossRef]
29. Huang, B.-N.; Hwang, M.J.; Yang, C.W. Causal relationship between energy consumption and GDP growth revisited: A dynamic panel data approach. *Ecol. Econ.* **2008**, *67*, 41–54. [CrossRef]

30. Saidi, K.; Hammami, S. The impact of energy consumption and CO_2 emissions on economic growth: Fresh evidence from dynamic simultaneous-equations models. *Sustain. Cities Soc.* **2015**, *14*, 178–186. [CrossRef]
31. Lee, C.-C. The causality relationship between energy consumption and GDP in G-11 countries revisited. *Energy Policy* **2006**, *34*, 1086–1093. [CrossRef]
32. Apergis, N.; Payne, J.E. Energy consumption and economic growth: Evidence from the Commonwealth of Independent States. *Energy Econ.* **2009**, *31*, 641–647. [CrossRef]
33. Jalil, A.; Mahmud, S.F. Environment Kuznets curve for CO_2 emissions: A cointegration analysis for China. *Energy Policy* **2009**, *37*, 5167–5172. [CrossRef]
34. Waheed, R.; Sarwar, S.; Wei, C. The survey of economic growth, energy consumption and carbon emission. *Energy Rep.* **2019**, *5*, 1103–1115. [CrossRef]
35. Pandey, S.; Dogan, E.; Taskin, D. Production-based and consumption-based approaches for the energy-growth-environment nexus: Evidence from Asian countries. *Sustain. Prod. Consum.* **2020**, *23*, 274–281. [CrossRef]
36. Considine, T.J. The impacts of weather variations on energy demand and carbon emissions. *Resour. Energy Econ.* **2000**, *22*, 295–314. [CrossRef]
37. Jin, T.; Kim, J. Coal Consumption and Economic Growth: Panel Cointegration and Causality Evidence from OECD and Non-OECD Countries. *Sustainability* **2018**, *10*, 660. [CrossRef]
38. Chang, C.P.; Hao, Y. Environmental performance, corruption and economic growth: Global evidence using a new data set. *Appl. Econ.* **2017**, *49*, 498–514. [CrossRef]
39. Niu, J.; Chang, C.P.; Yang, X.Y.; Wang, J.S. The long-run relationships between energy efficiency and environmental performance: Global evidence. *Energy Environ.* **2017**, *28*, 706–724. [CrossRef]
40. Fankhauser, S.; Tol, R.S.J. On climate change and economic growth. *Resour. Energy Econ.* **2005**, *27*, 1–17. [CrossRef]
41. Zhao, X.; Gerety, M.; Kuminoff, N.V. Revisiting the temperature-economic growth relationship using global subnational data. *J. Environ. Manag.* **2018**, *223*, 537–544. [CrossRef] [PubMed]
42. Rezai, A.; Taylor, L.; Foley, D. Economic Growth, Income Distribution, and Climate Change. *Ecol. Econ.* **2018**, *146*, 164–172. [CrossRef]
43. Magazzino, C.; Mutascu, M.; Sarkodie, S.A.; Adedoyin, F.F.; Owusu, P.A. Heterogeneous effects of temperature and emissions on economic productivity across climate regimes. *Sci. Total Environ.* **2021**, *775*, 145893. [CrossRef]
44. Chen, W.H.; Lei, Y. The impacts of renewable energy and technological innovation on environment-energy-growth nexus: New evidence from a panel quantile regression. *Renew. Energy* **2018**, *123*, 1–14. [CrossRef]
45. Zakari, A.; Tawiah, V.; Oyewo, B.; Alvarado, R. The impact of corruption on green innovation: The case of OECD and non-OECD countries. *J. Environ. Plan. Manag.* **2022**, *65*, 1–33. [CrossRef]
46. Maneejuk, P.; Yamaka, W. An analysis of the impacts of telecommunications technology and innovation on economic growth. *Telecommun. Policy* **2020**, *44*, 102038. [CrossRef]
47. Assi, A.F.; Zhakanova Isiksal, A.; Tursoy, T. Renewable energy consumption, financial development, environmental pollution, and innovations in the ASEAN + 3 group: Evidence from (P-ARDL) model. *Renew. Energy* **2021**, *165*, 689–700. [CrossRef]
48. Thompson, M. Social capital, innovation and economic growth. *J. Behav. Exp. Econ.* **2018**, *73*, 46–52. [CrossRef]
49. Adak, M. Technological Progress, Innovation and Economic Growth; the Case of Turkey. *Procedia Soc. Behav. Sci.* **2015**, *195*, 776–782. [CrossRef]
50. Töbelmann, D.; Wendler, T. The impact of environmental innovation on carbon dioxide emissions. *J. Clean. Prod.* **2020**, *244*, 118787. [CrossRef]
51. Nguyen, T.T.; Pham, T.A.T.; Tram, H.T.X. Role of information and communication technologies and innovation in driving carbon emissions and economic growth in selected G-20 countries. *J. Environ. Manag.* **2020**, *261*, 110162. [CrossRef]
52. Erdogan, S.; Yildirim, S.; Yildirim, D.C.; Gedikli, A. The effects of innovation on sectoral carbon emissions: Evidence from G20 countries. *J. Environ. Manag.* **2020**, *267*, 110637. [CrossRef]
53. Carrión-Flores, C.E.; Innes, R. Environmental innovation and environmental performance. *J. Environ. Econ. Manag.* **2010**, *59*, 27–42. [CrossRef]
54. Sun, H.; Edziah, B.K.; Kporsu, A.K.; Sarkodie, S.A.; Taghizadeh-Hesary, F. Energy efficiency: The role of technological innovation and knowledge spillover. *Technol. Forecast. Soc. Change* **2021**, *167*, 120659. [CrossRef]
55. Wurlod, J.-D.; Noailly, J. The impact of green innovation on energy intensity: An empirical analysis for 14 industrial sectors in OECD countries. *Energy Econ.* **2018**, *71*, 47–61. [CrossRef]
56. Wang, D.; Tian, S.; Fang, L.; Xu, Y. A functional index model for dynamically evaluating China's energy security. *Energy Policy* **2020**, *147*, 111706. [CrossRef]
57. Xu, X.L.; Chen, Y.J. A comprehensive model to analyze straw recycling logistics costs for sustainable development: Evidence from biomass power generation. *Environ. Prog. Sustain. Energy* **2020**, *39*, e13394. [CrossRef]
58. Khan, I.; Hou, F.; Zakari, A.; Tawiah, V.K. The dynamic links among energy transitions, energy consumption, and sustainable economic growth: A novel framework for IEA countries. *Energy* **2021**, *222*, 119935. [CrossRef]
59. Zhou, G.; Zhu, J.; Luo, S. The impact of fintech innovation on green growth in China: Mediating effect of green finance. *Ecol. Econ.* **2022**, *193*, 107308. [CrossRef]

60. Xu, X.L.; Shen, T.; Zhang, X.; Chen, H.H. The role of innovation investment and executive incentive on financial sustainability in tech-capital-labor intensive energy company: Moderate effect. *Energy Rep.* **2020**, *6*, 2667–2675. [CrossRef]
61. Liu, S.J.; Xu, X.L. The pollution halo effect of technology spillover and pollution haven effect of economic growth in agricultural foreign trade: Two sides of the same coin? *Environ. Sci. Pollut. Res. Int.* **2021**, *28*, 20157–20173. [CrossRef] [PubMed]
62. Xu, X.L.; Chen, H.H.; Li, Y.; Chen, Q.X. The role of equity balance and executive stock ownership in the innovation efficiency of renewable energy enterprises. *J. Renew. Sustain. Energy* **2019**, *11*, 055901. [CrossRef]
63. Arellano, M.; Bond, S. Some tests of specification for panel data: Monte Carlo evidence and an application to employment equations. *Rev. Econ. Stud.* **1991**, *58*, 277–297. [CrossRef]
64. Arellano, M.; Bover, O. Another look at the instrumental variable estimation of error-components models. *J. Econom.* **1995**, *68*, 29–51. [CrossRef]
65. Blundell, R.; Bond, S. Initial conditions and moment restrictions in dynamic panel data models. *J. Econom.* **1998**, *87*, 115–143. [CrossRef]
66. Saidi, K.; Hammami, S. Economic growth, energy consumption and carbone dioxide emissions: Recent evidence from panel data analysis for 58 countries. *Qual. Quant.* **2015**, *50*, 361–383. [CrossRef]
67. Sekrafi, H.; Sghaier, A. Examining the Relationship Between Corruption, Economic Growth, Environmental Degradation, and Energy Consumption: A Panel Analysis in MENA Region. *J. Knowl. Econ.* **2016**, *9*, 963–979. [CrossRef]
68. Bhattacharya, M.; Awaworyi Churchill, S.; Paramati, S.R. The dynamic impact of renewable energy and institutions on economic output and CO_2 emissions across regions. *Renew. Energy* **2017**, *111*, 157–167. [CrossRef]
69. Farhadi, M.; Islam, M.R.; Moslehi, S. Economic Freedom and Productivity Growth in Resource-rich Economies. *World Dev.* **2015**, *72*, 109–126. [CrossRef]
70. Roodman, D. How to do xtabond2: An introduction to difference and system GMM in Stata. *Stata J.* **2009**, *9*, 86–136. [CrossRef]

energies

MDPI

Article

The Boundary of Porter Hypothesis: The Energy and Economic Impact of China's Carbon Neutrality Target in 2060

Shenhai Huang [1,2], Chao Du [1], Xian Jin [1], Daini Zhang [1], Shiyan Wen [3,*], Yu'an Wang [4], Zhenyu Cheng [5] and Zhijie Jia [6,*]

1 Jiaxing Hengchuang Electric Power Design & Institute Co., Ltd., Jiaxing 314100, China
2 School of Economics and Management, North China Electric Power University, Baoding 071003, China
3 School of Economics, Xi'an University of Finance and Economics, Xi'an 710003, China
4 School of Applied Economics, Renmin University of China, Beijing 100872, China
5 Shandong Academy of Social Sciences, Jinan 250000, China
6 School of Economics and Finance, Xi'an Jiaotong University, Xi'an 710049, China
* Correspondence: sywen_cn@163.com (S.W.); zjjia_cn@163.com or zhijie_jia@xjtu.edu.cn (Z.J.)

Abstract: The process of carbon neutrality does have economic costs; however, few studies have measured the cost and the economic neutral opportunities. This paper uses a dynamic computable general equilibrium (CGE) model to simulate China's carbon neutrality path from 2020 to 2060 and analyzes its economic impact. This paper innovatively adjusts the CGE modeling technology and simulates the boundary of the Porter hypothesis on the premise of economic neutrality. The results show that the carbon neutrality target may reduce the annual GDP growth rate by about 0.8% in 2020–2060. To make the carbon pricing method under the carbon neutrality framework meet the strong version of the Porter hypothesis (or economic neutrality), China must increase its annual total factor productivity by 0.56–0.57% in 2020–2060; this is hard to achieve. In addition, the study finds that China's 2030 carbon target has little impact on the economy, but the achievement of the 2060 carbon neutrality target will have a significant effect. Therefore, the paper believes that the key to carbon neutrality lies in the coexistence of technological innovation and carbon pricing to ensure that we can cope with global warming with the lowest cost and resistance.

Keywords: carbon neutrality; China; economic impact; computable general equilibrium model; carbon tax; carbon emission trading scheme

Citation: Huang, S.; Du, C.; Jin, X.; Zhang, D.; Wen, S.; Wang, Y.; Cheng, Z.; Jia, Z. The Boundary of Porter Hypothesis: The Energy and Economic Impact of China's Carbon Neutrality Target in 2060. *Energies* **2022**, *15*, 9162. https://doi.org/10.3390/en15239162

Academic Editors: Junpeng Zhu and Xinlong Xu

Received: 31 October 2022
Accepted: 1 December 2022
Published: 2 December 2022

1. Introduction

This paper studies the possible economic impact of China's carbon neutrality process and studies how many total factor productivity increases can make up for the economic losses; that is, under what conditions the strong version of the Porter hypothesis can be established. The strong version of the Porter hypothesis [1] means that reasonable and strict environmental regulation can stimulate enterprise innovation and hedge the costs caused by environmental regulation.

1.1. Background and Motivation

Many countries have announced their target goals on carbon neutrality. In the low-carbon development process, EU countries are at the forefront of the world both in technologies and management [2,3]. The EU proposes to achieve the goal of carbon neutrality by 2050, and China has announced the goal of carbon neutrality in 2060, indicating that nearly 1/3 of the world's emissions will go zero in 2060 [4,5]. However, no matter how we talk about carbon neutrality, we still need to consider the economics of carbon neutrality because there is indeed a strong trade-off between emission mitigation and economic development [6–8], regardless of many opposite hypotheses, such as the Porter hypothesis [9,10].

If the cost of emission mitigation is beyond expectation, governments may reduce their interest in reducing emissions.

Certainly, measuring the cost of emission mitigation, especially carbon neutrality, is essential for us human beings. The existing literature has estimated the marginal cost or efficiency cost of emission mitigation in many ways from different perspectives. Qin et al. (2019) [11] simulated the cost-effectiveness of China's green transition during the 12th five-year plan (2011–2015). Wang et al. (2016) [12] found that different regions have totally different abatement costs (measured by shadow price), and potential emissions (measured by the growth rates of emissions and economic outputs), highlighting the importance of specializing the carbon mitigation policies among the different regions. Cui et al. (2014) [13] applied a computable general equilibrium model to simulate the cost-saving effect of the emission trading scheme (ETS) and found that the carbon emissions trading only covered the pilots and that the unified carbon emissions trading market could reduce the total abatement costs by 4.50% and 23.67%.

However, only measuring the economic cost may not be enough. The economic cost varies by space and time. Using the directional distance function model, Wang et al. (2020) [14] measured the policy effects on CO_2 emission mitigation and abatement costs during the 13th Five-Year Plan (2016–2020) in China. They found high emission mitigation targets accompanied by high emission reduction costs. In the short term, the impact on different targets is not that obvious, but as time goes by, the effect increases. Uncertainties also affect the cost. Guo et al. (2019) [15] used a stochastic dynamic programming model to evaluate the impacts of uncertainties on the abatement planning process and found that uncertainties could increase the total abatement costs by around 5–7%.

Reducing the cost of the carbon mitigation policy is an essential topic of emission mitigation strategies [16]. Scholars have focused on the following methods: the clean development mechanism (CDM) [17], certified emission reduction (CER), carbon linkage, and low-carbon technologies. Wang et al. (2016) [18] analyzed the cost–benefit of waste-to-energy projects under China's clean development mechanism. They found that with or without the CDM, there is still a huge GHG reduction potential in solid waste management in China, which may reduce the cost of emission mitigation in China. Li et al. (2019) [19] found that the Chinese Certified Emission Reduction (CCER) scheme saves the national carbon trading system costs by applying a game theory. Zhang et al. (2019) [20] found that carbon linkage could reduce China's ETS pilots' carbon emission trading scheme's cost. Sun et al. (2018) [21] argued that most low-carbon technologies are cost-effective, with average annual cost savings of 71.43 billion CNY. Johansson et al. (2020) [22] found that the biofuels mandate in the United States reduced the emission reduction cost significantly, ranging up to 20 USD per ton. The Canadian case study finds a similar perspective [23].

Similar to the perspectives in the literature above [13], this paper also considers carbon pricing as a relatively cost-effective way to reduce emissions. Carbon pricing has proven effective in many regions and has been studied from many perspectives [24–27]. Among them, carbon emission trading schemes and carbon tax (CT) are two of the most popular mitigation strategies.

In 1990, the Netherlands began to levy a carbon tax: one of the earliest countries in the world to impose carbon taxes [28]. Sweden and Denmark also have strong and effective carbon tax policies [29–31]. The carbon tax policies of many countries, such as the United States, Australia, France, China, and Japan, are full of twists and turns. However, in most of these countries, a new kind of carbon pricing occurred: ETS. California seems to be the only state in the USA that has implemented a cap-and-trade scheme since 2013 [32]. EU-ETS is the first and the largest ETS in the world currently [33,34]. In 2010, the world's first city-level compulsory emission trading system was established in Tokyo, Japan. Then, Saitama Prefecture established the emission trading system in 2011. Saitama Prefecture's ETS is mainly a copy of the Tokyo ETS [35]. China's ETS pilot started in 2013 [36], and China's national ETS has already officially commenced on July 16, 2021. The revenue recycling scheme is one of the most concerning topics that may affect emission mitigation efficiency. For example, Liu and Lu (2015) [37] argued that the recycling scheme matters in the long-term effect of the carbon tax and the sectors' burden. Sun et al. (2021) [38] designed a recycling scheme to improve the emission mitigation effect and reduce the gross domestic product (GDP) loss.

Until now, there have been many studies focusing on the comparison of carbon pricing strategies [39]. However, it seems that there is little in the literature focusing on the economic losses of the strategies and no paper measuring how much of the productivity should be improved to neutralize the economic losses, or in other words, the boundary of the Porter hypothesis. It is the knowledge gap that this paper wants to fill. To fill the gap, the paper first measures the GDP loss of ETS and CT under carbon neutrality, then focuses on the changes by carbon neutrality and the boundary of the strong Porter hypothesis using CGE modeling technology.

1.2. Contributions and Paper Structure

Although existing papers focus on carbon neutrality, little in the literature studies the economic cost of carbon neutrality by applying carbon pricing [40–42]. Thus, this paper wants to fill the knowledge gap, for exploring the impact of carbon neutrality by emission trading and carbon tax from the perspective of GDP loss, energy structure, the compensation of the total factor productivity, and commodity price. The specific contributions and findings of the paper are shown below:

1. The paper finds that the cost of achieving carbon neutrality is in reducing the average annual growth rate in 2020–2060 by about 0.8%. The annual growth rate of the GDP will decrease from 1.2% to 1.8% in 2050–2060.
2. Carbon tax and carbon trading can significantly increase the share of renewable energy and make the energy system cleaner. Coal consumption in the counterfactual scenario will be cut in half compared with the benchmark, and the total energy demand will be reduced significantly because of the high actual energy prices.
3. If the whole society wants to make up for the loss of GDP, then in 2020–2060, society's average annual total factor productivity (TFP) must increase by 0.56–0.57% compared with the benchmark scenario. In other words, an additional 0.56–0.57% of the annual TFP growth could meet the strong Porter hypothesis.
4. The improvement of TFP can further stimulate the renewable energy structure and may reduce the producer's price of all kinds of goods. Therefore, technological progress may be the key to reducing the negative impact of achieving the carbon neutrality target.

The rest of the paper is organized as follows: Section 2 is the scenario design, which is the paper's exogenous assumptions. Section 3 introduces the paper's methodology, including the introduction of the computable general equilibrium (CGE) model used in the article, the dynamic method, and the data source. Section 4 presents the results of carbon neutrality impacts during 2020–2060 and discusses the results. Section 5 further assumes that the increase in TFP makes the process of carbon neutrality economically neutral (no impact on GDP), then presents and discusses the results. Section 6 discusses the key results found in the paper and compares them with other examples in the literature. Section 7 concludes the paper's results and proposes policy implications.

2. Scenario Design

The most important part of carbon neutrality is carbon emissions. This paper does not consider the impact of adding negative carbon emission technology at present because if negative carbon emission technology is to be added (such as forest carbon sequestration and carbon capture), more assumptions will be included in the model, which may affect the robustness of the model's conclusions.

We need to consider the remaining emissions by fossil energy consumption which are challenging to replace, such as air transportation, marine transportation, and energy chemicals, that will be captured by negative carbon emission technology. The paper calculates the emission share of these sectors from the total emissions in 2019 from the CEADs database (https://www.ceads.net.cn/, accessed on 1 November 2022) to be about 9%. Considering how other sectors may have positive emissions (increasing remaining emissions) and land transportation may be substituted by electric vehicles (reducing remaining emissions), the paper holds the assumption of 9% remaining. Thus, we have assumed a significant reduction in carbon emissions by 2060 caused by carbon neutrality measures such as ETS and CT, which is 91% lower than the baseline (BAU) scenario (Figure 1). Although negative carbon emission technology is not included in the model, we still need to assume the existence of negative carbon emission technology to ensure the reliability of the scenario simulation.

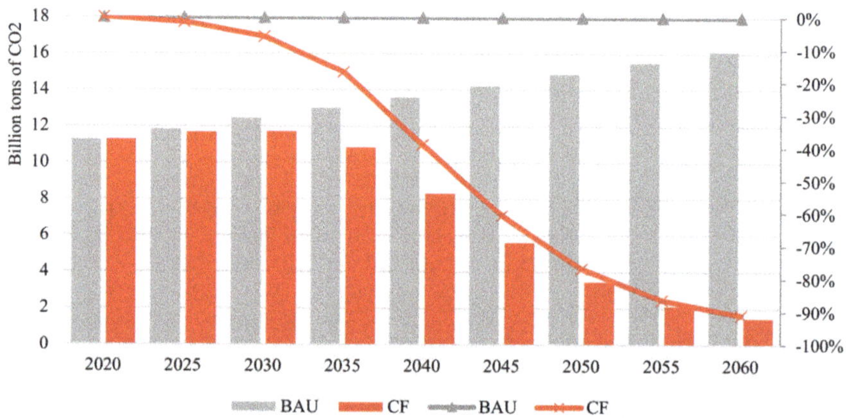

Figure 1. Energy-related CO_2 emissions during 2020-2060 in all scenarios. Notes: the bar chart shows the carbon emissions in 2020-2060 under different scenarios. The line chart shows the reduction rate of carbon emissions in the CF scenarios in each year compared with the BAU scenario. The CF scenarios include the ETS and the CT scenarios, which are illustrated in Table 1.

Table 1. The first scenario design.

Scenario Design	Descriptions
BAU	Assuming that there are no carbon pricing measures.
CT	Assuming that carbon tax is imposed in 2021 and full carbon tax coverage will be introduced in 2040.
ETS	Assuming that carbon emission trading is constructed in 2021 and full coverage of ETS will be introduced in 2040.

This paper assumes that China mainly uses carbon pricing to achieve the goal of carbon neutrality. The carbon tax is a long-discussed carbon pricing strategy for Chinese policymakers. Although a national carbon trading market was online in 2021, the current carbon trading market only covers the power generation industry, mainly because of the poor quality of data detection [43]. Therefore, a carbon tax may be a supplementary policy for China's carbon trading market or may even become the primary emission mitigation strategy.

In other words, at present, the carbon trading mechanism is relatively mainstream, and the positive stimulating effect of such a mechanism on enterprises seems to be more pronounced. Therefore, China may gradually improve the quality of monitoring, reporting, and verification (MRV) and cover more industries in the carbon trading system. Thus, ETS may also become the primary emission mitigation strategy in China.

Therefore, this paper first considers the construction of a CGE model based on the benchmark scenario, named the BAU scenario, and then constructs the CGE model based on the carbon tax or carbon trading scenario under the carbon neutrality framework and then compares the results of the carbon neutrality scenarios and the BAU scenario through similar analyzing methods for the experimental group and the control group. The scenario design is shown in Table 1. The paper assumes that the coverage of CT and ETS is in line with China's government plan; all the energy-intensive industries will be covered by carbon pricing [44]. Moreover, in 2040, carbon pricing will cover all kinds of enterprises.

3. Methodology

3.1. CGE Model

3.1.1. Why Do We Choose CGE Model?

Generally speaking, if we want to simulate an event that does not actually happen, and the event will lead to significant changes in the economic structure, the data-driven empirical evidence model will no longer be applicable, such as the econometric model (for example, the panel model, generalized method of moment model, time-series model), and machine learning (such as the BP neural network algorithm, genetic algorithm, and hybrid algorithm).

Therefore, when considering the simulation of 2060 carbon neutrality, we need to use the scenario analysis model, which is good at simulating counterfactual events. For example, the LEAP (long-range energy alternatives planning system) model, system dynamics, DSGE (dynamic stochastic general equilibrium) model, and the CGE model.

However, among these scenario analysis models, only the CGE model can describe the relationship between industries in detail because the CGE model is the model with the largest data demand (requiring the input-output table and other data, energy, and emission data of various departments), and it is also a model with relatively weak assumptions. The CGE model can simulate the behavior of maximizing the utility/profit of enterprises, residents, governments, and foreign manufacturers. Additionally, the model considers the mutual restriction relationship between different actors.

3.1.2. The Brief Introduction of CGE Model

The model is widely used to simulate various policies' macro impact [45–47]. This paper's CGE model is from the existing literature, and the exogenous parameters in the model are basically passed through several rounds of inspection [48–50], and the substitution elasticity is set based on a well-known CGE model [51,52]. The CGE model constructed in this paper includes more than 3200 endogenous variables and corresponding equations. It considers the behavior patterns of residents, enterprises, the government, and international firms. It is mainly based on a general equilibrium theory (advanced theory of game theory) and a large number of microeconomic theories (such as manufacturer behavior theory, resident consumption theory, etc). In order to couple the energy and environment block, we additionally considered the relevant theories of energy economics and environmental economics.

The applied model's name is the China Energy-Environment-Economy Analysis 2.0 (CEEEA) model, and it is a dynamic recursive model considering multi-sector and multi-households. The flow chart for establishing and simulating the CGE model is shown in Figure 2. It has five blocks:

1. Production block. This block describes the production behavior in all sectors. These behaviors are simulated by the constant elasticity of substitution (CES) production function considering the energy input, labor input, capital input, Leontief technology, and the intermediate inputs aggregation.
2. Trade block. The block expresses the import behavior of domestic consumers and the export behavior of domestic sectors. The former is simulated by the CES function, and the latter is simulated by constant elasticity of transformation (CET) technology.
3. Income and expenditure block. This block expresses the cash flow among four main economic entities: government, households, firms, and the foreign world.
4. Energy and environment block. The block describes the relationship between energy use in the energy balance table and energy input in the input-output table, the rela-

tionship between energy use and CO_2 emissions, and the carbon pricing strategies of the government.

5. Macroscopic closure and market-clearing block. This block is used to simulate the closure conditions and market-clearing assumptions of the whole economy. Based on the neoclassical macro-closure conditions, the model considers the clearing of commodity and factor markets and assumes that there is no factor redundancy or shortage.

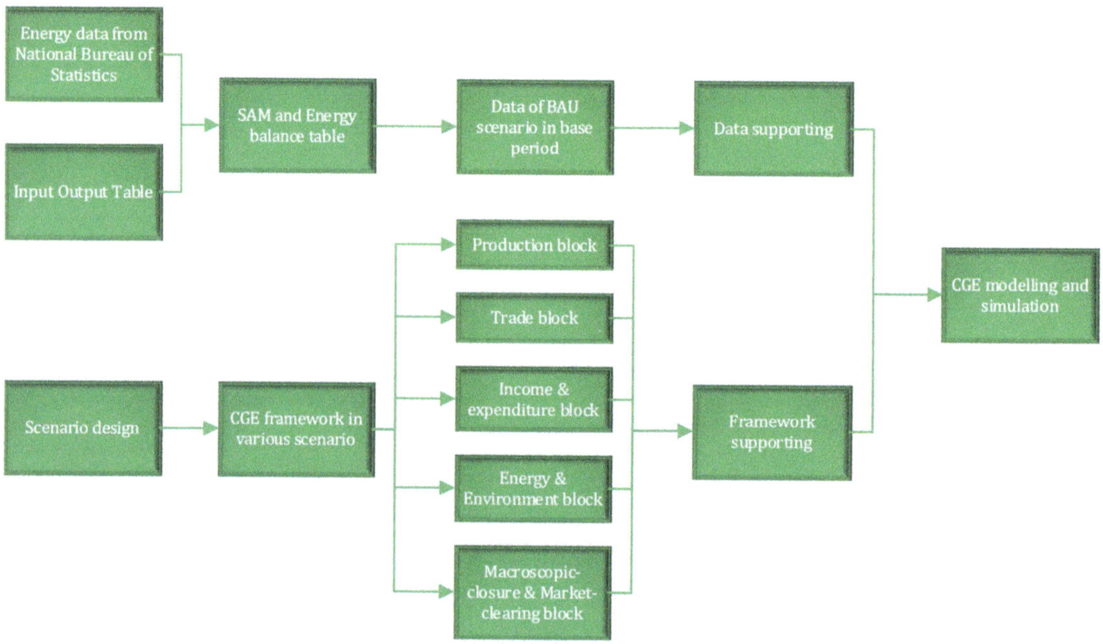

Figure 2. Flow chart of constructing and simulating the CGE model.

3.2. Dynamics

In this paper, the recursive method is used to make the CGE model dynamic. In general, the dynamic strategy under Neoclassical assumptions entails the growth of labor endowment, capital endowment, and technological progress. This paper makes fundamental assumptions about labor growth and calculates capital endowment using the perpetual inventory method. This paper simulates the total factor productivity (TFP) by setting exogenous GDP as an economic growth path and endogenous TFP first. Then, the paper uses TFP as the exogenous variable to simulate all scenarios, considering the steady technological progress.

Except for carbon pricing and leading to different carbon emission pathways, this paper assumes that all exogenous variables in the counterfactual (CF) scenario are the same as in the BAU scenario. The model additionally assumes the same total CO_2 emissions in the CT and ETS scenarios. The significance of this assumption is that it is assumed that the government's carbon pricing parameters are used to control the total amount of carbon emissions, and that the CT and ETS scenarios have the same carbon emissions making the scenario comparison more scientific.

3.3. Data Source

Most of the data in China's input-output table, the energy consumption data in China's energy statistical yearbook, and CO_2 emission data in the CEADs database are required in the model. Based on these data, the study compiles a social accounting matrix and energy balance table. The paper re-classifies the sectors, as presented in Table 2.

Table 2. Sector classification.

Abbreviation of the Sectors	Sectors
AGR	Primary industry
COL	Coal mining
COLP	Coal processing
O_G	Oil and gas exploitation
REFO	Refined oil
REFG	Refined gas
OMIN	Other mining's
LGT	Other mining industries
CMC	Chemicals
BMTL	Building material
STL	Steel
MTL_P	Metal products
MFT	Manufacturing
THP	Thermal power
HYP	Hydropower
WDP	Wind power
NCP	Nuclear power
SOP	Solar power
CST	Construction
TSPT	Transportation
SER	Services

At present, the latest input-output table has been updated to 2020, but we have not adopted it. The main reasons are: (1) The 2020 table is an extended table based on the 2017 table, and there may be a larger difference between the intermediate input value in the table and the actual situation. (2) The year 2020 witnessed the COVID-19 pandemic, so the data on transportation, tourism, and other industries cannot reflect the normal economic operation. Therefore, in order to ensure the reliability of the simulation, we conducted the simulation based on the table in 2018 (there was no input-output table in 2019). In addition, we believe that the epidemic will eventually pass and that society will gradually return to normal. Therefore, it is significantly better to simulate the relationship among industries, households, and government with data in 2018 than with data in 2020.

This paper obtains the physical quantity of energy consumption in various industries through China's energy statistical yearbook. However, the energy consumption data by industry in the China energy statistics yearbook is different from the industrial division in this paper, and the article integrates the sectors. For the sectors that need to be split, this paper separates them through the corresponding input-output coefficient.

Data on carbon emissions. Based on the calculated energy consumption data of re-classified sectors, IPCC's carbon emission calculation references and data, such as the average low calorific value, carbon content per unit calorific value, and carbon oxidation rate, are calculated in this paper through the carbon dioxide emission of these sectors.

Tax and resident income. The cash flow among the government, households, and firms is also simulated in this paper. So, factor income and direct tax are required and derived from the CEIC database (https://www.ceicdata.com/en, accessed on 1 November 2022).

It should be noted that if we have IOT from different regions and some relevant data, after sector adaptation and parameter calibration, the model can be used in any country with data support. However, due to the difference in the industry classification, productivity level, and trade relations in different countries, the results will be very different. Therefore, although the model in this paper can be applied to most countries, the research conclusion can only be considered unchanged.

4. Simulation Results

4.1. The Basic Situation in the BAU Scenario

The BAU scenario is the benchmark scenario of the paper; that is, almost all results are based on the comparison between the CT/ETS scenario and the BAU scenario. So, the paper needs to report the basic information about the BAU scenario first.

The CO_2 emissions in 2060 will be about 16.1 billion tons, increasing by 43% compared with 2020 emissions, which is illustrated in Figure 1. In terms of the primary energy structure, in the BAU scenario in 2060, China's primary fossil energy accounts for 32.0%, and the primary electricity (renewable energy) accounts for 68.0%. The energy structure is much better than the current situation, but there is still a big gap from in carbon neutrality goal.

Because this paper uses a long-term model, we do not consider using Keynesian macro closure conditions but neoclassical macro closure conditions. That is, the factor is fully utilized. Therefore, the main constraints of the whole model come from factor endowment. Thus, in the dynamic process, this paper considers the technological progress and changes in factor endowment (Section 3.2). The changes in capital and labor endowments and technological progress will lead to an increase in GDP. Without carbon constraints, the GDP in the BAU scenario will increase to 739 trillion CNY in 2060, with the primary sectors accounting for 9.3%, the secondary sectors accounting for 25.2%, and the tertiary sectors accounting for 65.5%. In the labor market, the labor population of the primary sectors accounts for 16.9%. Among them, the secondary sectors account for 16.0%, the tertiary sectors account for 65.5%, and the tertiary industry accounts for 67.1%.

4.2. Impacts on GDP

Figure 3 shows the impact of CT and ETs on the gross domestic product (GDP). The bar shows the GDP every five years, and the line shows the average annual growth rate of these five years. In the BAU scenario, China's GDP growth rate gradually decreases. The growth rate will be 5.5% in 2020–2025 but will reduce to 4.50% in 2055–2060. The BAU scenario's GDP settings are similar to several relevant studies. For instance, compared with Fang et al. (2015) [53], the path is pessimistic about the GDP growth from 2020 to 2040 but relatively optimistic from 2040 to 2060. However, in general, it is a fairly optimistic estimate compared with the other literature [54] because this paper considers that China is a developing country with a vast population and is implementing the rural revitalization strategy, so there is still much room for GDP growth in rural and backward areas.

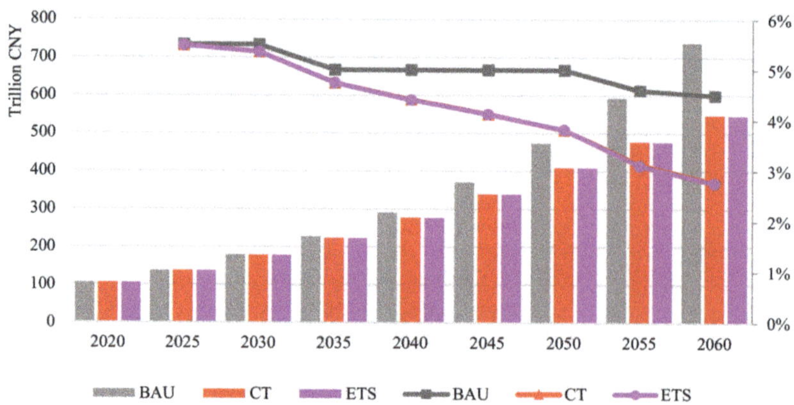

Figure 3. Impacts on GDP during 2020–2060.

Note that from 2021 to 2060, the average annual growth rate of the CT scenario was 4.23%, and in the ETS scenario, the average annual growth rate was 4.22%, indicating that the growth rate in the two counterfactual scenarios is similar to each other. As under the same emission mitigation path, there may be little differences between the economic impact of the full coverage carbon tax and carbon trading [39].

In the long run, it seems that there is no apparent difference in the impact of the carbon tax and carbon trading covering the same industry and emission mitigation on the total economy. In the early stage of the carbon neutrality process (2020–2040), the economic loss caused by the carbon pricing method will not be substantial. However, with the strict carbon emission reduction targets, economic losses gradually increased. From 2035 to 2040, the target of carbon neutrality will reduce economic growth by 0.6%; from 2055 to 2060, the economic growth will decrease by 1.7% in the final period of carbon neutrality. In summary, in the year 2020–2060, the average growth rate of the BAU scenario is about 5.0%, while the average growth rate under the carbon-neutral framework is around 4.2%: a decrease of about 0.8%.

4.3. Impacts on the Energy Mix

As renewable energy is the key to further energy supply [55], we need to focus on the energy mix in the long-term simulation. Figure 4 illustrates the impacts on the primary energy structure in 2060 and the energy mix in the 2020 BAU scenario. The electric power industry is the main source of carbon emissions in China, especially the coal-fired plant [56]. In the BAU scenario, the primary fossil energy has significantly reduced to 32.0% in 2060. A total of 21.0% of the primary energy is from coal consumption, and oil and gas account for 11.04%, while renewables account for 68.0%. Under China's current investment situation (dynamic investment preference of each scenario), China's renewable energy will significantly thrive and accounts for a large share. However, coal still accounts for about 1/5 of the primary power. Many examples in the literature also believe that the carbon pricing mechanism may increase the renewable energy share, consistent with relevant research [57,58].

In CT and ETS scenarios, coal consumption will be nearly cut in half in 2060. Moreover, the share of renewables will increase by 3.7–4.8%. The increase will be more significant in the ETS scenario by 4.8%. The percentage of oil and gas will increase by 6.0% and 4.8% in CT and ETS scenarios, respectively. However, the increasing share does not mean increasing consumption, as the total energy demand will significantly reduce under carbon neutrality scenarios. Under the carbon neutrality target, it seems that oil will be more difficult to remove than coal. The main reason for this may be that oil is more inclined to be used by the transportation industry and service industry, which is somehow irreplaceable, especially in air and water traffic.

4.4. Impacts on the Producer Price Index

Figure 5 depicts the impact on all sectors' producer price indexes in 2060. We found that energy-intensive industries are the most vulnerable sectors when achieving carbon neutrality targets. Energy processing sectors, such as the processing of coal, oil, and gas (COLP, REFO, and REFG), will be the first three most affected sectors. The results are similar to the relevant literature [59,60]. The prices in 2060 will increase by more than 300% compared with the 2060 BAU scenario. Steel and thermal power prices will increase by about 200% because coke (COLP sector) and raw coal (COL sector) are among the main upstream products of steel and thermal power. The rise in raw materials is a significant factor in the price rises of these sectors.

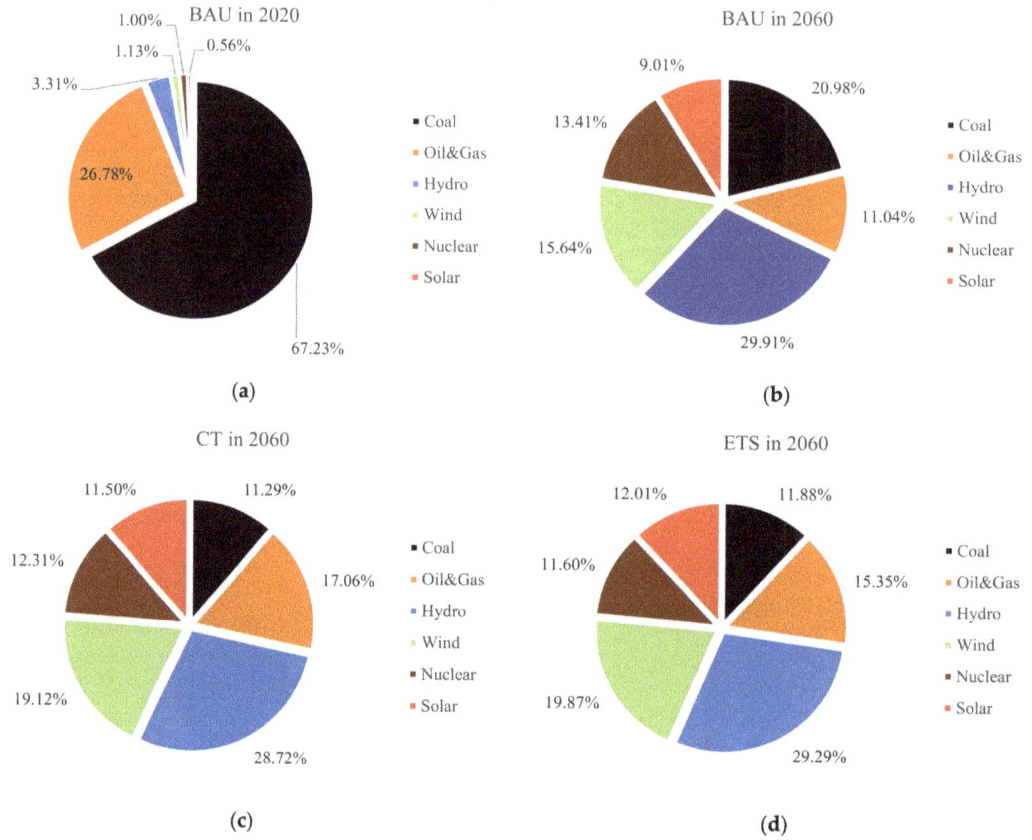

Figure 4. Impacts on energy structure in 2060. (**a**) BAU scenario in 2020; (**b**) BAU scenario in 2060; (**c**) CT scenario in 2060; (**d**) ETS scenario in 2060.

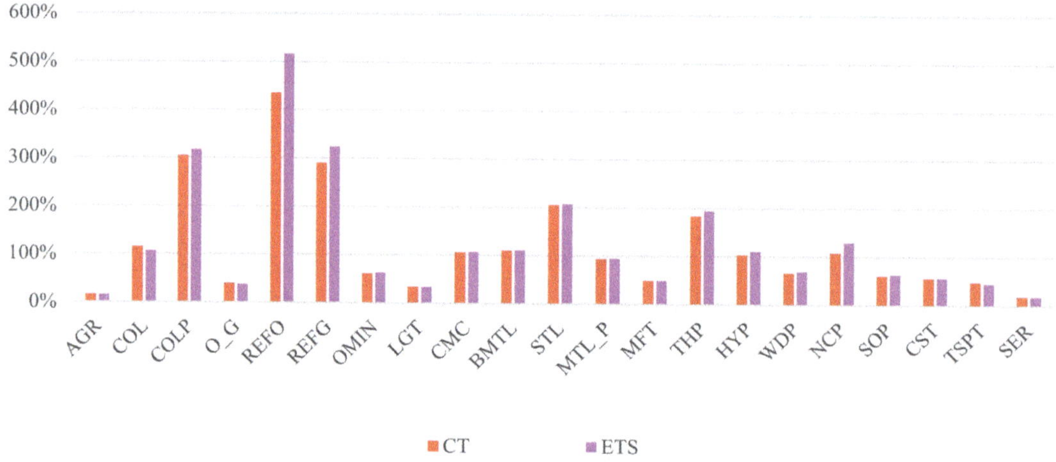

Figure 5. Impacts on producer price index in 2060.

In addition, we noticed that the price of low energy-intensive sectors, such as agriculture and services, may not be affected too much by the carbon neutrality target. The price of agriculture and services will increase by about 17.5% and 18.5%, respectively. The price rises in these industries may be caused by the combined effects of the price rise of other industries.

5. The Boundary of Porter Hypothesis under Carbon Neutrality

5.1. Scenario Design

The Porter hypothesis is an important issue related to carbon constraint. The weak version of the hypothesis describes how appropriate environmental regulations will stimulate technological innovation, while the strong version expresses that environmental regulation positively affects total factor productivity (TFP) or business performance by stimulating technological innovation. This paper focuses on TFP to test the boundary of the Porter hypothesis rather than green productivity, such as in many papers [61] because the definition of the strong version of the hypothesis is productivity.

The two types of carbon pricing models studied in this paper belong to environmental regulation. Therefore, the article wants to simulate the border of the Porter hypothesis in the CT and ETS scenarios. In the CGE model, the TFP of the sector is given exogenously. Therefore, the CGE model implies a critical assumption: the carbon pricing strategy will not affect the change in TFP. Thus, the model cannot use the CGE model to directly verify whether the Porter hypothesis is valid in a region. However, it can study the boundary of the Porter hypothesis through modeling technology: the changes in endogenous and exogenous variables. This section intends to discuss how much additional TFP is needed to increase and meet economic neutrality under carbon neutrality. In other words, what the study wants to know in this section is how much more TFP the enterprise needs to improve and meet the carbon neutrality target without reducing GDP.

Based on this idea, additional research and designs are carried out. We first add the endogenous TFP exchange rates to the scale factor in the CES production function in a value-added bundle and make GDP exogenous to be the same as BAU's GDP. Specifically, the modeling technology changes can be described in the CES production function and GDP calculation, as presented in Equations (1) and (2):

$$\mathrm{Y}_{it} = \mathrm{A}_{it}\left(\sum\nolimits_j \delta_{ij} Input_{ijt}^{\rho_i}\right)^{1/\rho_i} \tag{1}$$

$$\mathrm{GDP}_t = \sum\nolimits_i (XP_{it} + XG_{it} + XV_{it} + EX_{it} - IM_{it}) \tag{2}$$

where Y_{it} is the gross output in sector i and period t. The sector produces goods and services through the CES production function technology. A_{it} is the TFP in sector i and period t. δ_{ij} is the share parameter of input j in the production process by sector i, and ρ_i is the elasticity parameter. $Input_{ijt}$ is the total input of factor j in period t. GDP_t is the gross domestic product in period t, while XP_{it}, XG_{it}, XV_{it}, EX_{it}, and IM_{it} are household consumption, government consumption, investment, export, and import.

Usually, A_{it} is the exogenous variable and GDP_t is the endogenous variable in the model, which means that we performed the comparative analysis based on the same technology level, and we can analyze different external shocks on GDP or other endogenous variables. However, this section wants to explore the border of the strong version of the Porter hypothesis. So, the TFP should be the endogenous variable, and GDP should be controlled to be equal to BAU's GDP in other scenarios. Therefore, the paper changes the model by Equations (3) and (4):

$$\mathrm{Y}_{it}^{cf} = (1+\vartheta)\mathrm{A}_{it}\left(\sum\nolimits_j \delta_{ij} Input_{it\ ijt}^{cf\,\rho_i}\right)^{1/\rho_i} \tag{3}$$

$$\overline{\mathrm{GDP}_t} = \sum\nolimits_i \left(XP_{it}^{cf} + XG_{it}^{cf} + XV_{it}^{cf} + EX_{it}^{cf} - IM_{it}^{cf}\right) \tag{4}$$

where ϑ, which is an endogenous variable that catches the changes in TFP in the condition of the same GDP. The variables with superscript cf denote that they are endogenous variables in this section whose values are different from those in the previous section. $\overline{GDP_t}$ is an exogenous variable, which is the same in all scenarios in this section. Other settings are the same as in Section 4. The scenario design in this section is described in Table 3.

Table 3. The second scenario design.

Scenario Design	Descriptions
BAU	Assuming that there are no carbon pricing measures.
CT-TFP	Assuming that carbon tax is imposed in 2021 and full coverage of carbon tax will be introduced in 2040. The average TFP will be increased additionally to meet economic neutrality.
ETS-TFP	Assuming that carbon emission trading is constructed in 2021 and full coverage of ETS will be introduced in 2040. The average TFP will be increased additionally to meet economic neutrality.

5.2. Results

5.2.1. Additional TFP: Boundary of Strong Porter Hypothesis

Figure 6 depicts the additional TFP needed for economic neutrality during 2020–2060. The additional TFP required for carbon peak (2030) is not large, but after 2035, a higher TFP growth is needed every year to maintain GDP unchanged. For reaching the emission peak, only an additional 0.056% of total factor productivity per year is required to keep the GDP unchanged during 2020–2030. In contrast, 0.564–0.568% additional TFP is needed per year during 2020–2060 to keep the GDP unchanged for the carbon neutrality goal in 2060. It shows that the difficulty of carbon neutrality may be far more incredible than that of carbon peaking. To achieve the goal of carbon neutrality, we may need both policy guidance and technological change.

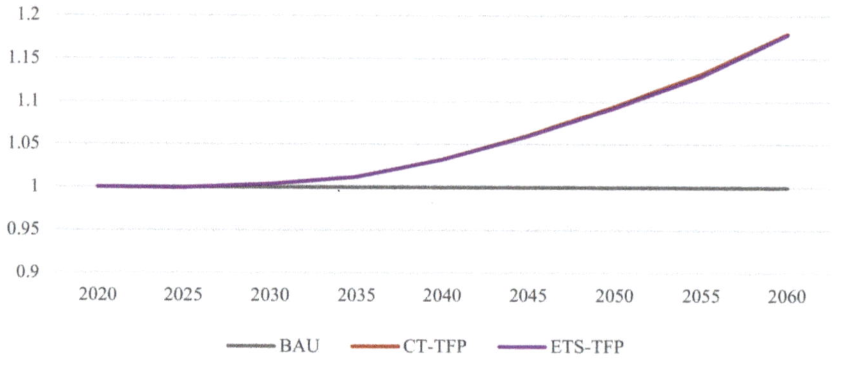

Figure 6. Additional TFP needed for economic neutrality during 2020–2060 (five-year smoothing index).

5.2.2. Impacts on the Energy Mix

Figure 7 illustrates the energy structure of all countermeasure scenarios. To better compare the impact of TFP improvement, we also put the CT scenario and the ETS scenario into Figure 7 for better comparative analysis. The results show that an increase in TFP will further increase the share of renewable energy. The share of renewables will increase by 3.7–3.9% in CT-TFP and ETS-TFP scenarios compared with that in CT and ETS scenarios. A possible reason is that technological progress leads to the increase in TFP, which leads to lower prices and higher demand. Simultaneously, due to the limitation of carbon emissions, more energy demand can only be satisfied by the growth of renewable energy. Thus, TFP growth may increase the share of renewable energy under the constraint of carbon emissions.

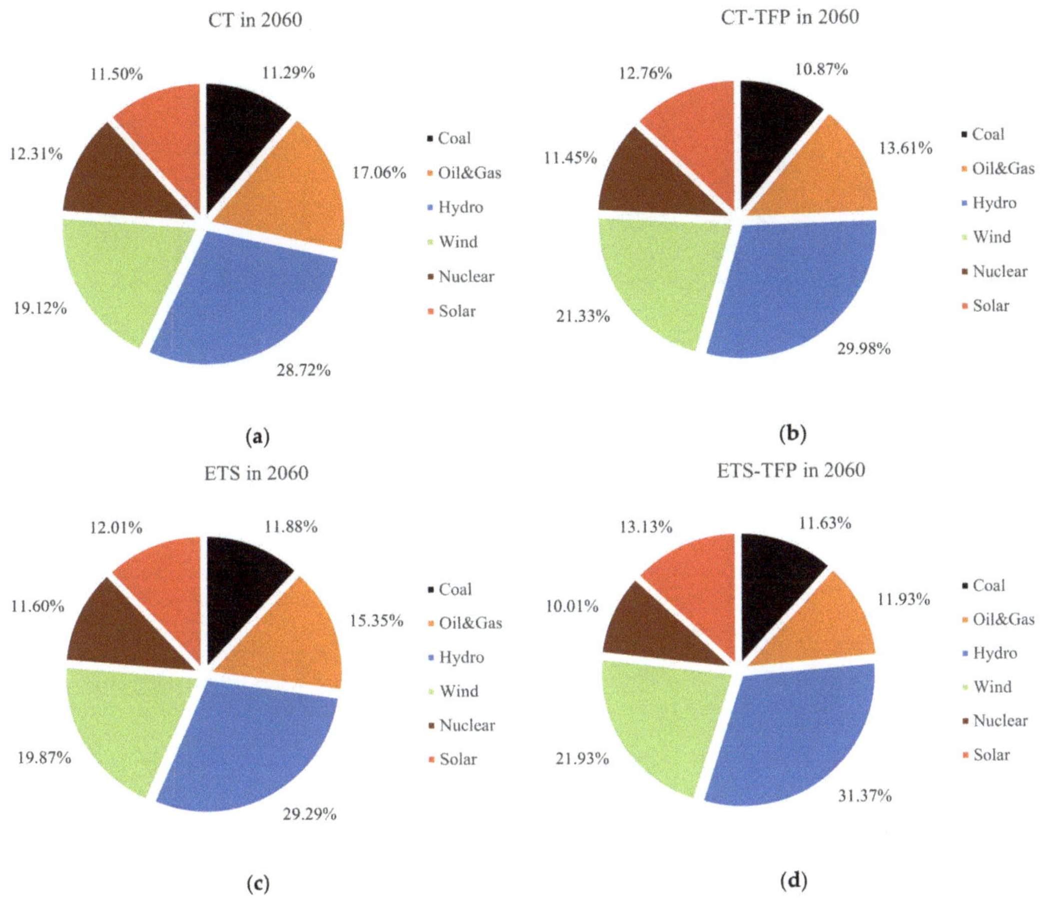

Figure 7. Impacts on energy structure in 2060. (**a**) CT scenario in 2060; (**b**) CT-TFP scenario in 2060; (**c**) ETS scenario in 2060; (**d**) ETS-TFP scenario in 2060.

5.2.3. Impacts on PPI

Figure 8 expresses the PPI changes. Due to the improvement of TFP, PPI changes in various products and services show apparent inconsistency. Specifically, the prices of energy-intensive commodities (such as refined oil and refined gas) have increased to a certain extent. Nevertheless, the PPI of non-energy-intensive enterprises, such as agriculture, light industry, and service industry, has decreased significantly.

The overall PPI is declining, but there is heterogeneity in the PPI of different industries. The main reason is that the increase in total factor productivity reduces the production cost of enterprises, so the overall price will decrease. However, due to the constraints of carbon emissions under the carbon neutrality target, carbon pricing will increase due to the increased energy demand, which will increase the product prices of energy-intensive enterprises.

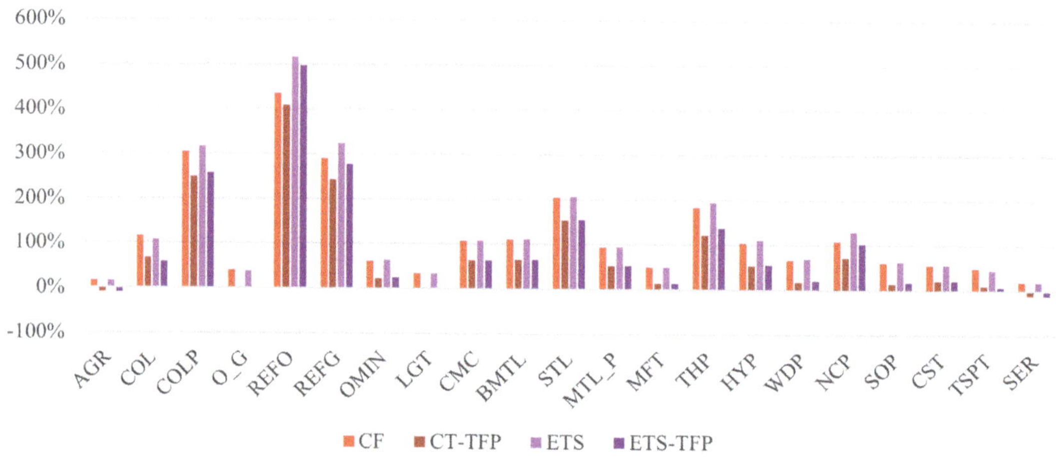

Figure 8. Impacts on producer price index in 2060.

6. Discussions

The results of this paper show that the economic cost of carbon neutralization is very large. If we want to make up for this part of the economic loss, we need an additional TFP growth of about 0.564–0.568% per year.

How difficult is the 0.564–0.568% annual additional TFP growth? According to Park (2012) [62], the TFP growth in most Asian countries (such as China, Japan, Korea, Thailand, Pakistan, and India) will range from 0.95% to 2.66% during 2010–2030. Additionally, the US's TFP growth ranged from 0%–3% during 2000–2020 [63]. Although we could not find the literature about the long-term TFP growth projection, it is certain that China's TFP growth will no longer be higher than China is used to, and the annual growth rate may range from 1% to 2.5%. Thus, the additional 0.568% TFP growth means that the TFP growth should increase by 22.72% to 56.8%.

The paper also finds other examples in the literature identifying Porter's hypothesis from which to take references. Zhao and Sun (2016) [64] argued that flexible control policies meet the weak Porter hypothesis using 2007–2012 enterprise-level data. Lin and Chen (2020) [65] supported the strong Porter hypothesis in the non-ferrous metal industry using province-industry level data in China. Zhou et al. (2021) [66] found that the weak version hypothesis for China's revised environmental protection law does not hold using the listed company's data. Lanoie et al. (2008) [67] found an average of 3% TFP growth in the Quebec manufacturing sector brought by environmental regulation. Their study has data for six years, so the annual additional TFP growth is about 0.5%, which is similar to this study; however, rather than for the manufacturer, this study is for the whole society. Other examples in the literature on testing the hypothesis also prove how hard it can be made in the context of carbon neutrality.

7. Conclusions, Policy Implications, and Limitations

7.1. Conclusions and Policy Implications

The Chinese government announced the goal of achieving carbon neutrality by 2060. This paper analyzes the impact of carbon pricing (carbon tax and carbon trading) on the economy and energy through the CGE model. In addition, according to the strong Porter hypothesis, this paper constructs new scenarios to explore the additional TFP value under the assumption of GDP neutrality and carbon neutrality, which provides a marginal contribution to the current literature.

This paper simulates carbon pricing (carbon tax and carbon trading) to achieve carbon neutrality in 2060. It was found that achieving the carbon neutrality target would reduce

China's annual economic growth by about 0.6% during 2020–2060. Carbon pricing can significantly reduce the share of fossil energy consumption and reduce overall energy consumption but also partially increase the irreplaceable energy share (such as water and air transportation and oil consumption). The process of carbon neutrality will significantly increase the price of energy-intensive products, such as energy-processing products and steel, and such a process will hardly have a significant impact on agriculture and services.

According to the strong Porter hypothesis, environmental regulations may lead to technological innovation, thus improving enterprise productivity. In addition, this paper simulates the additional total factor productivity needed to recover the economic loss caused by the carbon neutrality target. The results show that TFP needs to increase by about 0.056% every year from 2020 to 2030 to make up for the economic losses caused by the 2030 carbon peak. However, if we want to make up for the economic losses caused by carbon neutrality in 2060, TFP needs to be increased by about 0.568% annually in 2020–2060. From this point of view, the impact of carbon neutrality on the economy may be far greater than that of carbon peaking, and the economic cost should be carefully considered. By referring to the other literature, this paper believes that additional 0.568% annual TFP growth is hard to achieve, not to mention the cost of increasing productivity.

The growth of TFP will lead to improved production efficiency, and the prices of most commodities will be reduced by varying degrees, especially the prices of non-energy-intensive commodities. However, the prices of energy-intensive commodities will rise due to the dual effects of TFP and carbon neutrality. The main reason for this is that the increase in TFP will make the factor demand rise, but due to carbon constraints, the supply of energy-intensive products is restrained. Therefore, in the case of rising demand and limited supply, the price of energy-intensive products will rise relatively. Similarly, the growth of TFP will increase the share of renewable energy, mainly because under the carbon constraint, the additional energy demand must be provided by renewable energy rather than fossil energy. At the same time, the paper also finds that the price of renewable generation increases in TFP-related scenarios, which also shows that the demand for renewable energy will increase with the increase in energy demand and the limitation of thermal power generation.

These conclusions have a specific significance for our scientific understanding of carbon neutrality. TFP in 2020–2030 only needs an additional 0.056% to make up for the economic loss caused by the peak of carbon emissions; however, we need an annual TFP increase of 0.568% to make up for the economic loss caused by carbon neutrality, and the GDP growth loss is about 0.6% every year. Therefore, in the process of carbon neutrality, encouraging enterprise innovation and improving efficiency may be the key to reducing economic loss and welfare loss.

7.2. Limitations

The boundary of the Porter hypothesis in this paper assumes that all industries should increase the same level of TFP growth to meet the economic neutrality goals, and it is not the real case. However, we cannot know the actual change value of TFP in different industries under the carbon neutrality target. Therefore, we need to understand this boundary as the average boundary of additional TFP increases in the total society.

Another potential bias of this paper is that the model does not consider the cost of technological progress. Although the proportion of R&D investment in the total social cost is low, it may increase under carbon neutrality. Therefore, technological progress is not free. This paper has no appropriate reference to describe the cost of technological progress. Thus, the paper may underestimate the boundaries of strong Porter's hypothesis to some extent.

Author Contributions: Conceptualization, S.H. and Z.J.; methodology, Z.J.; formal analysis, C.D., X.J. and D.Z.; investigation, Y.W.; data curation, S.W.; writing—original draft preparation, S.H. and S.W.; writing—review and editing, Z.C.; visualization, S.W.; supervision, Z.J.; funding acquisition, S.H. All authors have read and agreed to the published version of the manuscript.

Funding: This research received support by the provincial management industrial unit of State Grid Zhejiang Electric Power Co., Ltd. (HC-SJ-F01-2022337-BY), the Ministry of Education of China (No. 22YJC790047), the Fundamental Research Funds for the Central Universities (Humanities and Social Sciences) (No. SK2022026), and the Shaanxi Federation of Social Sciences (No. 2022ND0312).

Data Availability Statement: The data available are in a publicly accessible repository that does not issue DOIs.

Conflicts of Interest: The authors declare no conflict of interest.

References

1. Porter, M.E. America' Green Strategy. *Sci. Am.* **1991**, *264*, 168. [CrossRef]
2. Borowski, P.F. Management of Energy Enterprises in Zero-Emission Conditions: Bamboo as an Innovative Biomass for the Production of Green Energy by Power Plants. *Energies* **2022**, *15*, 1928. [CrossRef]
3. Borowski, P.F. Zonal and Nodal Models of Energy Market in European Union. *Energies* **2020**, *13*, 4182. [CrossRef]
4. Fernández-Braña, A.; Feijoo, G.; Dias-Ferreira, C. Turning Waste Management into a Carbon Neutral Activity: Practical Demonstration in a Medium-Sized European City. *Sci. Total Environ.* **2020**, *728*, 138843. [CrossRef] [PubMed]
5. Shi, X.; Zheng, Y.; Lei, Y.; Xue, W.; Yan, G.; Liu, X.; Cai, B.; Tong, D.; Wang, J. Air Quality Benefits of Achieving Carbon Neutrality in China. *Sci. Total Environ.* **2021**, *795*, 148784. [CrossRef]
6. Wang, F.; Li, Y.; Zhang, W.; He, P.; Jiang, L.; Cai, B.; Zhang, J.; Zhang, P.; Pan, H.; Jiang, H. China's Trade-Off Between Economic Benefits and Sulfur Dioxide Emissions in Changing Global Trade. *Earth's Future* **2020**, *8*, e2019EF001354. [CrossRef]
7. Fleischhacker, A.; Lettner, G.; Schwabeneder, D.; Auer, H. Portfolio Optimization of Energy Communities to Meet Reductions in Costs and Emissions. *Energy* **2019**, *173*, 1092–1105. [CrossRef]
8. Yang, Z.; Shao, S.; Yang, L. Unintended Consequences of Carbon Regulation on the Performance of SOEs in China: The Role of Technical Efficiency. *Energy Econ.* **2021**, *94*, 105072. [CrossRef]
9. Ranocchia, C.; Lambertini, L. Porter Hypothesis vs Pollution Haven Hypothesis: Can There Be Environmental Policies Getting Two Eggs in One Basket? *Environ. Resour. Econ.* **2021**, *78*, 177–199. [CrossRef]
10. Wang, Y.; Sun, X.; Guo, X. Environmental Regulation and Green Productivity Growth: Empirical Evidence on the Porter Hypothesis from OECD Industrial Sectors. *Energy Policy* **2019**, *132*, 611–619. [CrossRef]
11. Qin, C.; Wang, J.; Ge, C.; Su, J. Simulating the Cost-Effectiveness of China's Green Transition Based on Emission Reduction Targets during the 12th Five-Year Plan Period. *J. Clean. Prod.* **2019**, *208*, 19–34. [CrossRef]
12. Wang, S.; Chu, C.; Chen, G.; Peng, Z.; Li, F. Efficiency and Reduction Cost of Carbon Emissions in China: A Non-Radial Directional Distance Function Method. *J. Clean. Prod.* **2016**, *113*, 624–634. [CrossRef]
13. Cui, L.-B.; Fan, Y.; Zhu, L.; Bi, Q.-H. How Will the Emissions Trading Scheme Save Cost for Achieving China's 2020 Carbon Intensity Reduction Target? *Appl. Energy* **2014**, *136*, 1043–1052. [CrossRef]
14. Wang, Z.; Chen, H.; Huo, R.; Wang, B.; Zhang, B. Marginal Abatement Cost under the Constraint of Carbon Emission Reduction Targets: An Empirical Analysis for Different Regions in China. *J. Clean. Prod.* **2020**, *249*, 119362. [CrossRef]
15. Guo, J.-X.; Tan, X.; Gu, B.; Qu, X. The Impacts of Uncertainties on the Carbon Mitigation Design: Perspective from Abatement Cost and Emission Rate. *J. Clean. Prod.* **2019**, *232*, 213–223. [CrossRef]
16. Lin, B.; Jia, Z. Economic, Energy and Environmental Impact of Coal-to-Electricity Policy in China: A Dynamic Recursive CGE Study. *Sci. Total Environ.* **2020**, *698*, 134241. [CrossRef]
17. Rahman, S.M.; Kirkman, G.A. Costs of Certified Emission Reductions under the Clean Development Mechanism of the Kyoto Protocol. *Energy Econ.* **2015**, *47*, 129–141. [CrossRef]
18. Wang, Y.; Geng, S.; Zhao, P.; Du, H.; He, Y.; Crittenden, J. Cost–Benefit Analysis of GHG Emission Reduction in Waste to Energy Projects of China under Clean Development Mechanism. *Resour. Conserv. Recycl.* **2016**, *109*, 90–95. [CrossRef]
19. Li, L.; Ye, F.; Li, Y.; Chang, C.-T. How Will the Chinese Certified Emission Reduction Scheme Save Cost for the National Carbon Trading System? *J. Environ. Manag.* **2019**, *244*, 99–109. [CrossRef]
20. Zhang, W.; Zhang, N.; Yu, Y. Carbon Mitigation Effects and Potential Cost Savings from Carbon Emissions Trading in China's Regional Industry. *Technol. Forecast. Soc. Change* **2019**, *141*, 1–11. [CrossRef]
21. Sun, D.-Q.; Yi, B.-W.; Xu, J.-H.; Zhao, W.-Z.; Zhang, G.-S.; Lu, Y.-F. Assessment of CO_2 Emission Reduction Potentials in the Chinese Oil and Gas Extraction Industry: From a Technical and Cost-Effective Perspective. *J. Clean. Prod.* **2018**, *201*, 1101–1110. [CrossRef]
22. Johansson, R.; Meyer, S.; Whistance, J.; Thompson, W.; Debnath, D. Greenhouse Gas Emission Reduction and Cost from the United States Biofuels Mandate. *Renew. Sustain. Energy Rev.* **2020**, *119*, 109513. [CrossRef]
23. Wang, H.; Zhang, S.; Bi, X.; Clift, R. Greenhouse Gas Emission Reduction Potential and Cost of Bioenergy in British Columbia, Canada. *Energy Policy* **2020**, *138*, 111285. [CrossRef]
24. Green, J.F. Does Carbon Pricing Reduce Emissions? A Review of Ex-Post Analyses. *Environ. Res. Lett.* **2021**. [CrossRef]
25. Huang, Y.; Dai, X.; Wang, Q.; Zhou, D. A Hybrid Model for Carbon Price Forecasting Using GARCH and Long Short-Term Memory Network. *Appl. Energy* **2021**, *285*, 116485. [CrossRef]

26. Chen, Z.; Su, Y.; Wang, X.; Wu, Y. The Price Discrimination and Environmental Effectiveness in Carbon Emission Trading Schemes: A Theoretical Approach. *J. Clean. Prod.* **2021**, *283*, 125196. [CrossRef]
27. Zhang, W.-W.; Zhao, B.; Ding, D.; Sharp, B.; Gu, Y.; Xu, S.-C.; Xing, J.; Wang, S.-X.; Liou, K.-N.; Rao, L.-L. Co-Benefits of Subnationally Differentiated Carbon Pricing Policies in China: Alleviation of Heavy PM2.5 Pollution and Improvement in Environmental Equity. *Energy Policy* **2021**, *149*, 112060. [CrossRef]
28. Kok, R. Six Years of CO 2 -Based Tax Incentives for New Passenger Cars in The Netherlands: Impacts on Purchasing Behavior Trends and CO 2 Effectiveness. *Transp. Res. Part A Policy Pract.* **2015**, *77*, 137–153. [CrossRef]
29. Andersson, J.J. Carbon Taxes and CO_2 Emissions: Sweden as a Case Study. *Am. Econ. J. Econ. Policy* **2019**, *11*, 1–30. [CrossRef]
30. Tirkaso, W.T.; Gren, I.-M. Road Fuel Demand and Regional Effects of Carbon Taxes in Sweden. *Energy Policy* **2020**, *144*, 111648. [CrossRef]
31. Hájek, M.; Zimmermannová, J.; Helman, K.; Rozenský, L. Analysis of Carbon Tax Efficiency in Energy Industries of Selected EU Countries. *Energy Policy* **2019**, *134*, 110955. [CrossRef]
32. Tapia Granados, J.A.; Spash, C.L. Policies to Reduce CO_2 Emissions: Fallacies and Evidence from the United States and California. *Environ. Sci. Policy* **2019**, *94*, 262–266. [CrossRef]
33. Bruninx, K.; Ovaere, M.; Delarue, E. The Long-Term Impact of the Market Stability Reserve on the EU Emission Trading System. *Energy Econ.* **2020**, *89*, 104746. [CrossRef]
34. Soliman, A.M.; Nasir, M.A. Association between the Energy and Emission Prices: An Analysis of EU Emission Trading System. *Resour. Policy* **2019**, *61*, 369–374. [CrossRef]
35. Nguyen, D.H.; Chapman, A.; Farabi-Asl, H. Nation-Wide Emission Trading Model for Economically Feasible Carbon Reduction in Japan. *Appl. Energy* **2019**, *255*, 113869. [CrossRef]
36. Qi, S.-Z.; Zhou, C.-B.; Li, K.; Tang, S.-Y. Influence of a Pilot Carbon Trading Policy on Enterprises' Low-Carbon Innovation in China. *Clim. Policy* **2021**, *21*, 318–336. [CrossRef]
37. Liu, Y.; Lu, Y. The Economic Impact of Different Carbon Tax Revenue Recycling Schemes in China: A Model-Based Scenario Analysis. *Appl. Energy* **2015**, *141*, 96–105. [CrossRef]
38. Sun, Y.; Mao, X.; Yin, X.; Liu, G.; Zhang, J.; Zhao, Y. Optimizing Carbon Tax Rates and Revenue Recycling Schemes: Model Development, and a Case Study for the Bohai Bay Area, China. *J. Clean. Prod.* **2021**, *296*, 126519. [CrossRef]
39. Jia, Z.; Lin, B. Rethinking the Choice of Carbon Tax and Carbon Trading in China. *Technol. Forecast. Soc. Change* **2020**, *159*, 120187. [CrossRef]
40. Schreyer, F.; Luderer, G.; Rodrigues, R.; Pietzcker, R.C.; Baumstark, L.; Sugiyama, M.; Brecha, R.J.; Ueckerdt, F. Common but Differentiated Leadership: Strategies and Challenges for Carbon Neutrality by 2050 across Industrialized Economies. *Environ. Res. Lett.* **2020**, *15*, 114016. [CrossRef]
41. Griffiths, S.; Sovacool, B.K. Rethinking the Future Low-Carbon City: Carbon Neutrality, Green Design, and Sustainability Tensions in the Making of Masdar City. *Energy Res. Soc. Sci.* **2020**, *62*, 101368. [CrossRef]
42. Salvia, M.; Reckien, D.; Pietrapertosa, F.; Eckersley, P.; Spyridaki, N.-A.; Krook-Riekkola, A.; Olazabal, M.; De Gregorio Hurtado, S.; Simoes, S.G.; Geneletti, D.; et al. Will Climate Mitigation Ambitions Lead to Carbon Neutrality? An Analysis of the Local-Level Plans of 327 Cities in the EU. *Renew. Sustain. Energy Rev.* **2021**, *135*, 110253. [CrossRef]
43. Lin, B.; Jia, Z. What Will China's Carbon Emission Trading Market Affect with Only Electricity Sector Involvement? A CGE Based Study. *Energy Econ.* **2019**, *78*, 301–311. [CrossRef]
44. National Development and Reform Commission (NDRC) Notice on the Key Tasks for Launching the National Carbon Emissions Trading Market. Available online: https://www.ndrc.gov.cn/xxgk/zcfb/tz/201601/t20160122_963576.html (accessed on 1 January 2020).
45. Sun, Y.; Zhi, Y.; Zhao, Y. Indirect Effects of Carbon Taxes on Water Conservation: A Water Footprint Analysis for China. *J. Environ. Manag.* **2021**, *279*, 111747. [CrossRef] [PubMed]
46. Wang, Z.; Nong, D.; Countryman, A.M.; Corbett, J.J.; Warziniack, T. Potential Impacts of Ballast Water Regulations on International Trade, Shipping Patterns, and the Global Economy: An Integrated Transportation and Economic Modeling Assessment. *J. Environ. Manag.* **2020**, *275*, 110892. [CrossRef] [PubMed]
47. Su, Q.; Dai, H.; Lin, Y.; Chen, H.; Karthikeyan, R. Modeling the Carbon-Energy-Water Nexus in a Rapidly Urbanizing Catchment: A General Equilibrium Assessment. *J. Environ. Manag.* **2018**, *225*, 93–103. [CrossRef]
48. Jia, Z. What Kind of Enterprises and Residents Bear More Responsibilities in Carbon Trading? A Step-by-Step Analysis Based on the CGE Model. *Environ. Impact Assess. Rev.* **2023**, *98*, 106950. [CrossRef]
49. Jia, Z.; Lin, B. CEEEA2.0 Model: A Dynamic CGE Model for Energy-Environment-Economy Analysis with Available Data and Code. *Energy Econ.* **2022**, *112*, 106117. [CrossRef]
50. Jia, Z.; Wen, S.; Liu, Y. China's Urban-Rural Inequality Caused by Carbon Neutrality: A Perspective from Carbon Footprint and Decomposed Social Welfare. *Energy Econ.* **2022**, *113*, 106193. [CrossRef]
51. Lou, F. *Theory and Application of Chinese Economy-Energy-Environment-Tax Dynamic Computable General Equilibrium Model*, 5th ed.; China Social Sciences Press: Beijing, China, 2015; (In Chinese). ISBN 978-7-5161-6423-5.
52. Fujimori, S.; Masui, T.; Matsuoka, Y. *AIM/CGE [Basic] Manual*; Center for Social and Environmental Systems Research, NIES: Frankfurt am Main, Germany, 2012.

53. Fang, X.; Roe, T.L.; Smith, R.B.W. Water Shortages, Intersectoral Water Allocation and Economic Growth: The Case of China. *China Agric. Econ. Rev.* **2015**, *7*, 2–26. [CrossRef]
54. Cao, J.; Dai, H.; Li, S.; Guo, C.; Ho, M.; Cai, W.; He, J.; Huang, H.; Li, J.; Liu, Y.; et al. The General Equilibrium Impacts of Carbon Tax Policy in China: A Multi-Model Comparison. *Energy Econ.* **2021**, *99*, 105284. [CrossRef]
55. Chen, Y.; Shao, S.; Fan, M.; Tian, Z.; Yang, L. One Man's Loss Is Another's Gain: Does Clean Energy Development Reduce CO_2 Emissions in China? Evidence Based on the Spatial Durbin Model. *Energy Econ.* **2022**, *107*, 105852. [CrossRef]
56. Yao, X.; Lei, H.; Yang, L.; Shao, S.; Ahmed, D.; Ismaail, M.G.A. Low-Carbon Transformation of the Regional Electric Power Supply Structure in China: A Scenario Analysis Based on a Bottom-up Model with Resource Endowment Constraints. *Resour. Conserv. Recycl.* **2021**, *167*, 105315. [CrossRef]
57. Azam, A.; Rafiq, M.; Shafique, M.; Yuan, J. An Empirical Analysis of the Non-Linear Effects of Natural Gas, Nuclear Energy, Renewable Energy and ICT-Trade in Leading CO_2 Emitter Countries: Policy towards CO_2 Mitigation and Economic Sustainability. *J. Environ. Manag.* **2021**, *286*, 112232. [CrossRef] [PubMed]
58. Khan, S.A.R.; Yu, Z.; Belhadi, A.; Mardani, A. Investigating the Effects of Renewable Energy on International Trade and Environmental Quality. *J. Environ. Manag.* **2020**, *272*, 111089. [CrossRef]
59. Huang, Z.; Du, X. Toward Green Development? Impact of the Carbon Emissions Trading System on Local Governments' Land Supply in Energy-Intensive Industries in China. *Sci. Total Environ.* **2020**, *738*, 139769. [CrossRef]
60. Zhu, L.; Zhang, X.B.; Li, Y.; Wang, X.; Guo, J. Can an Emission Trading Scheme Promote the Withdrawal of Outdated Capacity in Energy-Intensive Sectors? A Case Study on China's Iron and Steel Industry. *Energy Econ.* **2017**, *63*, 332–347. [CrossRef]
61. Zhang, D. Green Credit Regulation, Induced R&D and Green Productivity: Revisiting the Porter Hypothesis. *Int. Rev. Financ. Anal.* **2021**, *75*, 101723. [CrossRef]
62. Park, J. Total Factor Productivity Growth for 12 Asian Economies: The Past and the Future. *Jpn. World Econ.* **2012**, *24*, 114–127. [CrossRef]
63. Crouzet, N.; Eberly, J. Intangibles, Markups, and the Measurement of Productivity Growth. *J. Monet. Econ.* **2021**, *124*, S92–S109. [CrossRef]
64. Zhao, X.; Sun, B. The Influence of Chinese Environmental Regulation on Corporation Innovation and Competitiveness. *J. Clean. Prod.* **2016**, *112*, 1528–1536. [CrossRef]
65. Lin, B.; Chen, X. Environmental Regulation and Energy-Environmental Performance—Empirical Evidence from China's Non-Ferrous Metals Industry. *J. Environ. Manag.* **2020**, *269*, 110722. [CrossRef] [PubMed]
66. Zhou, D.; Qiu, Y.; Wang, M. Does Environmental Regulation Promote Enterprise Profitability? Evidence from the Implementation of China's Newly Revised Environmental Protection Law. *Econ. Model.* **2021**, *102*, 105585. [CrossRef]
67. Lanoie, P.; Patry, M.; Lajeunesse, R. Environmental Regulation and Productivity: Testing the Porter Hypothesis. *J. Product. Anal.* **2008**, *30*, 121–128. [CrossRef]

energies

MDPI

Article

Industrial Policy and Technological Innovation of New Energy Vehicle Industry in China

Hongchen Li [1,†], Huijun Qi [1,†], Hongjian Cao [1,2,*] and Li Yuan [1,2]

1 Hunan Key Laboratory of Macroeconomic Big Data Mining and Its Application, School of Business, Hunan Normal University, Changsha 410081, China
2 Center for Large Country Economic Research, Hunan Normal University, Changsha 410081, China
* Correspondence: caohj@hunnu.edu.cn; Tel.: +86-18907316398
† These authors contributed equally to this work.

Abstract: Promoting the development of new energy vehicles is one of the important measures to ensure energy security and deal with global warming. Technological innovation is an inexhaustible driving force for the development of the new energy vehicle industry. This study considered listed enterprises in China's new energy vehicle industry as research samples and used the fixed effect model to study the impact of government subsidies on the quantity and quality of technological innovation in the new energy vehicle industry. The empirical results show that government subsidies have a significant positive impact on the quantity of technological innovation in the new energy vehicle industry; however, government subsidies have no significant impact on the quality of technological innovation. Government subsidies increase the quantity of technological innovation in the new energy vehicle industry by increasing R&D investment, mitigating financing constraints, and improving the external attention of enterprises. Compared to downstream enterprises in the industrial chain, government subsidies have a better incentive effect on the technological innovation of upstream enterprises, which increases the number of patents and enhances the quality of utility model patents. Government subsidies have a better effect on promoting the quantity of technological innovation in large enterprises.

Keywords: new energy vehicle industry; technological innovation; industrial policy; government subsidy; innovation quality

Citation: Li, H.; Qi, H.; Cao, H.; Yuan, L. Industrial Policy and Technological Innovation of New Energy Vehicle Industry in China. *Energies* **2022**, *15*, 9264. https://doi.org/10.3390/en15249264

Academic Editor: Bert Scholtens

Received: 4 October 2022
Accepted: 4 December 2022
Published: 7 December 2022

1. Introduction

The new energy vehicle industry is a strategic emerging industry that China focuses on developing. Over the past 40 years, China's economic development has progressed significantly. The Chinese government has constantly adjusted economic policies to achieve coordinated development between the economy and environment. Developing strategic emerging industries is the key to achieving high-quality development in China. The new energy vehicle industry has been listed as one of the seven key strategic emerging industries in the file issued by the State Council named *Decision on Speeding up the Cultivation and Development of Strategic Emerging Industry* in 2010. Today, the development of the new energy vehicle industry can ensure China's energy security and is an important means of low-carbon consumption under carbon peaking and carbon neutrality goals (China aims to reach a CO_2 emissions peak before 2030 and achieve carbon neutrality before 2060 to accelerate the world's transition to green and low-carbon development).

Developing a new energy vehicle industry is vitally important to energy security for China, since it has relatively insufficient oil resources and a low crude oil reserve, but has a very high demand on oil consumption as China has become the world's second largest oil consumer in recent years due to economic growth. The country depends significantly on imported oil from foreign countries, and China's dependence on imported oil has been rising from 5% in the early 1990s to as high as 70% by 2020. The increasing dependence on imported oil reflects the great risk of China's energy security. New energy vehicles use electricity and clean energy fuels as driving forces that can reduce China's demand on oil resources, and further ensure China's energy security.

Developing the new energy vehicle industry can help China achieve carbon peaking and carbon neutrality goals. In September 2020, the Chinese government proposed the goals of reaching peak carbon emissions in 2030 and carbon neutrality in 2060. China's transportation, thermal power generation, and steel industries are the most important industries for carbon dioxide emission. Automobile carbon emissions account for up to three-quarters of the total carbon emissions in the field of transportation [1]. The use of new energy vehicles has enormous advantages in reducing carbon emissions. The well-to-wheel greenhouse gas emission intensities of battery electric vehicles (BEVs) is 22–293 g CO_2eq/km, while that of gasoline internal combustion engine vehicles (ICEVs) is 227–245 g CO_2eq/km [2]. Developing new energy vehicles is a fast and effective way to achieve the goal of carbon peaking and carbon neutrality in China.

The development of a new energy vehicle industry cannot do without technological innovation. Many studies have shown that to promote industry development and achieve low-carbon transformation, it is necessary to gradually shift from factor driven to innovation driven. Zhao et al. [3] analyzed the R&D incentive mechanism of China's photovoltaic industry based on the system dynamics model, and believed that technological progress in the photovoltaic industry could reduce carbon emissions. Wu et al. [4] studied the listed companies of new energy vehicles in China as a sample, and found that the firm's technological capability is an important factor to promote industrial development and increase R&D investment. These studies all show that technological innovation is a driving force for industrial development.

The new energy vehicle industry and its technological innovation have strong positive externalities. Technological innovation has a long cycle and causes great uncertainty. At the same time, the benefits generated by innovation are difficult to be fully owned by private individuals [5], which is prone to "free riding" behavior, thus inhibiting the R&D momentum of micro subjects [6]. Therefore, the Chinese government has paid much attention to the guidance of industrial policies in the new energy vehicle industry and its technological innovation.

The Chinese government has started to provide policy guidance for the new energy vehicle industry since the beginning of the 21st century. Before 2009, there were few supporting policies for China's new energy vehicle industry, and these mainly focused on planning from the production-side. From 2009–2015, China's new energy industry policy has focused on consumption. Meanwhile, the central government and local governments have launched industrial policies intensively to stimulate the technical breakthrough of key links such as new energy vehicle drive systems, battery management systems, and vehicle integration, and also began to encourage the construction of new energy vehicle supporting facilities. Since 2015, the government's industrial policy has begun to attach importance to the combination of promising governments and effective markets. *The Notice on the Financial Support Policy for the Promotion and Application of New Energy Vehicles from 2016–2020* in 2015 indicated that the subsidy would gradually decline after 2016. "The dual credit policy" issued in 2017 represents the industrial policy's impact on leading the new energy vehicle industry transit from being policy driven to market driven, the gradual withdrawal of subsidy policy, and the function of the market mechanism (In September 2017, the Ministry of Industry and Information Technology and other five departments jointly issued *the measures for the parallel management of average fuel consumption and new energy vehicle credits of passenger vehicle enterprises* (hereinafter referred to as *"the double credits policy"*), which was implemented on 1 April 2018. "The double credits policy" set up two credits for the average fuel consumption of automobile manufacturers and new energy vehicles, and established a credit trading mechanism. "The double credits policy" is an assessment system. The assessment indicators are the average fuel consumption credits and new energy vehicle credits. The purpose is to promote enterprises to develop new energy vehicles to alleviate the energy and environmental pressure.) The subsidy policy for energy vehicles after 2018 has increasingly higher standards for key technical indicators such as the energy density of power battery systems, vehicle energy consumption, and endurance to stimulate the innovation vitality of enterprises and improve the product quality [7].

Government subsidies are the most common industrial policy tool in China's new energy vehicle industry. However, industrial policies such as government R&D subsidies may lead to distortions in resource allocation and incentives, resulting in negative effects [8]. Therefore, the policy's impact on technological innovation remains controversial. Some scholars have shown that government subsidies have a positive impact on the technological innovation of enterprises. Hottenrott and Lopes-Bento [9] used the Belgian Community Innovation Survey data and found that public R&D support had a significant incentive effect on enterprise innovation output. Huergo and Moreno [10] used Spanish company data and found that obtaining any type of direct assistance significantly increased the possibility of carrying out R&D activities. However, some believe that government subsidies have had a negative impact on the enterprises' technological innovation. Wallsten [11] found that the Small Business Innovation Research (SBIR) program funding in the United States had a significant negative effect on enterprise R&D expenditure. Link and Scott [12] also found that the commercialization probability of the R&D achievements funded by the SBIR program was very low, while other studies have shown that the impact of government subsidy on technological innovation is uncertain. Marino et al. [13] used the data of French companies from 1993 to 2009, and based on the DID model, found that public subsidies had neither an incentive effect nor crowding out effect on private R&D expenditure. Montmartin and Herrera [14] used a database of 25 OECD countries and found that there was a nonlinear relationship between R&D subsidiaries and financial investments implemented within a country and private R&D.

The effects of government subsidies on the new energy vehicle industry are also controversial. Some scholars hold a positive attitude toward the effect of government subsidies on the new energy vehicle industry. Using data from 32 European countries, Münzel et al. [15] found a significant positive correlation between financial incentives and plug-in electric vehicle (PEV) adoption. Xing et al. [16] found that federal income tax credits from the United States could increase the sales of electric vehicles. Breetz and Salon [17] found that government subsidies could significantly improve the cost competitiveness of new energy vehicles by studying 14 cities in the United States. Jiao et al. [18] and Wang and Li [19] believe that government subsidies could significantly promote the expansion of China's new energy vehicle market. Gao and Hu [20] found that the subsidy policy for new energy vehicles played a significant role in promoting enterprise performance through two mechanisms: enterprise size and patent behavior. Some scholars also hold a negative attitude toward the implementation effect of government subsidies for the new energy vehicle industry. Zhang et al. [21] found that in Beijing, the license plate lottery policy was better than the subsidy policy in promoting electric vehicles. Sheldon and Dua [22,23] explored the impact and cost-effectiveness of electric vehicle subsidies by using the data of U.S. new car buyers and Chinese new vehicle consumers. Both research results showed that the cost of the subsidies was too high and the subsidy target should be determined according to the policy objectives.

Reviewing the existing studies, scholars have used data from various countries to conduct extensive research on the impact of industrial policies on energy vehicles and their technological innovation. Relevant research includes the impact of industrial policies on the use and diffusion of new energy vehicles [24], the new energy vehicle industrial policies on environmental pollution [25], and industrial policies on the R&D and development strategies of new energy vehicle enterprises [26]. Only a few studies have examined the impact of industrial policies on technological innovation in the new energy vehicle industry [27,28]. So far, whether the existing industrial policies have really improved the technological innovation level of the new energy vehicle industry is open to debate.

Furthermore, the impact of government industrial policies on the quantity and quality of technological innovation in the new energy vehicle industry has not been compared and analyzed yet. This study examined the impact of government subsidies on technological innovation in the new energy vehicle industry from the dimensions of the quantity and quality of technological innovation and explored how government subsidies impact the effect of policies to enrich the research in related fields.

The marginal contributions of this study are as follows. First, the different effects of government subsidies on the quantity and quality of technological innovation in the new energy vehicle industry were investigated; it was found that government subsidies could significantly promote the quantity of technological innovation but could not improve the quality of technological innovation. Second, it was found through empirical study that the industrial policy could increase the number of innovations in the new energy vehicle industry through three mechanisms: improving the attention of enterprises, increasing the R&D investment, and mitigating the financing constraints. Third, we examined the differences in the impact of industrial policies on technological innovation in different links of the industrial chain, which provides valuable insights for industrial policies to promote technological innovation in the new energy vehicle industry.

The rest of this paper is organized as follows. Section 2 introduces the empirical method and data. Section 3 presents the benchmark regression and robustness test results. Section 4 discusses the mechanism test and a series of heterogeneity analyses. Section 5 concludes the study and puts forward policy implications.

2. Method and Data

2.1. Model Design

This paper used the two-way fixed effect model for estimation, which can make the estimation result control some individual heterogeneity that will not change over time and is difficult to observe as well as reduce the problem of missing variables. The following is the benchmark model of this paper:

$$innov\_n_{it} = \beta_0 + \beta_1 sub_{it} + \beta_2 X_{it} + firm_i + year_t + \varepsilon_{it} \quad (1)$$

$$innov\_q_{it} = \beta_0 + \beta_1 sub_{it} + \beta_2 X_{it} + firm_i + year_t + \varepsilon_{it} \quad (2)$$

where *innov_n* represents the quantity of technological innovation including *tpatent* (the number of total patents); *ipatent* (invention patents); and *upatent* (utility model patents). *innov_q* represents the quality of technological innovation including *width* (patent quality); *iwidth* (invention patent quality); and *uwidth* (utility model patent quality). *sub* represents the government subsidy; X represents a series of control variables including capital structure (*lev*), profitability (*roa*), enterprise size (*size*), proportion of fixed assets (*ppe*), proportion of independent directors (*dir*), enterprise age (*age*), enterprise growth ability (*gov*), and enterprise human capital (*hc*). *firm* is the enterprise fixed effect; *year* is the year fixed effect; ε is the random disturbance term.

2.2. Definition of the Variable

(1) Quantity and quality of technological innovations

Patents are generally considered good indicators of technological innovation. Compared to utility model and design patents, invention patents have higher requirements and are more innovative and breakthrough. Utility model patents represent the improvement in existing technology by enterprises to a certain extent, but the improvement is relatively small. The patent does not contain any technological innovations. Therefore, this study selected the number of invention patent applications and utility model patent applications to measure the quantity of technological innovation.

This study used patent knowledge width to measure the quality of technological innovation. The patent knowledge width can measure the quality of patents based on the complexity and universality of the knowledge contained in patents. Therefore, this study used the practices of Zhang and Zheng [29] and Aghion et al. [30] to measure the quality of each patent using the knowledge width method, which can measure the complexity of knowledge contained in a patent. In this study, the formula for calculating the width of patent knowledge is $width = 1 - \sum x^2$; x is the proportion of each group of patent IPC classification numbers. We then added the patent knowledge width to the enterprise level according to the average value. The greater the knowledge width value of a patent, the wider the knowledge involved in the patent, and therefore the higher the quality of innovation.

(2) Government subsidies

In this study, the data of government subsidy were processed logarithmically.

(3) Control variables

In order to exclude the influence of other factors on the regression model and estimation results, referring to the research of Chen et al. [31], this study selected some variables related to the nature and capabilities of the enterprise for control including: capital structure (*lev*), profitability (*roa*), enterprise size (*size*), proportion of fixed assets (*ppe*), proportion of independent directors (*dir*), enterprise age (*age*), enterprise growth ability (*gov*), and enterprise human capital (*hc*) (see Table 1 for the definition of variables).

Table 1. Definition of the variables.

Variable Classification	Variable	Symbol	Definition
Dependent variable	Quantity of technological innovation	*tpatent*	*ln*(Number of invention patent applications + number of utility model patent applications + 1)
		ipatent	*ln*(Number of invention patent applications + 1)
		upatent	*ln*(Number of utility model patent applications + 1)
	Quality of technological innovation	*width*	Quality of patent
		iwidth	Quality of invention patent
		uwidth	Quality of utility mode patent
Independent variable	Government subsidy	*sub*	*ln*(Government subsidy)
Control variable	Capital structure	*lev*	Asset liability ratio
	Profitability	*roa*	Net interest rate of total assets
	Enterprise size	*size*	*ln*(Number of employees)
	Proportion of fixed assets	*ppe*	Net fixed assets/total assets
	Proportion of independent directors	*dir*	Number of independent directors/numbers of board of directors
	Enterprise age	*age*	*ln*(Year—establishment year + 1)
	Enterprise growth ability	*gov*	Year on year growth rate of operating revenue
	Enterprise human capital	*hc*	Number of undergraduates/number of employees

2.3. *Data*

2.3.1. Data Source

This study used the listed enterprises in the field of new energy vehicle industry as the research object. China's support policies for the new energy vehicle industry began to grow rapidly after 2010. In this study, enterprises listed on energy vehicles from 2010 to 2019 were selected as the research samples, and the samples were processed as follows: (1) the samples with ST and ST* marks were excluded; and (2) eliminated samples with missing data. Finally, we obtained 242 new energy automobile enterprises, with a total of 1671 observations. To prevent interference from extreme values, all continuous variables were subjected to tailing reduction.

The patent data for this study came from the Incopat database. First, the name of each listed company and its subsidiaries was extracted from the Chinese Research Data Services (CNRDS) database. Then, the names of the listed companies and their subsidiaries were manually retrieved by patent applicants through the Incopat database, and the patent data of the listed companies and their affiliates were counted. Finally, the number and quality data of patents applied for by enterprises every year were obtained through sorting and statistics. Government subsidies and other financial data received by the enterprises were obtained from the Wind and CSMAR databases.

2.3.2. Descriptive Statistics

Table 2 presents the descriptive statistics for the main variables. The average number of enterprise patents (tpatent) was 3.559, the average number of invention patents (ipatent) was 2.628, and the average number of utility model patents (upatent) was 3.017. It can be seen that new energy vehicle enterprises apply for more utility model patents. The maximum values of the invention patents, utility model patents, and total patents were 6.443, 6.724, and 7.407, respectively. The standard deviation was also relatively large, which indicates that there is a large gap in the level of technological innovation between different enterprises. The average quality of patents (width) was 0.219, the average quality of invention patents (iwidth) was 0.245, and quality of utility model patents (uwidth) was 0.18. It can be seen that the average quality of the utility model patents was significantly lower than that of the invention patents. The maximum patent quality, invention patent quality, and utility model patent quality were 0.723, 0.75, and 0.57, respectively. The maximum value of the government subsidies was 20.88, the minimum value was 12.68, and the standard deviation was 1.537. It can be seen that there were certain differences in the government subsidies received by enterprises.

Table 2. Descriptive statistics.

Variable	N	Mean	SD	Min	Max
tpatent	1671	3.559	1.582	0	7.407
ipatent	1671	2.628	1.498	0	6.443
upatent	1671	3.017	1.612	0	6.724
width	1671	0.219	0.133	0	0.723
iwidth	1671	0.245	0.164	0	0.75
uwidth	1671	0.18	0.12	0	0.57
sub	1671	16.53	1.537	12.68	20.88
lev	1671	0.417	0.187	0.0681	0.933
roa	1671	4.88	5.916	−22.11	18.1
size	1671	7.9	1.15	5.733	11.46
ppe	1671	0.204	0.104	0.0157	0.541
dir	1671	0.37	0.0486	0.333	0.556
age	1671	2.841	0.294	2.079	3.526
gov	1671	17.32	28.56	−41.29	130
hc	1446	18.45	13.02	3.16	73.28

In order to see the development of technological innovation in China's new energy vehicle industry in detail, Table 3 presents the annual mean value of the dependent variables. As shown in Table 3, from 2010 to 2019, the number of invention and utility model patents of the listed companies in the new energy vehicle industry maintained a steady upward trend. In 2010, the average annual number of patents of enterprises was 40.99, which nearly quadrupled to 159.32 in 2019, with an average annual growth rate of 16.28%. Overall, in 2010, the annual average patent quality of the listed companies in the new energy vehicle industry was 0.195, and the patent quality increased to 0.275 in 2019. The patent quality fluctuated from 2010 to 2015 and improved rapidly after 2015. In terms of the patent type, the quality of invention patents was significantly higher than that of the utility model patents.

Table 3. Descriptive statistics of the dependent variables by year (mean value).

Year	*Tpatent*	*Ipatent*	*Upatent*	*Width*	*Iwidth*	*Uwidth*
2010	40.99	13.61	27.38	0.194	0.194	0.151
2011	69.63	25.36	44.27	0.176	0.2	0.126
2012	76.98	29.05	47.94	0.191	0.207	0.148
2013	97.87	36.46	61.41	0.177	0.195	0.136
2014	106.7	40.99	65.71	0.187	0.213	0.159
2015	109.8	42.61	67.14	0.188	0.197	0.156
2016	132.6	55.37	77.28	0.228	0.248	0.174
2017	140.9	59.28	81.66	0.244	0.282	0.202
2018	160.6	71.1	89.45	0.247	0.29	0.206
2019	159.3	69.4	89.93	0.275	0.314	0.247

3. Empirical Results and Analysis

3.1. Analysis of Benchmark Regression Results

First, we discuss the impact of government subsidies on the number of new energy vehicle patents. It can be seen from columns (1)–(3) in Table 4 that the coefficients of government subsidies were significantly positive at the level of 1%, and the coefficients were 0.173, 0.147, and 0.162, respectively. The number of enterprise patent applications increased by 0.173%. The more subsidies the government gives to enterprises, the more invention patents and utility model patents the enterprises apply for.

Table 4. The regression results of government subsidies on the quantity and quality of technological innovation.

	(1)	(2)	(3)	(4)	(5)	(6)
	Tpatent	*Ipatent*	*Upatent*	*Width*	*Iwidth*	*Uwidth*
sub	0.173 ***	0.147 ***	0.162 ***	0.00003	−0.00295	0.00211
	(4.540)	(4.483)	(4.284)	(0.00791)	(−0.428)	(0.443)
Constant	−4.693 *	−6.205 **	−4.382	0.577 *	0.194	0.186
	(−1.849)	(−2.414)	(−1.632)	(1.958)	(0.564)	(0.749)
CONTROLS	YES	YES	YES	YES	YES	YES
COMPANY FE	YES	YES	YES	YES	YES	YES
YEAR FE	YES	YES	YES	YES	YES	YES
N	1446	1446	1446	1446	1446	1446
R^2	0.465	0.425	0.443	0.120	0.133	0.138

Note: Robust standard errors are given in parentheses. * $p < 0.1$, ** $p < 0.05$, *** $p < 0.01$. CONTROLS includes capital structure (*lev*), profitability (*roa*), enterprise size (*size*), proportion of fixed assets (*ppe*), proportion of independent directors (*dir*), enterprise age (*age*), enterprise growth ability (*gov*), and human capital (*hc*).

Second, we discuss the impact of government subsidies on the NEV patent quality of new energy vehicles. As shown in columns (4)–(6) in Table 4, the coefficients of sub were relatively small and not significant, indicating that government subsidy had no significant impact on the quality of patents. The reason is that enterprises encouraged by industrial policies will significantly increase their patent applications in order to obtain more government subsidies. However, due to many uncertain risks in the process of early research and development, some enterprises prefer to carry out low-quality technological innovation with relatively short cycles and low investment than high-quality technological innovation to reduce the costs and risks [32–35].

3.2. Endogenous Test

Considering the endogeneity problem of reverse causality may exist between government subsidies and the amount of technological innovation of enterprises, that is, the higher the level of the technological innovation of enterprises, the easier it is for them to meet the standards for granting subsidies and obtain more government subsidies. This study selected the mean value of government subsidies in the new energy vehicle industry lagging behind the first phase and government subsidies lagging behind the second phase as instrumental variables for the two-stage least squares estimation [36]. Reasons for selecting tool variables are as follows. (1) When applying for government subsidies, enterprises are likely to refer to the subsidy amount applied by other enterprises in the same industry in the previous period to ensure that the maximum subsidy amount can be obtained on the basis of successful application; and (2) if the enterprise can obtain government support in the early stage, it may send a positive signal to the government, which is conducive to the enterprise applying again for government subsidies. Table 5 shows that government subsidies had a significant effect on the total number of patents, invention patents, and utility model patents at the 5% level. The impact of government subsidies on patent quality was still insignificant. These results are consistent with the benchmark regression results.

Table 5. Regression results of the endogenous test.

	(1)	(2)	(3)	(4)	(5)	(6)
	Tpatent	*Ipatent*	*Upatent*	*Width*	*Iwidth*	*Uwidth*
sub	0.257 *	0.226 **	0.290 *	−0.0187	−0.000443	−0.0105
	(1.819)	(2.033)	(1.886)	(−1.057)	(−0.0175)	(−0.573)
CONTROLS	YES	YES	YES	YES	YES	YES
COMPANY FE	YES	YES	YES	YES	YES	YES
YEAR FE	YES	YES	YES	YES	YES	YES
Hansen-J P	0.8786	0.5249	0.9651	0.4827	0.9791	0.5214
N	1117	1117	1117	1117	1117	1117
R^2	0.408	0.376	0.378	0.127	0.145	0.130

Note: z-statistics are given in parentheses * $p < 0.1$, ** $p < 0.05$, *** $p < 0.01$. CONTROLS includes capital structure (*lev*), profitability (*roa*), enterprise size (*size*), proportion of fixed assets (*ppe*), proportion of independent directors (*dir*), enterprise age (*age*), enterprise growth ability (*gov*), and human capital (*hc*).

3.3. Robustness Test

3.3.1. Replace Dependent Variables

This study replaced the quantitative index of technological innovation with the number of patent applications provided by the CNRDS database (*cpt, cpi, cpu*). The patent quality index was replaced by the number of cited patents (*cited*) and patent claims (*claim*). The regression results are presented in Table 6. It can be seen that after replacing the measurement indicators, the regression results are consistent with the benchmark regression results.

Table 6. Robustness test: Replace the dependent variables.

	(1)	(2)	(3)	(4)	(5)
	cpt	*cpi*	*cpu*	*Cited*	*Claim*
sub	0.159 ***	0.122 ***	0.149 ***	0.0476	0.0894
	(5.151)	(3.969)	(4.631)	(0.734)	(1.383)
Constant	−4.182 *	−5.951 **	−3.090	4.525	2.781
	(−1.693)	(−2.327)	(−1.208)	(0.922)	(0.720)
CONTROLS	YES	YES	YES	YES	YES
COMPANY FE	YES	YES	YES	YES	YES
YEAR FE	YES	YES	YES	YES	YES
N	1446	1446	1446	1374	1374
R^2	0.410	0.335	0.455	0.467	0.281

Note: Robust standard errors are given in parentheses. * $p < 0.1$, ** $p < 0.05$, *** $p < 0.01$. CONTROLS includes capital structure (*lev*), profitability (*roa*), enterprise size (*size*), proportion of fixed assets (*ppe*), proportion of independent directors (*dir*), enterprise age (*age*), enterprise growth ability (*gov*), and human capital (*hc*).

3.3.2. Subsample Regression

Considering that some enterprises may enter the new energy vehicle industry in a certain period, this paper verified the time when each enterprise entered the industry by consulting the annual report of the enterprise, and then selected the sub-sample after the enterprise entered the new energy vehicle industry for regression. The regression results are shown in Table 7. We can see that the regression results are consistent with the benchmark regression results.

Table 7. Robustness test: Subsample.

	(1)	(2)	(3)	(4)	(5)	(6)
	Tpatent	*Ipatent*	*Upatent*	*Width*	*Iwidth*	*Uwidth*
sub	0.154 ***	0.131 ***	0.145 ***	0.00142	−0.00320	0.00405
	(4.187)	(4.081)	(3.855)	(0.275)	(−0.433)	(0.804)
Constant	−4.157	−4.821 **	−4.354	0.667 **	0.269	0.195
	(−1.592)	(−2.012)	(−1.527)	(2.050)	(0.729)	(0.683)
CONTROLS	YES	YES	YES	YES	YES	YES
COMPANY FE	YES	YES	YES	YES	YES	YES
YEAR FE	YES	YES	YES	YES	YES	YES
N	1350	1350	1350	1350	1350	1350
R^2	0.443	0.407	0.420	0.120	0.130	0.141

Note: Robust standard errors are given in parentheses. * $p < 0.1$, ** $p < 0.05$, *** $p < 0.01$. CONTROLS includes capital structure (*lev*), profitability (*roa*), enterprise size (*size*), proportion of fixed assets (*ppe*), proportion of independent directors (*dir*), enterprise age (*age*), enterprise growth ability (*gov*), and human capital (*hc*).

3.3.3. Panel Tobit Model

Because the patent quantity and quality data were non-negative, and had the characteristics of truncated data [37], this study used the panel Tobit model to estimate, and the results are shown in Table 8. The results show that the coefficients of sub on the quantity of technological innovation were all significantly positive, and the coefficients on the quality of technological innovation were still not significant, which is consistent with the benchmark regression results.

Table 8. Robustness test: Panel Tobit model.

	(1)	(2)	(3)	(4)	(5)	(6)
	Tpatent	*Ipatent*	*Upatent*	*Width*	*Iwidth*	*Uwidth*
sub	0.211 ***	0.199 ***	0.193 ***	0.00253	−0.000928	0.00443
	(7.950)	(7.220)	(6.744)	(0.685)	(−0.189)	(1.187)
Constant	−4.646 ***	−5.863 ***	−4.945 ***	0.338 **	0.383 **	0.151
	(−4.532)	(−5.595)	(−4.450)	(2.448)	(2.103)	(1.086)
sigma_u	0.867 ***	0.869 ***	0.969 ***	0.0866 ***	0.112 ***	0.0658 ***
	(18.44)	(18.07)	(18.44)	(15.83)	(15.29)	(12.83)
sigma_e	0.711 ***	0.730 ***	0.761 ***	0.104 ***	0.137 ***	0.110 ***
	(47.61)	(46.60)	(46.24)	(46.31)	(44.77)	(44.48)
CONTROLS	YES	YES	YES	YES	YES	YES
COMPANY FE	YES	YES	YES	YES	YES	YES
YEAR FE	YES	YES	YES	YES	YES	YES
N	1446	1446	1446	1446	1446	1446

Note: z-statistics are given in parentheses. * $p < 0.1$, ** $p < 0.05$, *** $p < 0.01$. CONTROLS includes capital structure (*lev*), profitability (*roa*), enterprise size (*size*), proportion of fixed assets (*ppe*), proportion of independent directors (*dir*), enterprise age (*age*), enterprise growth ability (*gov*), and human capital (*hc*).

3.3.4. System-GMM Model

Considering that the level of technological innovation in the previous period may have an impact on the level of technological innovation in the current period, this paper introduced the independent variable lagging behind the first phase and used the system-GMM model to estimate. As shown in Table 9, it was found that the quantity and quality of technological innovation in the previous period had a significant positive impact on the quantity and quality of technological innovation in the current period. The government subsidies had a significant positive impact on the quantity of technological innovation, but had no significant impact on the quality of technological innovation, which is consistent with the benchmark regression results.

Table 9. Robustness test: System-GMM model.

	(1)	(2)	(3)	(4)	(5)	(6)
	Tpatent	*Ipatent*	*Upatent*	*Width*	*Iwidth*	*Uwidth*
L.tpatent	0.549 ***					
	(7.313)					
L.ipatent		0.507 ***				
		(8.150)				
L.upatent			0.529 ***			
			(6.411)			
L.width				0.205 ***		
				(3.511)		
L.iwidth					0.224 ***	
					(4.103)	
L.uwidth						0.113 **
						(2.110)
sub	0.183 **	0.238 ***	0.156 *	0.000848	−0.0117	0.000674
	(2.374)	(4.138)	(1.846)	(0.112)	(−1.116)	(0.0719)
Constant	−4.646 ***	−5.863 ***	−4.945 ***	0.338 **	0.383 **	0.151
	(−4.532)	(−5.595)	(−4.450)	(2.448)	(2.103)	(1.086)
CONTROLS	YES	YES	YES	YES	YES	YES
COMPANY FE	YES	YES	YES	YES	YES	YES
YEAR FE	YES	YES	YES	YES	YES	YES
AR(1)	0.000	0.000	0.000	0.000	0.000	0.000
AR(2)	0.259	0.228	0.133	0.546	0.785	0.142
Hansen *p* value	0.264	0.460	0.321	0.676	0.675	0.255
N	1446	1446	1446	1446	1446	1446

Note: z-statistics are given in parentheses. * $p < 0.1$, ** $p < 0.05$, *** $p < 0.01$. CONTROLS includes capital structure (*lev*), profitability (*roa*), enterprise size (*size*), proportion of fixed assets (*ppe*), proportion of independent directors (*dir*), enterprise age (*age*), enterprise growth ability (*gov*), and human capital (*hc*).

4. Mechanism Test and Heterogeneity Analysis

4.1. Mechanism Test

In order to empirically test the mechanism of government subsidies affecting the amount of technological innovation of new energy vehicle enterprises, this paper designed the following mechanism test econometric models:

$$attention_{it} = \beta_0 + \beta_1 sub_{it} + \beta_2 X_{it} + firm_i + year_t + \varepsilon_{it} \tag{3}$$

$$R\&D_{it} = \beta_0 + \beta_1 sub_{it} + \beta_2 X_{it} + firm_i + year_t + \varepsilon_{it} \tag{4}$$

$$fund_{it} = \beta_0 + \beta_1 sub_{it} + \beta_2 X_{it} + firm_i + year_t + \varepsilon_{it} \tag{5}$$

where $attention$ represents the degree of external attention of an enterprise; $R\&D$ represents an enterprise's R&D capital investments; $fund$ represents the external financing of an enterprise. The other symbols have the same meaning as in Models (1) and (2).

In this paper, the logarithm of the number of analysts who make profit forecasts for enterprises every year was taken as the proxy variable that enterprises are concerned by the outside world [38]. The logarithm of an enterprise's annual R&D expenditure was used to measure $R\&D$. $fund$ was measured by the ratio of net cash flow from financing activities to total assets [39].

4.1.1. Improving the External Attention of Enterprises

Table 10 presents the regression results of the mechanistic tests. It can be seen from column (1) that government subsidies can significantly improve the number of analysts who pay attention to the subsidized enterprises. The more government subsidies the enterprises receive, the more attention they will receive from the outside world, which will increase the possibility of enterprises integrating various external innovation resources [38]. Therefore, government subsidies can promote technological innovation by increasing the attention of the enterprises.

Table 10. Mechanism test.

	(1)	(2)	(3)
	Attention	*R&D*	*Fund*
sub	0.162 ***	0.0688 **	0.00962 *
	(2.910)	(2.306)	(1.775)
Constant	−6.235 *	10.94 ***	0.195
	(−1.727)	(5.637)	(0.757)
CONTROLS	YES	YES	YES
COMPANY FE	YES	YES	YES
YEAR FE	YES	YES	YES
N	1138	1412	1446
R^2	0.190	0.662	0.072

Note: Robust standard errors are given in parentheses. * $p < 0.1$, ** $p < 0.05$, *** $p < 0.01$. CONTROLS includes capital structure (*lev*), profitability (*roa*), enterprise size (*size*), proportion of fixed assets (*ppe*), proportion of independent directors (*dir*), enterprise age (*age*), enterprise growth ability (*gov*), and human capital (*hc*).

4.1.2. Increasing R&D Capital Investment

Column (2) in Table 10 indicates that the coefficient of sub was 0.0688, which was significant at the 5% level, indicating that government subsidies play an important role in promoting the enterprises' R&D capital investment. Many studies have proven that R&D capital investment can significantly improve the enterprises' innovation output [40]. Therefore, government subsidies promote technological innovation by increasing enterprise R&D capital investment.

4.1.3. Mitigating Financing Constraints

Column (3) in Table 10 shows that the coefficient of sub was 0.00962, which was significant at the 10% level, indicating that government subsidies are conducive to enterprises obtaining more external financing. This alleviates enterprise financing constraints and promotes an increase in the quantity of technological innovation [41].

In general, these three mechanisms were significant. In comparison, the role of improving the attention of enterprises was greater, while the role of alleviating financing was relatively small.

4.2. Heterogeneity Analysis

4.2.1. Industrial Chain Perspective

This study divided enterprises into upstream, midstream, and downstream industries according to the industrial chain. Table 11 shows the regression results of the impact of government subsidies on the quantity of the enterprises' technological innovation in all links of the industrial chain. The results show that increasing government subsidies can increase the technological innovation across the entire industrial chain.

To test whether there was a difference in the significant impact of government subsidies on upstream, midstream, and downstream enterprises, we conducted an inter-group coefficient difference test. First, we set the dummy variables *chain*1, *chain*2, and *chain*3. The dummy variable *chain*1 is 1 when the enterprise belongs to the upstream, otherwise it is 0. The dummy variable *chain*2 is 1 when the enterprise belongs to the midstream, otherwise it is 0. The dummy variable *chain*3 is 1 when the enterprise belongs downstream, otherwise it is 0. Second, we multiplied the three dummy variables with the main independent variable (*sub*) to form the interactive terms *sub_chain*1, *sub_chain*2, and *sub_chain*3. Third, we set the three interactive terms in Model (1) for regression. The results are presented in Table 12. When government subsidies increased, Panel A shows that upstream enterprises applied for more utility model patents than the other links. Panel B shows that although the quantity of technological innovation in midstream enterprises increased, it was significantly less than the upstream and downstream enterprises. Panel C shows that compared to other links, downstream enterprises applied for more invention patents. Overall, government subsidies can promote technological innovation across the entire industrial chain. The number of invention patents of downstream enterprises increased the most, while the number of utility model patents of the upstream enterprises increased the most.

Table 11. Heterogeneity analysis: Industrial chain perspective (quantity of technological innovation).

	(1)	(2)	(3)
	Tpatent	*Ipatent*	*Upatent*
Panel A: upstream			
sub	0.210 **	0.161 **	0.241 ***
	(2.650)	(2.360)	(3.034)
Constant	−10.64 *	−11.17 **	−13.15 *
	(−1.857)	(−2.225)	(−1.910)
N	247	247	247
R^2	0.539	0.474	0.543
Panel B: midstream			
sub	0.109 **	0.0805 *	0.107 **
	(2.360)	(1.786)	(2.189)
Constant	−1.281	−4.747	0.0820
	(−0.452)	(−1.581)	(0.0287)
N	877	877	877
R^2	0.477	0.424	0.456
Panel C: downstream			
sub	0.233 ***	0.226 ***	0.203 ***
	(3.051)	(3.528)	(2.795)
Constant	−7.045	−5.189	−7.213
	(−1.429)	(−0.983)	(−1.379)
N	322	322	322
R^2	0.509	0.526	0.441
CONTROLS	YES	YES	YES
COMPANY FE	YES	YES	YES
YEAR FE	YES	YES	YES

Note: Robust standard errors are given in parentheses. * $p < 0.1$, ** $p < 0.05$, *** $p < 0.01$. CONTROLS includes capital structure (*lev*), profitability (*roa*), enterprise size (*size*), proportion of fixed assets (*ppe*), proportion of independent directors (*dir*), enterprise age (*age*), enterprise growth ability (*gov*), and human capital (*hc*).

Table 12. Heterogeneity analysis: Comparison of inter-group coefficients of the industrial chain perspective (quantity of technological innovation).

	(1)	(2)	(3)
	Tpatent	*Ipatent*	*Upatent*
Panel A			
sub	0.161 ***	0.149 ***	0.134 ***
	(3.854)	(3.988)	(3.256)
sub_chain1	0.0609	−0.00870	0.146 *
	(0.663)	(−0.116)	(1.709)
Constant	−4.674 *	−6.208 **	−4.335
	(−1.850)	(−2.409)	(−1.640)
N	1446	1446	1446
R^2	0.465	0.425	0.446
Panel B			
sub	0.235 ***	0.209 ***	0.222 ***
	(4.136)	(4.450)	(4.219)
sub_chain2	−0.128 *	−0.128 **	−0.125 *
	(−1.753)	(−1.992)	(−1.769)
Constant	−4.423 *	−5.936 **	−4.118
	(−1.737)	(−2.302)	(−1.531)
N	1446	1446	1446
R^2	0.468	0.429	0.447
Panel C			
sub	0.136 ***	0.0930 **	0.147 ***
	(3.229)	(2.553)	(3.327)
sub_chain3	0.116	0.169 **	0.0461
	(1.322)	(2.299)	(0.559)
Constant	−4.486 *	−5.903 **	−4.299
	(−1.752)	(−2.286)	(−1.589)
N	1446	1446	1446
R^2	0.467	0.430	0.443
CONTROLS	YES	YES	YES
COMPANY FE	YES	YES	YES
YEAR FE	YES	YES	YES

Note: Robust standard errors are given in parentheses. * $p < 0.1$, ** $p < 0.05$, *** $p < 0.01$. CONTROLS includes capital structure (*lev*), profitability (*roa*), enterprise size (*size*), proportion of fixed assets (*ppe*), proportion of independent directors (*dir*), enterprise age (*age*), enterprise growth ability (*gov*), and human capital (*hc*).

Table 13 shows the impact of government subsidies on the quality of the enterprises' technological innovation in all links of the industrial chain. The results showed that government subsidies had a significant positive impact only on the quality of the utility model patents of upstream enterprises. This shows that government subsidies can improve the quality of the technological innovation of upstream enterprises, but are limited in terms of the quality of utility model patents, and have no impact on the quality of invention patents.

Table 13. Heterogeneity analysis: Industrial chain perspective (quality of technological innovation).

	(1)	(2)	(3)
	Width	*Iwidth*	*Uwidth*
Panel A: upstream			
sub	0.00522	0.00176	0.0205 **
	(0.713)	(0.261)	(2.045)
Constant	0.778	−0.351	−1.651 **
	(0.598)	(−0.296)	(−2.299)
N	247	247	247
R^2	0.130	0.161	0.228
Panel B: midstream			
sub	−0.00179	−0.00815	−0.00340
	(−0.298)	(−0.908)	(−0.502)
Constant	0.332	−0.0793	0.495 *
	(0.957)	(−0.177)	(1.804)
N	877	877	877
R^2	0.117	0.131	0.125
Panel C: downstream			
sub	−0.00355	−0.00545	−0.00108
	(−0.352)	(−0.370)	(−0.140)
Constant	0.685	0.715	0.403
	(1.580)	(1.400)	(0.829)
N	322	322	322
R^2	0.247	0.220	0.233
CONTROLS	YES	YES	YES
COMPANY FE	YES	YES	YES
YEAR FE	YES	YES	YES

Note: Robust standard errors are given in parentheses. * $p < 0.1$, ** $p < 0.05$, *** $p < 0.01$. CONTROLS includes capital structure (*lev*), profitability (*roa*), enterprise size (*size*), proportion of fixed assets (*ppe*), proportion of independent directors (*dir*), enterprise age (*age*), enterprise growth ability (*gov*), and human capital (*hc*).

4.2.2. Enterprise Size Perspective

Enterprises are divided into large and small enterprises according to their scale. Table 14 shows the impact of government subsidies on the technological innovation of enterprises of different sizes. Panels A and B show the regression results for large and small enterprises. The results show that an increase in government subsidies had a significantly positive impact on the quantity of technological innovation in large and small enterprises. To further analyze the difference between the two significant effects, an intergroup coefficient test was conducted. We set the dummy variable *size*1. When the enterprise is large-scale, it is 1; otherwise, it is 0. Then, the dummy variable *size*1 was multiplied by the main independent variable (*sub*) to form the interaction term *sub_size*1, and Model (1) was added for regression (see Panel C for the results). The results show that compared to small enterprises, large enterprises can apply for more invention patents after receiving government subsidies. There are two possible reasons for this finding. First, large enterprises have a wider and better internal division of labor and a stronger ability to use R&D networks and knowledge spillovers, which is more conducive to technological innovation. Second, from the perspective of enterprise strategic objectives, large enterprises pay more attention to long-term returns, so they have a stronger willingness to engage in technological innovation activities.

Table 15 shows the impact of government subsidies on the technological innovation quality of enterprises of different sizes. The results show that government subsidies had no significant impact on the quality of the technological innovation of enterprises of different sizes. This proves that the result of benchmark regression is robust. Overall, increasing government subsidies is significantly effective in increasing the technological innovation of large and small enterprises, especially for large-scale enterprises. However, the enterprise size heterogeneity does not significantly affect the effect of government subsidies on the technological innovation quality.

Table 14. Heterogeneity analysis: Enterprise size perspective (quantity of technological innovation).

	(1)	(2)	(3)
	Tpatent	*Ipatent*	*Upatent*
Patent A: large			
sub	0.225 ***	0.193 ***	0.218 ***
	(4.098)	(3.340)	(4.028)
Constant	−2.016	−5.175	−0.399
	(−0.551)	(−1.280)	(−0.108)
N	728	728	728
R^2	0.393	0.373	0.357
Panel B: small			
sub	0.136 ***	0.117 ***	0.136 ***
	(3.286)	(3.347)	(3.316)
Constant	−1.325	−1.843	−3.152
	(−0.383)	(−0.552)	(−0.791)
N	718	718	718
R^2	0.353	0.268	0.359
Panel C: coefficient compare			
sub	0.198 ***	0.134 ***	0.206 ***
	(4.297)	(3.548)	(4.555)
sub_size1	0.0374	0.118 **	−0.000342
	(0.642)	(2.360)	(−0.00581)
Constant	−2.098	−3.057	−2.145
	(−0.880)	(−1.238)	(−0.847)
N	1446	1446	1446
R^2	0.447	0.403	0.426
CONTROLS	YES	YES	YES
COMPANY FE	YES	YES	YES
YEAR FE	YES	YES	YES

Note: Robust standard errors are given in parentheses. * $p < 0.1$, ** $p < 0.05$, *** $p < 0.01$. CONTROLS in all models includes capital structure (*lev*), profitability (*roa*), proportion of fixed assets (*ppe*), proportion of independent directors (*dir*), enterprise age (*age*), enterprise growth ability (*gov*), and enterprise human capital (*hc*).

Table 15. Heterogeneity analysis: Enterprise size perspective (quality of technological innovation).

	(1)	(2)	(3)	(4)	(5)	(6)
		Large			Small	
	Width	*Iwidth*	*Uwidth*	*Width*	*Iwidth*	*Uwidth*
sub	−0.000816	2.47×10^{-5}	−0.000319	−0.00354	−0.00888	0.00372
	(−0.169)	(0.00346)	(−0.0574)	(−0.437)	(−0.800)	(0.491)
Constant	0.146	0.0563	0.409	1.027 **	0.668	0.281
	(0.614)	(0.133)	(1.645)	(2.125)	(1.376)	(0.620)
N	728	728	728	718	718	718
R^2	0.218	0.149	0.186	0.094	0.148	0.133
CONTROLS	YES	YES	YES	YES	YES	YES
COMPANY FE	YES	YES	YES	YES	YES	YES
YEAR FE	YES	YES	YES	YES	YES	YES

Note: Robust standard errors are given in parentheses. * $p < 0.1$, ** $p < 0.05$, *** $p < 0.01$. CONTROLS in all models includes capital structure (*lev*), profitability (*roa*), proportion of fixed assets (*ppe*), proportion of independent directors (*dir*), enterprise age (*age*), enterprise growth ability (*gov*), and enterprise human capital (*hc*).

5. Conclusions and Policy Implications

Against the background of carbon peaking and carbon neutrality goals, the new energy vehicle industry is a strategic emerging industry with huge social and economic benefits, and needs reasonable guidance through industrial policies. Technological innovation is the fundamental driving force to promote the high-quality development of the new energy vehicle industry and is an important way to advance the low-carbon transformation of energy. Taking the listed enterprises in the new energy vehicle industry from 2010 to 2019 as the research sample, this study compared the patent data of the Incopat database with the financial data of enterprises in the Wind and CSMAR databases. The quantity of technological innovation was measured by the number of patents, and the quality of technological innovation was measured by the width of patent knowledge. The fixed effects model was used to empirically test the impact of government subsidies on the quantity and quality of technological innovation and the internal impact mechanism.

The conclusions are as follows. (1) The government subsidy only encouraged the quantity of technological innovation in the new energy vehicle industry, but had no incentive effect on the quality of technological innovation. (2) There are three mechanisms for government subsidies to promote the quantity of technological innovations in the new energy automobile industry. First, as an approval from the government, the government subsidy can increase the enterprises' credibility in the market. Second, government subsidies can encourage enterprises to increase their R&D capital investment. Third, government subsidies can mitigate the financing constraints in technological innovation. (3) Government subsidies can only materially improve the quality of the utility model patents of upstream enterprises.

The policy implications of this study are as follows. (1) Optimizing the government subsidy policy system for the new energy vehicle industry. First, the selection mechanism for subsidy objects should be improved. We should strengthen the fairness and openness of the selection of subsidy objects and provide a competitive environment for enterprises. Second, the subsidy effect evaluation mechanism should be improved, the traditional innovation evaluation system based on the number of innovations should be abandoned, and the inspection of enterprise innovation quality should be strengthened. (2) Formulating differentiated subsidy incentive policies. When the government grants subsidies, it is necessary to fully consider the heterogeneity factors such as the location of the industrial chain and the size of enterprises. The government should promote differentiated incentive policies according to local conditions and improve the allocation efficiency of government subsidy funds. The government should further improve the incentive effect of technological innovation for midstream and downstream enterprises in the new energy vehicle industry chain. For upstream enterprises of the new energy vehicle industrial chain, the incentives of invention patents and high-quality innovation should be emphasized. At the same time, the government should enhance the incentive effect of industrial policies on the technological innovation output of small- and medium-sized enterprises.

There are still some limitations in this study. As above-mentioned, the purpose, object and strength of the industrial policies of the new energy vehicle industry are different. This study failed to distinguish the heterogeneity of the effects of the supply and demand side on subsidy policies. There are many kinds of government subsidies related to China's new energy vehicle industry, but it is difficult to obtain the details of each subsidy received by enterprises from the financial data of listed companies. This is the difficulty and direction of future research.

Author Contributions: Conceptualization, H.L. and H.Q.; Methodology, H.C. and H.Q.; Software, H.Q. and H.L.; Formal analysis, H.L. and H.Q.; Resources, H.C.; Data curation, H.Q. and H.L.; Writing-original draft preparation, H.Q. and H.L.; Writing-review and editing, H.L. and H.C.; Supervision, H.C. and L.Y. All authors have read and agreed to the published version of the manuscript.

Funding: This research was funded by the National Social Science Foundation of China (Grant No. 22BJY059), the Natural Science Foundation of Hunan Province (Grant No. 2021JJ3045), and the Postgraduate Scientific Research Innovation Project of Hunan Province (Grant No. QL20210122).

Data Availability Statement: Not applicable.

Conflicts of Interest: The authors declare no conflict of interest.

References

1. Mao, X.; Yang, S.; Liu, Q.; Tu, J.; Jaccard, M. Achieving CO_2 Emission Reduction and the Co-Benefits of Local Air Pollution Abatement in the Transportation Sector of China. *Environ. Sci. Policy* **2012**, *21*, 1–13. [CrossRef]
2. Gan, Y.; Lu, Z.; He, X.; Hao, C.; Wang, Y.; Cai, H.; Wang, M.; Elgowainy, A.; Przesmitzki, S.; Bouchard, J. Provincial Greenhouse Gas Emissions of Gasoline and Plug-in Electric Vehicles in China: Comparison from the Consumption-Based Electricity Perspective. *Environ. Sci. Technol.* **2021**, *55*, 6944–6956. [CrossRef]
3. Zhao, X.G.; Wang, W.; Wu, L. A dynamic analysis of research and development incentive on China's photovoltaic industry based on system dynamics model. *Energy* **2021**, *233*, 121141. [CrossRef]
4. Wu, Y.; Gu, F.; Ji, Y.; Guo, J.; Fan, Y. Technological Capability, Eco-Innovation Performance, and Cooperative R&D Strategy in New Energy Vehicle Industry: Evidence from Listed Companies in China. *J. Clean. Prod.* **2020**, *261*, 121157. [CrossRef]
5. Arrow, K.J. The Economic Implications of Learning by Doing. *Rev. Econ. Stud.* **1962**, *29*, 155–173. [CrossRef]
6. Hsu, P.-H.; Tian, X.; Xu, Y. Financial Development and Innovation: Cross-Country Evidence. *J. Financ. Econ.* **2014**, *112*, 116–135. [CrossRef]
7. Li, J.Z.; Ku, Y.Y.; Li, L.Y.; Liu, C.L.; Deng, X.D. Optimal Channel Strategy for Obtaining New Energy Vehicle Credits under Dual Credit Policy: Purchase, Self-Produce, or both? *J. Clean. Prod.* **2022**, *342*, 130852. [CrossRef]
8. Acemoglu, D.; Akcigit, U.; Bloom, N.; Kerr, W.R. Innovation, Reallocation, and Growth. *Am. Econ. Rev.* **2018**, *108*, 3450–3491. [CrossRef]
9. Hottenrott, H.; Lopes-Bento, C. R&D Collaboration and SMEs: The Effectiveness of Targeted Public R&D Support Schemes. *Res. Policy* **2014**, *43*, 1055–1066.
10. Huergo, E.; Moreno, L. Subsidies or loans? Evaluating the Impact of R&D Support Programmes. *Res. Policy* **2017**, *46*, 1198–1214.
11. Wallsten, S.J. The Effects of Government-Industry R&D Programs on Private R&D: The Case of the Small Business Innovation Research Program. *Rand J. Econ.* **2000**, *31*, 82–100.
12. Link, A.N.; Scott, J.T. Private Investor Participation and Commercialization Rates for Government-Sponsored Research and Development: Would a Prediction Market Improve the Performance of the SBIR Programme? *Economica* **2009**, *76*, 264–281. [CrossRef]
13. Marino, M.; Lhuillery, S.; Parrotta, P.; Sala, D. Additionality or Crowding-Out? An Overall Evaluation of Public R&D Subsidy on Private R&D Expenditure. *Res. Policy* **2016**, *45*, 1715–1730.
14. Montmartin, B.; Herrera, M. Internal and External Effects of R&D Subsidies and Fiscal Incentives: Empirical Evidence Using Spatial Dynamic Panel Model. *Res. Policy* **2015**, *44*, 1065–1079.
15. Münzel, C.; Plötz, P.; Sprei, F.; Gnann, T. How large is the effect of financial incentives on electric vehicle sales?—A global review and European analysis. *Energy Econ.* **2019**, *84*, 104493. [CrossRef]
16. Xing, J.; Leard, B.; Li, S. What does an electric vehicle replace? *J. Environ. Econ. Manag.* **2021**, *107*, 102432. [CrossRef]
17. Breetz, H.L.; Salon, D. Do Electric Vehicles Need Subsidies? Ownership Costs for Conventional, Hybrid, and Electric Vehicles in 14 US Cities. *Energy Policy* **2018**, *120*, 238–249. [CrossRef]
18. Jiao, Y.; Yu, L.; Wang, J.; Wu, D.; Tang, Y. Diffusion of New Energy Vehicles under Incentive Policies of China: Moderating Role of Market Characteristic. *J. Clean. Prod.* **2022**, *353*, 131660. [CrossRef]
19. Wang, D.; Li, Y. Measuring the Policy Effectiveness of China's New-energy Vehicle Industry and its Differential Impact on Supply and Demand Markets. *Sustainability* **2022**, *14*, 8215. [CrossRef]
20. Gao, W.; Hu, X.Y. New Energy Vehicle Policy Effect: Does Scale or Innovation Serve as an Intermediary? *Sci. Res. Manag.* **2020**, *41*, 32–44. (In Chinese)
21. Zhang, X.; Bai, X.; Zhong, H. Electric Vehicle Adoption in License Plate-Controlled Big Cities: Evidence from Beijing. *J. Clean. Prod.* **2018**, *202*, 191–196. [CrossRef]
22. Sheldon, T.L.; Dua, R. Measuring the Cost-effectiveness of Electric Vehicle Subsidies. *Energy Econ.* **2019**, *84*, 104545. [CrossRef]
23. Sheldon, T.L.; Dua, R. Effectiveness of China's plug-in electric vehicle subsidy. *Energy Econ.* **2020**, *88*, 104773. [CrossRef]
24. Yu, P.; Zhang, J.; Yang, D.F.; Lin, X.; Xu, T.Y. The Evolution of China's New Energy Vehicle Industry from the Perspective of a Technology–Market–Policy Framework. *Sustainability* **2019**, *11*, 1711. [CrossRef]
25. Zhang, L.; Wang, L.; Chai, J. Influence of New Energy Vehicle Subsidy Policy on Emission Reduction of Atmospheric Pollutants: A Case Study of Beijing, China. *J. Clean. Prod.* **2020**, *275*, 124069. [CrossRef]
26. Zuo, W.; Li, Y.; Wang, Y. Research on the Optimization of New Energy Vehicle Industry Research and Development Subsidy about Generic Technology Based on the Three-Way Decisions. *J. Clean. Prod.* **2019**, *212*, 46–55. [CrossRef]
27. Fang, S.; Xue, X.; Yin, G.; Fang, H.; Li, J.; Zhang, Y. Evaluation and Improvement of Technological Innovation Efficiency of New Energy Vehicle Enterprises in China Based on DEA-Tobit Model. *Sustainability* **2020**, *12*, 7509. [CrossRef]
28. Yang, T.; Xing, C.; Li, X. Evaluation and Analysis of New-Energy Vehicle Industry Policies in the Context of Technical Innovation in China. *J. Clean. Prod.* **2021**, *281*, 125126. [CrossRef]
29. Zhang, J.; Zheng, W.P. Has Catch-up Strategy of Innovation Inhibited the Quality of China's Patents? *Econ. Res. J.* **2018**, *53*, 28–41. (In Chinese)
30. Aghion, P.; Akcigit, U.; Bergeaud, A.; Blundell, R.; Hemous, D. Innovation and Top Income Inequality. *Rev. Econ. Stud.* **2019**, *86*, 1–45. [CrossRef]

31. Chen, H.; Na, C.H.; Yu, T.M.Z.; Han, X.F. Internal Control and R&D Subsidy Performance. *J. Manag. World.* **2018**, *34*, 149–164. (In Chinese)
32. Kim, M.; Lee, S.Y. The Effects of Government Financial Support on Business Innovation in South Korea. *Asian. J. Technol. Innov.* **2011**, *19*, 67–83. [CrossRef]
33. Tian, X. Wang, T. Tolerance for Failure and Corporate Innovation. *Rev. Financ. Stud.* **2014**, *27*, 211–255. [CrossRef]
34. Li, W.J.; Zheng, M.N. Is it Substantive Innovation or Strategic Innovation? Impact of Macroeconomic Policies on Micro-enterprises' Innovation. *Econ. Res. J.* **2016**, *51*, 60–73. (In Chinese)
35. Zhang, Y.; Na, S.; Niu, J.; Jiang, B. The Influencing Factors, Regional Difference and Temporal Variation of Industrial Technology Innovation: Evidence with the FOA-GRNN Model. *Sustainability* **2018**, *10*, 187. [CrossRef]
36. Yu, J.; Shi, X.; Guo, D.; Yang, L. Economic Policy Uncertainty (EPU) and Firm Carbon Emissions: Evidence Using a China Provincial EPU Index. *Energy Econ.* **2020**, *94*, 105071. [CrossRef]
37. Faleye, O.; Kovacs, T.; Venkateswaran, A. Do Better-Connected CEOs Innovate More. *J. Financ. Quant. Anal.* **2014**, *49*, 1201–1225. [CrossRef]
38. Guo, Y. Signal Transmission Mechanism of Government Innovation Subsidy and Enterprise Innovation. *China Ind. Econ.* **2018**, *9*, 98–116. (In Chinese)
39. Li, H.D.; Tang, Y.J.; Zuo, J.J. On the Selection of Innovation Financing Resource: Based on the Financing Structure and the Innovation of Listed-company in China. *J. Financ. Res.* **2013**, *34*, 170–183. (In Chinese)
40. Hall, B.H.; Lotti, F.; Mairesse, J. Evidence on the Impact of R&D and ICT Investments on Innovation and Productivity in Italian Firms. *Econ. Innov. New Technol.* **2013**, *22*, 300–328.
41. Meuleman, M.; De Maeseneire, W. Do R&D Subsidies Affect SMEs' Access to External Financing? *Res. Policy* **2012**, *41*, 580–591.

energies

MDPI

Article

Risk Assessment of a Coupled Natural Gas and Electricity Market Considering Dual Interactions: A System Dynamics Model

Lin Wang and Yuping Xing *

Glorious Sun School of Business and Management, Donghua University, Shanghai 200051, China
* Correspondence: ypxing@dhu.edu.cn

Abstract: Because reliance on gas for electricity generation rises over time, the natural gas and electricity markets are highly connected. However, both of them are susceptible to various risk factors that endanger energy security. The intricate interactions among multiple risks and between the two markets render risk assessment more challenging than for individual markets. Taking a systematic perspective, this study first undertook a thorough analysis of the evolution mechanism that indicated the key risk factors and dual interactions, with real-world illustrative examples. Subsequently, a system dynamics model was constructed for understanding the causal feedback structures embedded in the operation of a coupled natural gas–electricity market in the face of risks. Quantitative experiments were conducted by using data from China's Energy Statistical Yearbook, China's Statistical Yearbook and other reliable sources to assess the effects of individual risks, depict the evolutionary behavior of coupled markets and compare the risk response strategies. The findings revealed the evolution of dominant risk factors and the aggregated effects of multiple risks in multiple markets, suggesting the need to comprehensively monitor dynamic risks. Moreover, risk factors can propagate from one market to another via interactions, yet it depends on multiple aspects such as the severity of the risk and the intensity of the interactions. Demand compression and emergency natural gas supply behave differently throughout the market's recovery, necessitating a balance between short-term and long-term risk response strategies.

Keywords: risk assessment; electricity market; natural gas market; system dynamics

Citation: Wang, L.; Xing, Y. Risk Assessment of a Coupled Natural Gas and Electricity Market Considering Dual Interactions: A System Dynamics Model. *Energies* **2023**, *16*, 223. https://doi.org/10.3390/en16010223

Academic Editors: Junpeng Zhu and Xinlong Xu

Received: 31 October 2022
Revised: 19 December 2022
Accepted: 21 December 2022
Published: 25 December 2022

1. Introduction

Driven by ever-increasing climate change, a worldwide consensus has been reached on the urgency of transitioning global energy towards a green future [1,2]. Common agreements such as the Kyoto Protocol and the Copenhagen Accord have been ratified by hundreds of governments, forcing the replacement of coal-fired electricity generation with cleaner and more reliable energy resources [3,4]. In this regard, being the cleanest fossil fuel with a relatively high efficiency, natural gas has become an indispensable option for supplying electricity demand [5]. The International Energy Agency (IEA) predicted in its 2017 report that, by 2040, natural gas would surpass oil as the second-largest fuel in the global fuel mix, accounting for one quarter of the world's energy demand [6,7]. Despite the fact that reliance on gas for electricity has kept rising over the years, unanticipated risks may occur and disrupt the gas-to-electricity progression. For instance, since natural gas is distributed unevenly across the globe, certain countries rely heavily on the international supply to satisfy their gas consumption needs [6]. However, due to human attacks, economic disputes or geopolitical issues, the supply is subject to significant uncertainties and fluctuations [8]. Examples include Russia's suspension of natural gas shipments to Ukraine in 2018 and the rupture of the Nord Stream pipelines in 2022, which resulted in a severe shortage of gas in European countries [9]. In the most recent quarterly report of the gas market [10], the IEA warned of an impending supply crisis and revised its 2022 gas demand forecast downward. Additionally, the electricity market may be vulnerable to a variety of hazards, such as natural disasters and technical issues, which can hinder the functioning of energy markets [11]. Hence, to ensure energy security, a comprehensive risk assessment approach is essential.

Alongside the diverse risks stemming from distinct sectors, complex interactions render risk assessments of the coupled two markets more challenging than those of individual energy markets. On the one hand, interactions may exist between multiple risks, e.g., excessively cold weather may cause a surge in electricity demand while enhancing the likelihood of a gas pipeline failure [12]. On the other hand, interactions between the two energy sectors enable the transfer of risks from one market to the other, e.g., the surge in electricity demand may drive an increase in natural gas demand, thus further widening the demand–supply imbalance [3]. Markets may adapt to the disruptions and dynamically evolve, e.g., a reduction in gas supply may result in a subsequent rise in gas prices and discourage the use of natural gas, which may exacerbate the insufficiency of electricity. The dynamic behaviors of numerous variables constitute feedback loops with various time delays, and these are too complex for decision-makers to grasp [3,13]. In light of these conditions, this study aimed to explore the impact of potential risks on the overall gas–electricity market from a holistic perspective, taking the diverse interactions into account.

In the current literature on energy security, many scholars have analyzed the risks encountered in the natural gas and electricity markets. Regarding the natural gas market, Chen et al. [6] established a worldwide gas trading network and examined the structural risks using data from the gas import trade in 2015. Considering the risk of supply shortages, Ding et al. [8] evaluated the resilience capabilities of China's natural gas system by integrating a system dynamics model and a resilience curve. Dong and Kong [14] investigated the impact of risks affecting gas imports on the Chinese economy by analyzing three categories of risk, i.e., exporting countries, transportation and foreign dependency. Egging and Holz [15] focused on three scenarios in a stochastic natural gas model and investigated the infrastructure investments under various risks based on the data from Europe, North America and China. Some research has highlighted the inherent vulnerabilities of the market. Using a natural gas pipeline in Zhuhai, China, Liu et al. [16] developed a simulation model for assessing the risks to gas pipelines by considering the probability of failure, the consequences of an accident and individual risks. Chen et al. [17] investigated the supply security of a gas pipeline network with stochastic demand. Zarei et al. [18] used FMEA to study the dynamic safety of a gas station and revealed that human error was the leading cause of system failure. Regarding the electricity market, Ahmad et al. [13] reviewed the studies that applied system dynamics in electricity sector modelling and highlighted the microworld models facilitating the trade and risk analysis in electricity markets. Salman and Li [19] proposed a framework for assessing multihazard risks in electric power systems exposed to seismic and hurricane threats, which could be used for disaster preparedness, mitigation and response planning. Based on the core elements of risk identification, measurement, assessment, evaluation, control and monitoring, Tummala and Mak [20] developed a risk management framework to improve the operations and maintenance of electricity transmission systems. Chiaradonna et al. [21] applied the stochastic activity network to construct a framework for quantitatively analyzing interactions between electricity generation and transmission infrastructures, so as to mitigate the losses induced by risks. Considering the context of the new economic normal, He et al. [22] applied system dynamics to a power consumption scenario for Tianjin to derive long-term energy demand predictions. Taking a systematic overview of the electricity market, the natural gas market and other energy markets, Burger et al. [23] investigated multiple categories of risks involved, as well as stochastic models for electricity and gas.

A recent emphasis has also been placed on the coupling between and interactions of the gas and electricity markets. Hibbard and Schatzki [24] reviewed multiple risk factors rising from the interdependence between electricity and natural gas markets and provided prominent strategies for mitigating the most significant risks. Different levels of interaction between gas and electricity systems were investigated in [25], and a two-stage stochastic programming approach was utilized to develop an integrated operational model for these systems with an unreliable power supply. Considering hourly real-time pricing in the gas and electricity markets, Tian et al. [26] explored the influence of gas market reform on the development of natural gas-fired units through a dynamic game-theoretic model. By applying a graph-theory-based technique, Beyza et al. [27] assessed structural robustness and the vulnerability of coupled gas and electricity systems by considering their interactions. Bao et al. [28] developed an integrated model to evaluate bidirectional cascade failures in an electricity–natural gas system by including coupling components such as gas-fired generators and electricity-driven gas compressors. Portante et al. [29] integrated two validated energy models (i.e., EPfast for electric power and NGfast for natural gas) to assess the propagation impact of risks and disruptions through interdependencies between the natural gas and electric power systems. Poljanšek et al. [30] constructed a probabilistic reliability model of the European gas and electricity transmission networks from a topological perspective, and the increased vulnerability resulting from market interdependencies could be observed from the results. Nazari-Heris et al. [31] exhaustively analyzed the interactions among electricity, gas and water systems, and improved the operation, economics and pollutant emissions of the integrated systems. Some studies considered both the dynamism and interactions involved in the coupled markets. Xiao et al. [1] analyzed the development pattern and constraints of China's natural gas power production, forecasting the natural gas prices of generation by using the market netback pricing approach. Esmaeili et al. [3] simulated the long-term impact of renewable energy resources' penetration on the natural gas–electricity market. Eusgeld et al. [32] constructed an integrated model to incorporate interdependencies between critical infrastructures and demonstrated the cascading effects of vulnerabilities and failures. Zhang et al. [33] coordinated the operations of power-to-gas units and generators in order to smooth the load curve of an integrated electricity and natural gas system.

These earlier studies established significant theoretical and methodological foundations for identifying the risk factors affecting the natural gas market or the electricity market, as well as the interactions between the two markets. However, the majority of them either addressed various risks in an individual market or concentrated on the impact of one specific risk event on interconnected markets, while the interactive behaviors of multiple risks and multiple markets still call for a comprehensive analysis. With the aim of observing the long-term behaviors of coupled natural gas and electricity markets under various interrelated risks, this study contributes to the research field by extending the risk assessment scenario to a more complex and dynamic setting, identifying prominent risks affecting the markets and constructing a quantitative model incorporating dual interactions between both risks and markets. System dynamics (SD) was introduced to support the assessment because of its advantages in integrating nonlinear interactions and modeling dynamic social systems [34].

The remaining sections of this article are structured as follows. The theoretical framework is presented in Section 2, depicting the dynamism and complexity of the coupled natural gas–electricity market. Section 3 proposes a system dynamics model with detailed descriptions of each component. The simulation experiments and results are presented in Section 4, followed by the conclusions.

2. Theoretical Framework

A thorough analysis of the mechanism of evolution can reveal the internal structure of the natural gas and electricity markets, their interactions and the associated risks to facilitate the risk assessment of coupled markets. The key risk factors and dual interactions are depicted, with illustrative examples from real-world accidents. These risks and interactions were determined on the basis of the relevant literature, by tracking news stories and through expert interviews.

2.1. Risk Evolution Mechanism

Taking a systematic perspective, a risk evolution mechanism of a coupled natural gas–electricity market is presented in Figure 1. The risk system, the natural gas system and the electricity system constitute a massive system-of-systems (SoS) [35,36], in which multiple components in each market, multiple risks and multiple interactions among the risk factors and market interactions interweave and evolve simultaneously. A key component of assessing the impact of specific risks on the overall SoS is to quantify their potential impact and aggregate the cascading disruptions induced by the interactions. To better understand the development and characteristics of the complex risk evolution process, each system and the interactive effects can be described as follows.

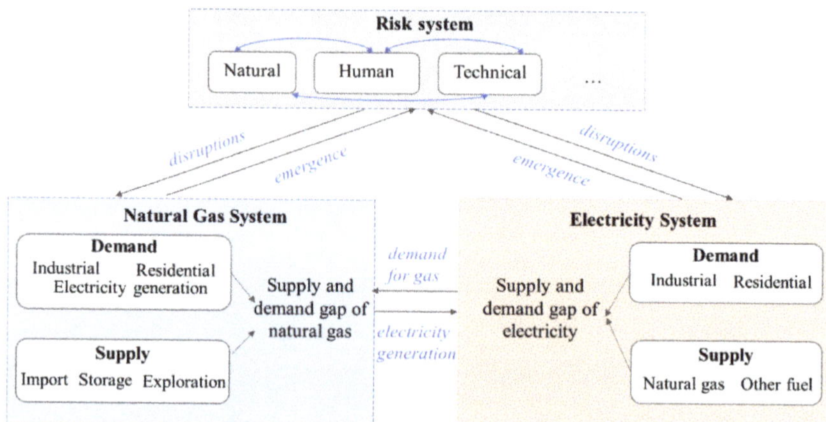

Figure 1. Mechanism of risk evolution in a coupled natural gas–electricity market.

Each market can be viewed as a collection of components functioning to achieve the common goal of satisfying market demand. In a natural gas market, gas utilization may be diverse, with electricity generation accounting for one of the major demands [1]. Natural gas is derived from multiple sources, including international trade and domestic production. An electricity market must satisfy the electricity demand of residents and industries. A diverse portfolio of fuels serves to supply the electricity, including natural gas, oil, coal, wind, etc. As natural gas is the primary focus of this study, the other categories of fuel were classified as "other fuels" in terms of the total proportion of electricity generation.

Nonetheless, the stochastic occurrence of unanticipated risks may result in market chaos [35]. These risk factors may emerge externally or internally, interact with each other and experience a rise or a decline during a certain period. The market's reactions may create counterintuitive side effects and cause the emergence of new risk factors [11]. Distinct bidirectional interactions among the facilities from different markets also act as the most prominent feature of the mechanism, as these are at the core of the complex risk assessment process. Through these interactions, excessive risks on one market may cause system inefficiency and then be transferred to another market. For instance, natural disasters or attacks may damage part of the gas production facilities, resulting in a severe shortage in the gas supply, which, in turn, would reduce the gas needed for supplying electricity and hinder the electricity system.

2.2. Prominent Risks and Their Impact

The long-term operations of natural gas and electricity markets suffer from various risks, such as natural disasters (e.g., hurricanes, earthquakes, extreme weather, etc.), hazards of human origin (e.g., terrorist attacks, cyberattacks, operational errors, etc.) and technical deficiencies (e.g., design defects, pipeline failures, corrosion, aging equipment, etc.) [11].

To quantify their consequences, the most prominent risk factors were identified based on evidence from multiple sources. After an in-depth investigation into the relevant literature [37–40], surveys [41] and online accident reports, a list of the risk factors affecting each market was extracted. Interviews were conducted to obtain experts' opinions regarding the most salient risks or alternative risks associated with the two energy markets. Table 1 provides a summary of the most significant risk factors. These risk factors are dynamic in nature, and each has its own evolutionary pattern and impact on the SoS.

Table 1. Prominent risk factors and their descriptions.

Systems	Risk Factors	Descriptions
Natural gas market	Pipeline defects	Corrosion, pipeline aging and other performance defects continually disrupt the natural gas supply.
	Import shortages	A sharp decline in cross-border trade or attacks posing a danger to pipelines induce severe import shortages.
	Extreme weather	Excessively low temperatures cause a demand peak in the natural gas market.
	Geopolitical risk	Risks associated with wars, terrorist acts and other geopolitical conflicts cause a supply shortage of the natural gas market
Electricity market	Infrastructural damage	Malfunctioning infrastructures results in ineffective electricity production and transmission.
	Electricity overload	Peaks in electricity demand widen the demand–supply gap and may even damage facilities.
	Extreme weather	Excessively low temperatures cause a demand peak in the electricity market.

In this risk assessment of the natural gas market, our emphasis was on the four common factors identified above, which include pipeline defects, import shortages, extreme weather and geopolitical risks [8,35,41,42]. First, according to a survey in [41], pipeline defects such as corrosion account for 38.5% of cases of pipeline failure. They create a continuous disruption of the pipelines' normal operations. Second, international trade is one of the main sources of natural gas supply [8]. With the escalation of international conflicts or deliberate attacks, countries face the uncertainties of sharp declines in cross-border natural gas trade. Due to a high dependence on imports, the decline can barely be compensated by domestic production, resulting in a severe supply shortfall. Third, extreme weather such as freezing stimulates the consumption of natural gas for heating, and thus a seasonal peak may occur in natural gas demand [8,35]. As the most proportion of global natural gas is supplied by specific countries, the geopolitical risks also convey much pressure on the natural gas market. Severe gas-supply shortages induced by geopolitical issues have been observed in a worldwide scope, including in Europe, Asia, America, the Eastern Mediterranean region, etc. [42–44].

In the electricity market, the three primary risk factors triggering system failures were also extracted. First, damage to the power grid's infrastructure hinders the system from maintaining a stable electricity supply [45]. The malfunction of infrastructure such as substations and transmission lines results in inefficient electricity production and transmission. Second, as residents' and industries' demand for electricity fluctuates with time, temperature and location, demand peaks that surpass the normal supply of electricity may occur [45]. This overload will further expand the demand–supply imbalance [35]. Third, various hazardous events such as hurricanes, earthquakes and freezing may inflict significant harm on the power grid [46]. In terms of frequency and consequences, among these natural hazards, extreme weather is also considered to be one of the main risks, which induces a rise in electricity demand with excessively low temperatures. Based on our analysis, extreme weather induces a demand peak in both the natural gas market and the electricity market.

2.3. *Dual Interactions*

Both the interactions among the risk factors and those between the markets constitute the structure of the coupled natural gas–electricity system and serve as the driving force for evolution. These interactions are highly dynamic and vary with stochastic events. For instance, a sudden gas rupture may impair the normal operation of the natural gas market, resulting in a gas supply shortage. In such a scenario, the electricity market may seek alternative energy sources to mitigate the demand–supply gap. As the ratio of gas to electricity declines, the interaction between the two systems decreases accordingly. To elucidate the initial data on the type and strength of the dual interactions, expert judgments were utilized. The cascading effects caused by the intricate interactions exert significant influences on the aggregated risks of the coupled markets.

- Interactions among risk factors. A risk interaction is defined as a relationship between one risk factor and another. Multiple risks may emerge concurrently. A typical illustrative example is the successive occurrence of an earthquake, a tsunami and a nuclear accident in Japan in March 2011 [35]. In the context of a coupled natural gas–electricity market, causal interactions exist. For instance, extreme weather and electricity overload are two interacted risk factors. There is a direct link between extreme weather and electricity overload, since it is possible for an extremely low temperature to produce a sudden rise in electricity demand. Owing to the interactions among risk factors, energy markets may be exposed to disruptions from combined direct and indirect risks.
- Interactions between markets. Interactions between the natural gas and electricity markets generated by the coupling of energy components can dramatically influence the strategic behavior of the SoS. Specifically, a failure or disruption in one market could propagate to the other through the coupling components, impairing the operation of numerous SoS facilities. Worse still, the propagation of disruptions may create negative feedback to the triggering system and result in even more severe ripple effects. For instance, a natural gas supply deficit or an interruption in electricity transmission may force the outage of electricity-driven compressors and further force gas generators offline, causing more severe disruptions. This work mainly considered the functional interactions involved in the coupled process of natural-gas-to-electricity generation.

3. Model Construction

To comprehensively investigate how multiple risks evolve and affect the coupled natural gas–electricity market considering complex interactions, system dynamics (SD) modeling was used for our analysis. SD is a simulation-based approach with the ability to quantitatively model dynamic and complex problems, offering decision-makers an intuitive interface for experimenting with numerous scenarios and revealing transparent results [34]. The basic principles of SD are that all outcomes of a system are determined by its unique inputs, and the behavior of a system originates from its structure [47,48]. Descriptions in the theoretical framework revealed the core structure of the systems, and physical activities of the coupled markets will be thoroughly investigated in this section, based on which the risk evolution processes, behaviors of multiple entities and state of the system can be characterized by continuously changing variables. These variables are interrelated, constituting feedback loops in response to system changes. Using differential equations, the peculiarity of variables and hypothesized relations can be quantified, which is in turn incorporated by SD software for the simulation.

Figure 2 depicts the SD model for a risk assessment of the coupled markets created using the software tool Vensim DSS. This model was adapted from well-established and verified models, including those developed by [1,8,22]. It portrays the general operation of the coupled two markets, and this model is flexible and can be structured with additional feedback loops representing specific ripple effects in other contexts. State variables such as the natural gas reserve capacity and the total natural gas supply are modeled as stocks. They are symbolized as containers or boxes, representing the accumulation of volume or capacity at a certain time. There might be inflows to or outflows from the stocks, which are symbolized by valves, inducing variations in the box per unit of time. Auxiliary variables, such as the import shortages and the electricity transmission efficiency, represent constant values or intermediate steps in calculations. The interactions between the variables are depicted as arrows, with "+" signifying a positive causal link and "−" signifying a negative link.

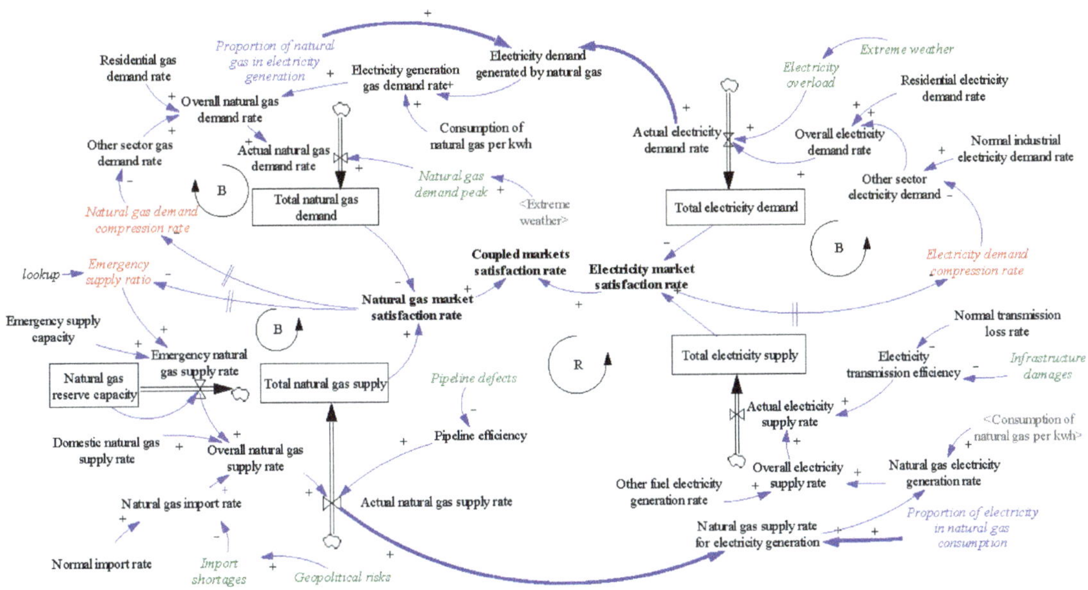

Figure 2. System dynamics model for assessing the risk of the coupled natural gas–electricity market (R: Reinforcing loop, B: Balancing loop).

We took China as an example, with China's Energy Statistical Yearbook, China's Statistical Yearbook and other reliable data sources for setting the variables and conducting experiments. As daily data are not accurately accessible for certain variables, we derived daily statistics based on the yearly data and seasonal peaks to preserve reasonableness [8]. The model's timescale was set from September to March (approximately 180 days), which was sufficient to accommodate for understanding a typical risk event, and the time step was set as 1 day. The mathematical settings of variables and links in Vensim DSS are illustrated in the following subsections.

3.1. The System's Boundaries and Structure

After analyzing the inherent interactions among the key risks and markets, this model determined the main variables, as well as their mutual influences and causal relationships. To describe the prominent variables and links in a structured way, the definitions and functions of the components within each market are presented in the subsequent subsections. Two bold arrows represent the interactions between the natural gas and electricity markets. Risk factors affecting each market may propagate to the coupled markets via these interactions. Different strategies can be adopted in response to the risks, including reducing demand and improving the emergency supplies [8], which can help the system to recover from risks or amplify the losses. As described below, four causal feedback loops can be easily observed.

- Loop number 1 (length = 5): Natural gas market satisfaction rate → emergency supply ratio → emergency natural gas supply rate → overall natural gas supply rate → actual natural gas supply rate;
- Loop number 2 (length = 6): Natural gas market satisfaction rate → natural gas demand compression rate → other sectors' gas demand rate → overall natural gas demand rate → actual natural gas demand rate → total natural gas demand;
- Loop number 3 (length = 6): Electricity market satisfaction rate → electricity demand compression rate → other sectors' electricity demand → overall electricity demand rate → actual electricity demand rate → total electricity demand;
- Loop number 4 (length = 20): Electricity market satisfaction rate → electricity demand compression rate → other sectors' electricity demand → overall electricity demand rate → actual electricity demand rate → electricity demand generated by natural gas → electricity generation gas demand rate → overall natural gas demand rate → actual natural gas demand rate → total natural gas demand → natural gas market satisfaction rate → emergency supply ratio → emergency natural gas supply rate → overall natural gas supply rate → actual natural gas supply rate → natural gas supply rate for electricity generation → natural gas electricity generation rate → overall electricity supply rate → actual electricity supply rate → total electricity supply.

Among the causal feedback loops, the notation "B" suggests a balancing loop that stabilizes the systems, while the notation "R" implies a reinforcing loop that amplifies the system's changes. These feedback loops foster the complex evolution of the coupled markets in the face of risks.

3.2. Natural Gas System

The supply subsystem and demand subsystem constitute the natural gas market, with the variable of the natural gas satisfaction rate being the indicator of market efficiency. Pipeline defects, import shortages and extreme weather are the risk factors affecting the market, and a decline in the satisfaction rate induced by these risks may trigger an increase in response strategies such as the natural gas compression rate and the emergency supply ratio.

In the supply subsystem, the different sources of natural gas were divided into domestic natural gas supply and international natural gas imports. According to the annual data derived from the available reports, the daily amount of domestic natural gas supply was set as 5.687 hundred million cubic meters per day (hMm^3/d) and the normal import rate as 4.603 hMm^3/d. However, import shortages may influence the actual import rate of natural gas. As defined in Equation (1), if the supply is cut or deliberate attacks occur at time 20, an excessively large decline emerges, and these events can barely be resolved within months. The supply of natural gas is transported to the end-users, during which, pipeline defects may occur, creating continual disturbances on the pipeline's efficiency, as denoted by Equation (2). If the natural gas supply declines, the emergency natural gas supply can supplement the supply shortage, which is constrained by the natural gas reserve capacity, the emergency supply capacity and the emergency supply ratio, as shown in Equation (3). According to real data and emergency policies, the natural gas reserve capacity was set as 261 hMm^3/d and the maximum daily emergency supply capacity as 2.058 hMm^3/d. The emergency supply ratio is highly dependent on the natural gas market's satisfaction rate; hence, a lookup function was used in Equation (4). When the satisfaction rate is lower than 95%, a proportion of emergency gas is supplied, and if the gap increases, the level of urgency rises. Full capacity is used if the market satisfaction rate is below 80%.

$$\text{Import shortages} = \text{IF THEN ELSE (Time plus} > 20, 0.9, 0) \tag{1}$$

$$\text{Pipeline defects} = \text{RANDOM NORMAL } (0.05, 0.25, 0.15, 0.05, 0.15) \tag{2}$$

$$\text{Emergency natural gas supply rate} = \text{IF THEN ELSE (Natural gas reserve capacity} > 0\text{, Emergency supply capacity} \times \text{Emergency supply ratio, 0),} \quad (3)$$

$$\text{Emergency supply ratio} = \text{lookup (Natural gas market satisfaction rate).} \quad (4)$$

In the demand subsystem, demands from the residential sector, the electricity sector and other sectors such as industry and transportation were considered. The daily residential gas demand rate was 3.6 hMm^3/d, with the other sectors' gas demand rate being equal to 5.16 hMm^3/d. The demand for electricity generation is dependent on the proportion of natural gas in electricity generation and the electricity demand generated by natural gas, as denoted in Equation (5). According to the data for 2021, natural gas accounts for 3% of electricity generation in China. Extreme weather in the winter may cause a natural gas demand peak. Normally, for every degree below 0 °C the temperature drops, the natural gas demand rises by 2%, as shown in Equation (6). The rise in demand may also cause dissatisfaction in the natural gas market, and a strategy of demand compression can be applied to mitigate the gap. As the residential sector always has the highest supply priority, we can assume that this compression occurs in other sectors such as in the industry. Equation (7) demonstrates that a high-level emergency triggers demand compression, and these actions experience a delay from the time when the market disruption occurred.

$$\text{Electricity generation gas demand rate} = \text{Consumption of natural gas per kwh} \times \text{Electricity demand generated by natural gas,} \quad (5)$$

$$\text{Natural gas demand peak} = \text{IF THEN ELSE (Extreme weather} \geq 0\text{, 0, 0.02} \times (-\text{Extreme weather)),} \quad (6)$$

$$\text{Natural gas demand compression rate} = \text{DELAY1 (IF THEN ELSE (Natural gas market satisfaction rate} \geq 0.8\text{, 0, 0.2), 10),} \quad (7)$$

3.3. Electricity System

The electricity market also consists of the electricity supply subsystem and the electricity demand subsystem, with the variable of the electricity satisfaction rate being the indicator of market efficiency. Infrastructural damages, electricity overload and extreme weather are the risk factors affecting the market, and a decline in the satisfaction rate induced by these risks may trigger response strategies such as electricity compression. As electricity cannot be stored, the additional supply under emergency circumstances was not considered to be a prominent risk response strategy in our experiments.

In the supply subsystem, multiple fuels can be adopted for generating electricity. We divided the sources into natural gas electricity generation and other fuels used for electricity generation, as natural gas was our main research focus. As defined in Equations (8) and (9), the natural gas electricity generation rate relies on the natural gas supply rate for electricity generation, which, in turn, depends on the actual natural gas supply rate from the natural gas market, as well as the proportion of electricity in natural gas consumption. According to the annual statistical data, the percentage of natural gas used for electricity generation was 16% of the total natural gas supply. However, during the transmission of electricity, damage to the infrastructure may also emerge and increase the rate of electricity transmission loss. Efforts would be made to repair the damaged infrastructure and recover the efficiency of electricity transmission to a normal level, as shown in Equations (10) and (11). China's Statistical Yearbook suggests a normal transmission loss rate of 5.26%.

$$\text{Natural gas electricity generation rate} = \text{Natural gas supply rate for electricity generation/Consumption of natural gas per kwh,} \quad (8)$$

$$\text{Natural gas supply rate for electricity generation} = \text{Actual natural gas supply rate} \times \text{Proportion of electricity in natural gas consumption,} \quad (9)$$

$$\text{Infrastructure damages} = \text{RANDOM NORMAL (0.3, 0.4, 0.3, 0.05, 0.3)} \times \text{(STEP (1, 70) + STEP (}-1\text{, 120)),} \quad (10)$$

$$\text{Electricity transmission efficiency} = 1 - \text{Infrastructure damages} - \text{Normal transmission loss rate}, \quad (11)$$

In the demand subsystem, the residential and nonresidential sectors, such as the industrial sector, were evaluated in Equation (12). The daily residential electricity demand rate was 32.17 hundred million kWh/d, while the daily normal industrial electricity demand rate equaled 195.52 hundred million kWh/d. Extreme weather in the winter may result in an electricity overload for heating or other purposes. Normally, when the temperature falls by 1 °C below 0, electricity consumption rises by 3%, as stated in Equation (13). The rise in demand may also produce dissatisfaction in the electricity market. Thus, a demand compression strategy might be utilized to close the gap. Since the residential sector always has the greatest supply priority, we also assumed that this compression would happen in other sectors. As illustrated in Equations (14) and (15), a high-level emergency generates demand compression, and these actions are delayed after the time when the market disruption occurred.

$$\text{Overall electricity demand rate} = \text{Other sector electricity demand} + \text{Residential electricity demand rate}, \quad (12)$$

$$\text{Electricity overload} = \text{IF THEN ELSE (Extreme weather} \geq 0, 0, 0.03 \times (-\text{Extreme weather})), \quad (13)$$

$$\text{Electricity demand compression rate} = \text{DELAY1 (IF THEN ELSE (Electricity market satisfaction rate} \geq 0.8, 0, 0.2), 10), \quad (14)$$

$$\text{Other sector electricity demand} = \text{Normal industrial electricity demand rate} \times (1 - \text{Electricity demand compression rate}), \quad (15)$$

4. Experiment and Results

After constructing the model, we first conducted tests to verify its authenticity, then the behaviors of coupled markets were simulated through multiple risk assessment experiments, namely, (1) the impact of individual risks on the coupled markets; (2) how the coupled natural gas–electricity market would evolve, considering the dual interactions; and (3) the possible market adaptation behaviors, considering whether the risks could be mitigated by different response strategies.

4.1. Authenticity Test

The validity of a model should be confirmed after the formulation of the model's structure and data collection. It is very critical to ensure that the model dynamically captures the relationships among the variables and that the model fits real-world conditions. First, since the variables were specified using open and reliable data sources and expert opinions, the equations were meaningful. A unit check using Vensim software was successful, and thus the model was deemed to be suitable and reasonable. We assumed two extreme scenarios and chose two representative variables to demonstrate if the coupled markets' actions were compatible with reality. Figure 3 displays the results when the systems are free from risks and when both the natural gas and the electricity markets collapse. We can see that in the absence of any risks, the markets' supply can meet demand at a good rate, and the emergency supply is not activated, keeping the natural gas reserve capacity constant. However, when both markets fail, we can see a rapid drop in the system's overall satisfaction rate. Additional natural gas will then be provided, depleting the natural gas reserves. The natural gas reserves are exhausted on day 126, and the system continues to collapse. The authenticity test results demonstrated that the model was both effective and valid.

4.2. Assessment of Individual Risks

This study first evaluated the influence of the prominent risk factors on the coupled markets when each risk factor functioned independently and the market did not apply a risk response strategy. Given that extreme weather affects both the natural gas and electricity systems, it disrupts the overall systems' performance by influencing demand peaks in natural gas and electricity overload. Moreover, as illustrated by practical conflicts, geopolitical risks often affect the natural gas market via its influence on the risk of import shortages. Hence, we reduced the initial list of risk factors to four tests. Figure 4 exhibits the input value of the risk factors described in the previous section, and Table 2 demonstrates their impacts on the coupled natural gas–electricity market with regard to the reduction in the satisfaction rate.

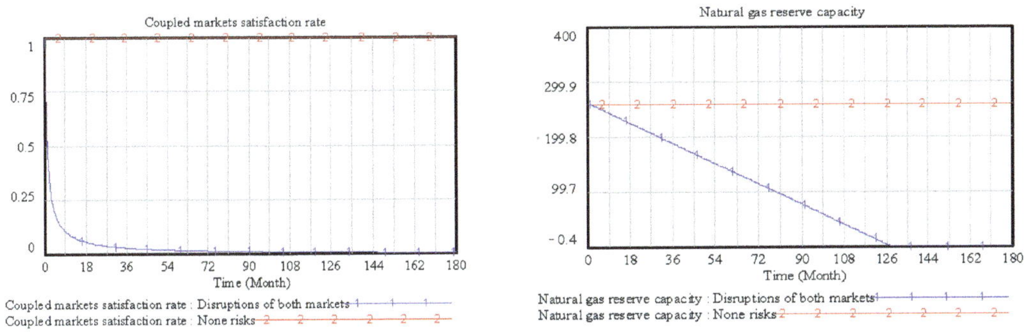

(a) (b)

Figure 3. Authenticity test results: (**a**) the behavior of the coupled markets' satisfaction rate under two extreme scenarios; (**b**) the behavior of natural gas reserves under two extreme scenarios.

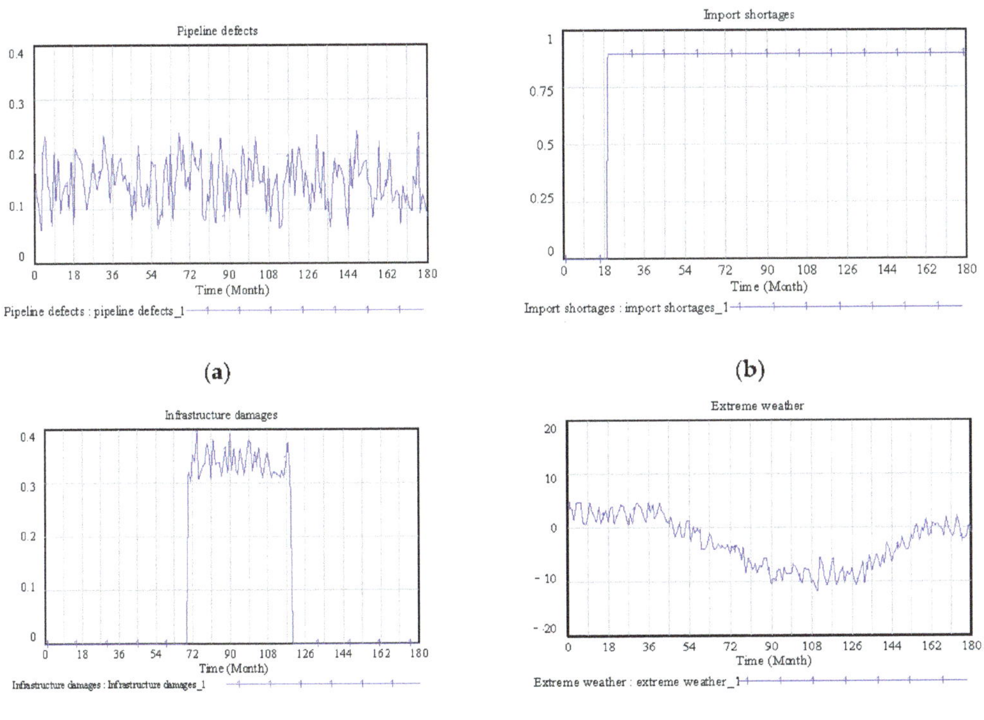

(a) (b) (c) (d)

Figure 4. Effects of the experimental risk factors: (**a**) pipeline defects; (**b**) import shortages; (**c**) infrastructural damages; (**d**) extreme weather.

Table 2. Market satisfaction losses induced by individual risks.

Risks	Coupled Markets	Natural Gas Market	Electricity Market
Pipeline defects	7.35%	14.69%	0%
Import shortages	17.88%	35.52%	0.28%
Infrastructural damages	4.53%	0%	9.06%
Extreme weather	8.63%	8.23%	9.03%

Regarding pipeline defects, it can be observed that defects such as corrosion and pipeline aging impose long-term and continual disruptions of the efficiency of natural gas transportation, which, in turn, affect the satisfaction of the natural gas market. According to the simulation's results, pipeline defects reduce the natural gas satisfaction rate by 14.69%. However, although it has an effect on the amount of natural gas supply that can be provided to generate electricity, its impact on the electricity market's satisfaction is invisible. This could be because the losses caused by this risk factor are relatively minor. Because only 16% of natural gas is used to generate electricity, and only 3% of the total electricity supply is provided by natural gas generation, such a minor disturbance can be considered negligible in an assessment of the electricity market.

Regarding import shortages, as previously indicated, if import shipments are cut off or an intentional attack is carried out on the key pipelines, this would result in a significant decrease in import supplies. In Figure 4, the natural gas market faces an import shortage starting on day 20, lowering the efficiency of imports to 10% of the initial level. Because of a lack of improvement in the global situation and the difficulty of reconstructing the pipeline, gas imports were still not recoverable within the experimental period (180 days). In light of the fact that natural gas imports constitute a proportion of China's overall natural gas supply, this significant import shortage would result in a loss of 35.52% in the satisfaction rate. Although natural gas supply accounts for a small fraction of the electricity market, the risk is transmitted to the overall markets through interactions as a consequence of the severe supply crisis. This results in a loss of 0.28% in the electricity satisfaction rate, and the loss rate of the coupled markets is around 17.88%.

If we consider damage to the infrastructure, despite the normal electricity transmission loss rate, the transmission infrastructure is frequently exposed to abrupt disruptions owing to exterior damage and other events, resulting in a considerable loss in transmission efficiency. Over a period of time, the infrastructure may be restored and returned to its normal transmission level. In Figure 4, the transmission infrastructure was severely damaged on day 70 and rebuilt on day 120. This reduced the satisfaction rate of the electrical market by 9.06%. However, because we ignored the impact of electricity supply on natural gas production, and the electricity market demand remained constant, the risk of infrastructural damage on the supply side of the electricity market was not propagated to the natural gas market.

In terms of extreme weather, from September to March, the coupled natural gas–electricity market first sees a decrease in temperature, followed by a rebound. In the simulation, the temperature began to oscillate downward after day 40 and stayed excessively low from day 90 to day 130, after which it rose upward to a warm situation. This risk factor increased demand in both the natural gas and electricity markets, which ultimately resulted in a decrease in the satisfaction rate of 8.23% and 9.03%, respectively. Note that in accordance with Table 2, even though in terms of the average loss in the market satisfaction rate of the coupled market, import shortages were ranked as the most significant risk factor because of the severity of their effects, extreme weather could not be ignored, as it influenced both markets.

4.3. Evolution of the Risk Behaviors Considering Dual Interactions

Section 4.2 quantitatively assessed the extent to which the risk factors, if functioning independently, would ultimately cause losses to the coupled natural gas–electric market.

If we consider the dynamism further, Figure 5 displays the evolutionary behavior of the coupled markets' key indicators over time when each of the four risk factors occur alone and when all the risk factors operate in collaboration. In particular, we focused on three indicator variables: the natural gas market's satisfaction rate, the electricity market's satisfaction rate and the coupled markets' satisfaction rate, which was calculated by combining the first two.

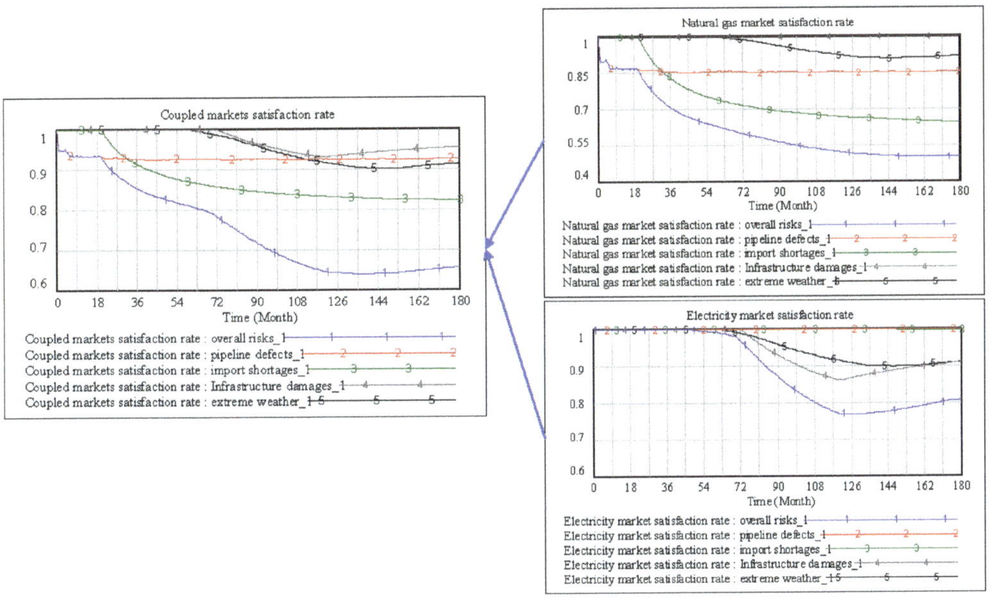

Figure 5. Dynamic behavior of the key variables, considering multiple risks and dual interactions.

First, the satisfaction rate of the natural gas and electricity markets were investigated separately. Consistent with the previous discussion, it could be observed that the natural gas market was primarily influenced by three risk factors: pipeline failures, import shortages and extreme weather. In contrast, the electricity market was heavily influenced by the risks of extreme weather and damage to the infrastructure, whereas pipeline failures and import shortages had negligible effects. For the natural gas market, pipeline failures lowered the market satisfaction rate, but the emergence of import shortages caused the market satisfaction rate to drop rapidly, surpassing pipeline failures as the most significant risk factor after day 32. The disruption caused by extreme weather shifted from growing to decreasing. However, even when the temperature had almost returned to the initial level, the market satisfaction rate did not return to normal. Since import shortages persisted as the most prevalent risk factor, the natural gas market satisfaction rate continued to decline, reaching the lowest level of 50.40% at the end of the simulation. For the electricity market, while infrastructural damage did not occur until day 70, after the emergence of extreme weather, it soon became the dominant risk factor owing to the substantial damage. After recovery of the damaged infrastructure on day 130, the satisfaction rate of the electricity market progressively recovered after day 157. The satisfaction rate of the electricity market showed a pattern of declining and then rising, with the lowest value occurring on day 121, at around 76.90%.

We then looked into the coupled markets' satisfaction rate. Considering the behavior of both markets and their interactions, we could observe that as time progressed, the dominant risk factors evolved, inducing more complexity for decision-makers attempting to make an adequate risk assessment. First, the different risk factors had their own evolutionary patterns. While pipeline defects and import shortages occurred abruptly and then persisted, infrastructural damages and extreme weather both showed a recovery trend, and the recovery pattern of extreme weather was even more obvious than that in each individual market. Second, various risk factors could be emphasized at different times. For instance, after the import shortages exceeded pipeline defects as the dominant risk factor, the disruptions induced by infrastructural damage also surpassed those induced by pipeline defects for some time. Third, the aggregated effects of minor risks may have serious effects on the coupled systems. While the disruptions from the most severe risk factor caused a decline of 17.88%, the satisfaction rate of the coupled market dropped by as much as 35.60% (e.g., on day 125). Hence, to better prevent crises, a holistic and dynamic perspective is essential when monitoring the performance of energy markets.

4.4. Risk Response Strategies and Their Effects

In the preceding section, the two most common risk response strategies for the markets were defined as demand compression and emergency supply. Similar tests were conducted in four scenarios: no risk response, emergency supply only, demand compression only and the two strategies together. As shown in Figure 6, both strategies were effective for responding to the overall risks, and the system's resilience was the highest when both strategies were applied collaboratively. However, situations may arise in which decision-makers have to choose between the two alternatives due to restricted resources and time.

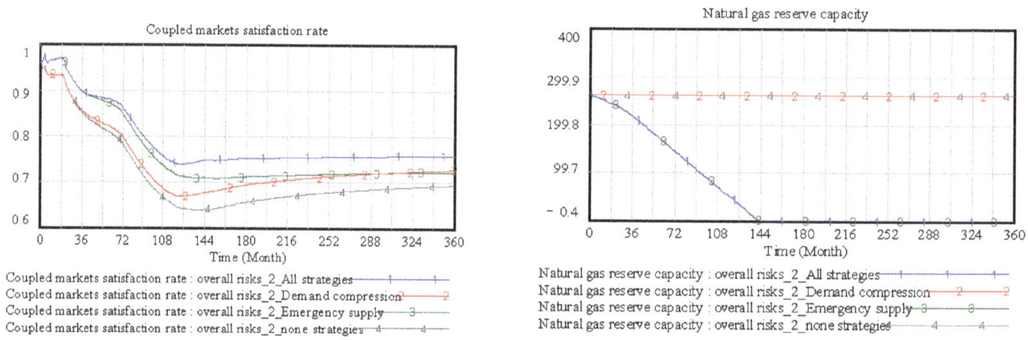

(a) (b)

Figure 6. Effects of the risk response strategies: (**a**) the behavior of the coupled markets' satisfaction rate for different risk response strategies; (**b**) the behavior of the capacity of natural gas reserves for different risk response strategies.

To properly investigate the long-term effects of different strategies, we extended the experimental duration from 180 days to 360 days. It can be observed that in the previous short-term experiments, the emergency supply of natural gas quickly brought the coupled market back to normal in the face of minor disturbances, and that even in a more severe risk scenario (e.g., when import shortages occurred), its response efficiency was also higher. At the end of day 180, the satisfaction rate of the coupled natural gas–electricity market using the emergency supply strategy was 71.28%, while the demand compression strategy achieved a rate of 69.49%. However, the improvements brought about by the emergency supply of natural gas stagnated with time, and demand compression became the superior strategy after day 297. This fact derives from the restrictions in the capacity of the natural gas reserves. While supplementation by additional natural gas is beneficial for mitigating the demand–supply gap induced by risks, the reserve capacity of a specific country is limited. Without boosting that capacity, this strategy will collapse if a crisis persists for an extended period of time.

5. Conclusions

By considering the complexity of multiple risks, the interactions among risks and market interactions, this study provided a comprehensive and transparent overview so that decision-makers could understand the evolving patterns of the risks influencing the coupled natural gas–electricity market. It first describes a list of the prominent risk factors and dual interactions based on a literature review and by tracking news about real-world accidents. Subsequently, a system dynamics model was constructed for the risk assessment. Four causal feedback loops were formulated that captured the dynamism and complexity embedded in the evolution of the coupled markets. Using China as an example, all variables were determined using China's Energy Statistical Yearbook, China's Statistical Yearbook, and other open and reliable data sources. After the construction of the model, three experiments were conducted, investigating the impact of each individual risk factor on the coupled market, the dynamic behaviors of the markets considering the dual interactions and a comparison of the two risk response strategies. The main findings are as follows.

- The dynamism and complexity all highly influence the results of the risk assessment. On the one hand, the dominant risk factors may evolve and change over time. The results in Table 2 demonstrated that among individual risks, the risk factor of import shortages ranked as the most severe one. It is in line with previous findings in the literature [8] concerning China's relatively high natural gas import dependency. However, the damage caused by this risk factor did not surpass that of pipeline defects until day 32, calling for a transition of the risk assessment's focus. On the other hand, the aggregated effects of multiple risks and multiple markets may induce a severe crisis even if the initial disturbances are minor. As illustrated in Figure 5, even though the decline in satisfaction rate caused by individual risks did not exceed 17.88%, the coupled market could see a decline of over 35% (e.g., on day 125).
- Risk factors can propagate from one market to another via interactions, yet they depend on multiple aspects such as the severity of the risks and the intensity of the interactions. In our experiments, given the fact that natural gas only accounts for about 3% of the electricity generation in China, the propagation effect was not obvious (please see Table 2, despite the extreme weather that affected both markets, only the abnormal shock of import shortages was observed propagating from the natural gas market to the electricity one). Compared with studies in other empirical backgrounds, however, the situation is different. For instance, in the European Union, where the share of electricity production from natural gas equals approximately 14%, the side-effects of shortages in the natural gas market supply on other energy markets have been partially observed [3,27]. This comparison provides transparent evidence to explain why the development of alternative energy sources is encouraged to improve the energy security, and to which extent it can save losses in the overall coupled market.
- Risk response strategies such as demand compression and emergency supply contribute to the recovery of the markets. Considering these two commonly used policies to tackle the natural gas and electricity insufficiency [8,25], our experiments revealed that they performed differently with the varying lengths of time. In the short term, an emergency supply will soon compensate the demand–supply gap, but this is always constrained by the country's reserve capacity. The demand compression strategy may create persistent improvements in the markets and thus perform better for long-term risk recovery. Note that as the expansion of reserve capacity calls for substantial investments, and the compression of the gas or electricity demand has the potential to influence the economy, the portfolio of risk response strategies should be further investigated through a financial analysis.

The following are policy recommendations based on the findings presented above: An isolated and static perspective of risk assessment is inevitably inaccurate; instead, monitoring the process and controlling the overall market are required to avert crises. Using the developed approach, decision-makers can identify when various disruptions may occur and which risk factors account for their occurrence and keep an eye on impending severe risks. For countries like China who have started embracing a new era of clean energy, determining the degree of long-term interactions between multiple energy markets is vital to guarantee energy security. In addition, among the risk response strategies, while the emergency supply strategy soon recovered the markets, the compression of demand had a longer enduring impact. Hence, decision-makers should strike a balance between the short-term and long-term effects of strategies, rather than adopting a myopic view.

The contribution of this study manifests in three aspects. First, it establishes an integrated framework for multiple stakeholders from different sectors to have a more systematic look at the underlying risks, with the objective of enhancing the overall performance of the coupled market. Second, the proposed model quantitatively captures both the stochastic nature of risks and the nonlinearity of interactions, offering a cost-effective and dynamic instrument that supports the whole risk assessment process through explicit experiments. Third, the visualization results in transparent graphics can help decision-makers to easily examine the evolutionary impact of risks and compare the consequences of various policies. Some limitations also exist that inspire future research. For instance, while the functional interactions are under investigation, geographical interactions may also contribute to the propagation of risks. It is possible to better characterize the complexity of relationships by using hybrid models that incorporate both geographical and functional information. Moreover, due to the complexity embedded, this study focuses primarily on how risks may result in a supply–demand imbalance and how various strategies will mitigate the gaps. Since resources in practice are often limited, when developing risk response strategies, multiple factors regarding the financial constraints and the carbon emissions can also be considered.

Author Contributions: Conceptualization and methodology, L.W.; validation and data curation, Y.X. All authors have read and agreed to the published version of the manuscript.

Funding: This research was funded by the MOE (Ministry of Education in China) Project of Humanities and Social Sciences (no. 22YJC630137), the Shanghai Yang Fan Program (no. 22YF1401600), and the Fundamental Research Funds for the Central Universities (no. 2232022E-04).

Data Availability Statement: Not applicable.

Conflicts of Interest: The authors declare no conflict of interest.

References

1. Xiao, B.; Niu, D.; Guo, X. Can natural gas-fired power generation break through the dilemma in China? A system dynamics analysis. *J. Clean. Prod.* **2016**, *137*, 1191–1204. [CrossRef]
2. Wójcik-Jurkiewicz, M.; Czarnecka, M.; Kinelski, G.; Sadowska, B.; Bilińska-Reformat, K. Determinants of Decarbonisation in the Transformation of the Energy Sector: The Case of Poland. *Energies* **2021**, *14*, 1217. [CrossRef]
3. Esmaeili, M.; Shafie-khah, M.; Catalão, P.S. A system dynamics approach to study the long-term interaction of the natural gas market and electricity market comprising high penetration of renewable energy resources. *Int. J. Elec. Power* **2022**, *139*, 108021. [CrossRef]
4. Ding, C.; Zhou, Y.; Ding, Q.; Li, K. Integrated Carbon-Capture-Based Low-Carbon Economic Dispatch of Power Systems Based on EEMD-LSTM-SVR Wind Power Forecasting. *Energies* **2022**, *15*, 1613. [CrossRef]
5. Chen, J.W.; Yu, T.; Xu, Y.; Cheng, X.H.; Yang, B.; Zhen, B.M. Fast analytical method for reliability evaluation of electricity-gas integrated energy system considering dispatch strategies. *Appl. Energy* **2019**, *242*, 260–272.
6. Chen, Z.; An, H.Z.; An, F.; Guan, Q.; Hao, X.Q. Structural risk evaluation of global gas trade by a network-based dynamics simulation model. *Energy* **2018**, *159*, 457–471. [CrossRef]
7. IEA. *World Energy Outlook 2017*; OECD: Paris, France, 2017.
8. Ding, Y.; Chen, S.; Zheng, Y.L.; Chai, S.L.; Nie, R. Resilience assessment of China's natural gas system under supply shortages: A system dynamics approach. *Energy* **2022**, *247*, 123518. [CrossRef]
9. Global Gas Markets to Remain Tight Next Year Amid Supply Squeeze, Energy Agency Says. Available online: https://www.reuters.com/business/energy/global-gas-markets-remain-tight-next-year-amid-supply-squeeze-iea-2022-10-03/ (accessed on 3 October 2022).
10. IEA. *Gas Market Report (Q4-2022) Including Global Gas Security Review 2022*; IEA: Paris, France, 2022.

11. Suo, W.L.; Wang, L.; Li, J.P. Probabilistic risk assessment for interdependent critical infrastructures: A scenario-driven dynamic stochastic model. *Reliab. Eng. Syst. Saf.* **2021**, *214*, 107730. [CrossRef]
12. Tang, Y.; Zhao, Y.; Li, W.Y.; Xie, K.G.; Yu, Y. Incorporating gas pipeline leakage failure modes in risk evaluation of electricity-gas integrated energy systems. *Int. J. Electr. Power Energy Syst.* **2023**, *144*, 1–9. [CrossRef]
13. Ahmad, S.; Mat Tahar, R.; Muhammad-Sukki, F.; Munir, A.B. Application of system dynamics approach in electricity sector modelling: A review. *Renew. Sust. Energy Rev.* **2016**, *56*, 29–37. [CrossRef]
14. Dong, X.; Kong, Z. The impact of China's natural gas import risks on the national economy. *J. Nat. Gas Sci. Eng.* **2016**, *36*, 97–107. [CrossRef]
15. Egging, R.; Holz, F. Risks in global natural gas markets: Investment, hedging and trade. *Energy Policy* **2016**, *94*, 468–479. [CrossRef]
16. Liu, A.; Chen, K.; Huang, X.F.; Li, D.D.; Zhang, X.C. Dynamic risk assessment model of buried gas pipelines based on system dynamics. *Reliab. Eng. Syst. Saf.* **2021**, *208*, 107326. [CrossRef]
17. Chen, Q.; Zuo, L.; Wu, C.; Cao, Y.; Bu, Y. Supply reliability assessment of a gas pipeline network under stochastic demands. *Reliab. Eng. Syst. Saf.* **2021**, *209*, 107482. [CrossRef]
18. Zarei, E.; Azadeh, A.; Khakzad, N.; Aliabadi, M.M.; Mohammadfam, I. Dynamic safety assessment of natural gas stations using Bayesian network. *J. Hazard. Mater.* **2017**, *321*, 830–840. [CrossRef]
19. Salman, A.M.; Li, Y. Multihazard Risk Assessment of Electric Power Systems. *J. Struct. Eng.* **2017**, *143*, 04016198. [CrossRef]
20. Tummala, R.V.; Mak, C.L. A risk management model for improving operation and maintenance activities in electricity transmission networks. *J. Oper. Res. Soc.* **2001**, *52*, 125–134. [CrossRef]
21. Chiaradonna, S.F.; Giandomenico, F.D.; Lollini Paolo, P. Definition, implementation, and application of a model-based framework for analyzing interdependencies in electric power systems. *Int. J. Crit. Infrastruct.* **2011**, *4*, 24–40. [CrossRef]
22. He, Y.; Jiao, J.; Chen, Q.; Ge, S.F.; Chang, Y.; Xu, Y. Urban long term electricity demand forecast method based on system dynamics of the new economic normal: The case of Tianjin. *Energy* **2017**, *133*, 9–22. [CrossRef]
23. Burger, M.; Graeber, B.; Schindlmayr, G. *Managing Energy Risk: An Integrated View on Power and Other Energy Markets*, 2nd ed.; Wiley: Hoboken, NJ, USA, 2014; pp. 101–151.
24. Hibbard, P.; Schatzki, T. The Interdependence of Electricity and Natural Gas: Current Factors and Future Prospects. *Electr. J.* **2012**, *25*, 6–17. [CrossRef]
25. Ordoudis, C.; Pinson, P.; Morales, J.M. An integrated market for electricity and natural gas systems with stochastic power producers. *Eur. J. Oper. Res.* **2019**, *272*, 642–654. [CrossRef]
26. Tian, R.; Zhang, Q.; Wang, G.; Li, H.; Chen, S.; Li, Y.; Tian, Y. Study on the promotion of natural gas-fired electricity with energy market reform in China using a dynamic game-theoretic model. *Appl. Energy* **2017**, *185*, 1832–1839. [CrossRef]
27. Beyza, J.; Ruiz-Paredes, H.F.; Garcia-Paricio, E.; Yust, J.M. Assessing the criticality of interdependent power and gas systems using complex networks and load flow techniques. *Phys. A* **2020**, *540*, 123169. [CrossRef]
28. Bao, Z.; Jiang, Z.; Wu, L. Evaluation of bi-directional cascading failure propagation in integrated electricity-natural gas system. *Int. J. Electr. Power Energy Syst.* **2020**, *121*, 106045. [CrossRef]
29. Portante, E.J.; Kavicky, J.A.; Craig, B.A.; Talaber, L.E.; Folga, S.M. Modeling electric power and natural gas system interdependencies. *J. Infrastruct. Syst.* **2017**, *23*, 04017035. [CrossRef]
30. Poljanšek, K.; Bono, F.; Gutiérrez, E. Seismic risk assessment of interdependent critical infrastructure systems: The case of European gas and electricity networks. *Earthq. Eng. Struct. Dyn.* **2012**, *41*, 61–79. [CrossRef]
31. Nazari-heris, M.; Jabari, F.; Mohammadi-ivatloo, B.; Asadi, S.; Habibnezhad, M. An updated review on multi-carrier energy systems with electricity, gas, and water energy sources. *J. Clean. Prod.* **2020**, *275*, 123136. [CrossRef]
32. Eusgeld, I.; Nan, C.; Dietz, S. System-of-systems approach for interdependent critical infrastructures. *Reliab. Eng. Syst.* **2011**, *96*, 679–686. [CrossRef]
33. Zhang, B.; Hu, W.; Li, J.; Cao, D.; Huang, R.; Huang, Q.; Chen, Z.; Blaabjerg, F. Dynamic energy conversion and management strategy for an integrated electricity and natural gas system with renewable energy: Deep reinforcement learning approach. *Energy Convers. Manag.* **2020**, *220*, 113063. [CrossRef]
34. Lin, W.; Kunc, M.; Li, J.P. Realizing value from project implementation under uncertainty: An exploratory study using system dynamics. *Int. J. Proj. Manag.* **2017**, *35*, 341–352.
35. Suo, W.L.; Zhang, J.; Sun, X.L. Risk assessment of critical infrastructures in a complex interdependent scenario: A four-stage hybrid decision support approach. *Saf. Sci.* **2019**, *120*, 692–705. [CrossRef]
36. Haimes, Y.Y. Risk modeling of interdependent complex systems of systems: Theory and practice. *Risk Anal.* **2018**, *38*, 84–98. [CrossRef] [PubMed]
37. Ianaloo, B.; Tansel, B.; Shames, K.; Jin, X.; Gan, A. A decision aid GIS-based risk assessment and vulnerability analysis approach for transportation and pipeline networks. *Saf. Sci.* **2016**, *84*, 57–66. [CrossRef]
38. Kong, J.J.; Simonovic, S.P. Probabilistic multiple hazard resilience model of an interdependent infrastructure systems. *Risk Anal.* **2019**, *39*, 1843–1863. [CrossRef] [PubMed]
39. Salman, A.M.; Li, Y. A probabilistic framework for multi-hazard risk mitigation for electric power transmission systems subjected to seismic and hurricane hazards. *Struct. Infrastruct. Eng.* **2018**, *14*, 1499–1519. [CrossRef]
40. Xian, H.; Jeong, C.E. Modeling the damage and recovery of interdependent critical infrastructure systems from natural hazards. *Reliab. Eng. Syst. Saf.* **2018**, *177*, 162–175.

41. Vairo, T.; Pontiggiab, M.; Fabiano, B. Critical aspects of natural gas pipelines risk assessments. A case-study application on buried layout. *Process Saf. Environ. Prot.* **2021**, *149*, 258–268. [CrossRef]
42. Tagliapietra, S.; Eri Enrico Mattei, F. Towards a new eastern Mediterranean energy corridor? natural gas developments between market opportunities and geopolitical risks. *FEEM Work Pap.* **2013**, *12*, 1–32. [CrossRef]
43. Burr, M. *The Geopolitical Risks of LNG*; Public Utilities Fortnightly: Arlington, VA, USA, 2005; pp. 29–33.
44. Cabalu, H. Indicators of Security of Natural Gas Supply in Asia. *Energy Policy* **2010**, *38*, 218–225. [CrossRef]
45. Rehman, O.; Ali, Y.; Sabir, M. Risk assessment and mitigation for electric power sectors: A developing country's perspective. *Int. J. Crit. Infrastruct. Prot.* **2022**, *36*, 100507. [CrossRef]
46. Duan, Y.; Sun, Y.; Zhang, Y.; Fan, X.Q.; Dong, Q.H.; Guo, S. Risk evaluation of electric power grid investment in China employing a hybrid novel MCDM method. *Mathematics* **2021**, *9*, 473. [CrossRef]
47. Sterman, J. *Business Dynamics: Modelling for a Complex World*; Mc-Graw Hill: New York, NY, USA, 2000; pp. 3–39.
48. Blair, C.; Gralla, E.; Wetmore, F.; Goentzel, J.; Peters, M. A Systems Framework for International Development: The Data-Layered Causal Loop Diagram. *Prod. Oper. Manag.* **2021**, *30*, 4374–4395. [CrossRef]

energies

MDPI

Article

Energy Literacy of Residents and Sustainable Tourism Interaction in Ethnic Tourism: A Study of the Longji Terraces in Guilin, China

Shaofeng Wu [1], Yanning Li [2], Changgan Fang [3,*] and Peng Ju [4,*]

1 College of Tourism, Hunan Normal University, Changsha 410081, China
2 School of Hospitality & Tourism Management, University of Surrey, Guildford GU2 7XH, UK
3 Scientific Research Office, Wuzhou University, Wuzhou 543002, China
4 Shenzhen Tourism College, Jinan University, Shenzhen 518053, China
* Correspondence: ff2986@163.com (C.F.); jupeng@sz.jnu.edu.cn (P.J.)

Abstract: Energy and environment form a nexus in which residents are the owners of tourism energy resources. Only a few studies have focused on the energy literacy of residents in ethnic tourism destinations and its impact on sustainable tourism. Using a qualitative research approach through field works and in-depth interviews in the Ping'an Village, Longji Terraces Scenic Area, this study explored the relationships between the energy literacy of residents and sustainable tourism in ethnic areas. The result showed that the energy literacy of the ethnic residents of Pingan village in terms of knowledge, attitude, and behavior has increased in line with the development of tourism, and both external and internal factors contribute to the improvement. Besides, the promotion of energy literacy among the residents not only has a positive impact on the tourists' behavior but also brings about effective improvements in the local energy use structure and infrastructure, thus contributing to the sustainable development of tourism. This research extends the understanding of energy literacy from the perspective of ethnic residents and changes in energy literacy in remote ethnic villages under tourism development. The results also deepen our understanding of such changes in the behavior of tourists and tourism destination sustainability and enrich the empirical research to promote energy conservation and sustainable tourism development in ethnic areas.

Keywords: Longji; ethnic areas; environment; ecological balance; tourism

Citation: Wu, S.; Li, Y.; Fang, C.; Ju, P. Energy Literacy of Residents and Sustainable Tourism Interaction in Ethnic Tourism: A Study of the Longji Terraces in Guilin, China. *Energies* **2023**, *16*, 259. https://doi.org/10.3390/en16010259

Academic Editors: Junpeng Zhu and Xinlong Xu

Received: 30 October 2022
Revised: 12 December 2022
Accepted: 21 December 2022
Published: 26 December 2022

1. Introduction

Energy is the basis for human survival and development and an essential resource for enhancing socioeconomic development. With the decline in global energy reserves, the issues of energy production and consumption have become a global concern. Although humans use science and technology to create a high quality of life, they also cause increasingly serious environmental pollution [1]. Energy and tourism are closely related; energy not only becomes an attraction but also a limitation to the tourism industry, where there is an increasing trend of energy consumption in tourism catering, accommodation, and transport [2]. There needs to be a coordinated effort between residents and tourists to effectively reduce the consumption of energy and emissions from tourism activities and protect the tourism landscape and ecological environment.

Energy literacy is an important concept related to the multifaceted phenomenon of energy consumption, and it has gained attention in many related fields and applications [3]. This concept can be explained as a person who is aware of the energy consumption of the appliances in their home, how to take the necessary measures to save energy, and the relationship between energy use and climate change [1–4]. Previous energy studies have focused on the concept of energy literacy [5,6], the relationship between energy knowledge, literacy and behavior, energy literacy scales, and factors that influence energy literacy [6,7];

however, few studies have discussed the relationship between energy and other social developments from a cross-disciplinary perspective. It is not surprising that the energy literacy of residents in ethnic tourist destinations remains unclear, as well as how energy literacy affects the development of tourist destinations.

Currently, tourism is an integrated industry in modern society that drives the flow and consumption of large numbers of tourists and logistics, and the energy behavior of its stakeholders is crucial for environmental sustainability [2]. Most of the energy-related tourism studies have discussed the benefits [8,9] and the constraints [10,11] that energy poses to tourism development, as well as the energy consumption and behavior in tourism [12,13]. Yet the relationship between tourism development and the energy literacy of residents remains unclear, particularly on how the energy literacy of ethnic village residents in remote areas has changed as a result of tourism development. Previous research on ethnic village tourism has mainly focused on the preservation of ethnic and cultural heritage [14], the participation of community residents in tourism [15], and the interaction between host and guest [16], with less attention paid to the energy environment of ethnic tourism destination. However, understanding how ethnic areas can escape poverty and gain knowledge of development for local economic and environmental sustainability are important for harmonious social development. Thus, this study focused on the Ping'an Zhuang Village in the Longji Terraces Scenic Area as an ethnic tourism site to explore the impact of the energy literacy of ethnic residents on sustainable tourism development by conducting field surveys and in-depth interviews and using qualitative research methods. Two questions answered in this study are as follows: How has tourism development brought about changes in residents' energy literacy? How does a change in energy literacy affect the sustainable development of local tourism? The results of this study can increase the understanding of the relationship between energy literacy and tourism development from a transdisciplinary perspective, enrich empirical research on energy literacy from the perspective of ethnic residents, and promote energy conservation and sustainable tourism development in ethnic areas.

2. Literature Review

2.1. Energy Literacy

Energy literacy is related to the perception and consumption of energy [5]. The earliest research focused on knowledge [17], while more recently, researchers have become increasingly aware of the importance of the willingness and actions of people [18]. When an individual is energy-aware, they know the energy production and consumption in everyday life, how to save energy in their home, how to adopt economic energy-efficient behaviors, and how their energy choice may be related to climate change [1]. Although the definition of energy literacy remains unclear, most authors have accepted that literacy should consist of three domains: knowledge, attitude (affect), and behavior [1,6,7]. Knowledge is a critical element in achieving energy literacy, comprising an understanding of energy efficiency, awareness of the environment, and the social impacts of energy production, distribution, and consumption [1,6,7]. Attitude refers to people's perceptions of the impact of energy issues on their lives and energy-related beliefs that are crucial for decision-making [1,6,7]. Though the affect may be instead of attitude in different studies, it means much the same as attitude, which refers to how you think of energy use [6]. Behavior includes personal attention to environmental issues and the need for energy efficiency, the responsibility each individual feels to be a citizen worldwide, and commitment to energy efficiency [1,6,7]. Previous studies have paid more attention to the relationship between energy knowledge, attitudes, and behavior. Hungerford and Volk found that knowledge contributes to behavior because it plays an important role in environmental protection decisions [19], whereas Alp et al. indicate that the energy attitudes of elementary school students significantly influence their environmental behavior [20]. Rioux proposed that neighborhood attachment as an affection variable is a critical element in the behavior of secondary school students. Energy knowledge of people, as well as their willingness and ability to act, are important for literacy [21]. De Waters and Powers also proposed that energy behavior and affect are more relevant than knowledge [6]. However, the traditional thinking that increasing knowledge could increase affect and result in behavioral changes has been gradually challenged by subsequent researchers. Numerous studies have reported that reasonable knowledge does not necessarily translate into sound behavior to save energy. For example, although Minnesota residents have already acquired some electricity-saving habits, they still show resistance to replacing equipment with more efficient ones and using public transport or more economical driving methods [1]. Similarly, although the Danes show a good energy knowledge level, they do not use more efficient equipment or renewable energy [22]. Chen et al. and Lee found that Taiwanese students achieved reasonable energy knowledge [23], but it did not seem to determine behavior change. University students in Portugal also showed a low level of concern and commitment to energy saving [24,25].

The scales and models of energy literacy also play an important role in energy literacy research. The foundation for developing a competitive scale for measuring energy literacy is still based on knowledge, attitude (affect), and behavior. In addition to the efforts of DeWaters to develop an energy literacy scale, Bodzin et al. developed two instruments to measure the energy literacy of middle-level education students: one related to knowledge and the other related to attitude and behavior [26]. Similarly, Kyriazi and Mavrikaki developed a scale to measure the environmental literacy of post-secondary students [27]. In general, the framework is defined as follows: (1) The knowledge dimensions refer to the influences of energy development, various types of daily energy use, basic rules of energy use, and the importance of efficient energy use. (2) The attitude dimensions refer to the influences of energy use in the daily lives of people, and the behavioral dimensions refer to the adoption of energy-efficient behaviors in daily life. Meanwhile, there are two main aspects of the discussion of energy literacy models: the education and the psychological model. The education model emphasizes the importance of acquiring energy, knowledge, and skills, which closely contribute to actions and behaviors. The psychological model encourages people to take responsibility for energy-saving actions. For example, Kollmuss and Agyeman proposed that internal and external factors are directly related to pro-environmental behavior [28].

The level of energy literacy is also related to personal characteristics and educational background; from the perspective of personal characteristics, gender is one of the most important factors. Studies have shown that although women have lower levels of energy-related financial literacy [29–31], they have higher levels of energy literacy than men [32]. Studies have also reported that the primary factors determining energy literacy are gender, going away from home to study, and the experience of energy poverty [33]. Meanwhile, age played an important role, and the results showed that children and elderly people presented lower levels of energy literacy [34]. Promoting the energy consumption habits of children has become a critical program; this includes increasing the frequency of energy-related curricular units, supplying a high level of parental education [35], and improving the practice of educational games [36,37]. Different educational backgrounds were also associated with different levels of energy literacy. As education programs can benefit students by achieving proficiency levels of energy literacy, students in different areas may have different opportunities to learn energy-saving knowledge and skills [38]. Differences also existed among students from different disciplines. For example, the results showed that students who majored in agriculture performed better than others [39]. Similarly, students of geography, earth and environmental sciences, marine sciences, engineering, and architecture obtained the broadest knowledge of energy issues from their education and are perceived to have a superior level of knowledge on the subject [40]. More recently, current research on energy literacy has been paying attention to more vulnerable energy users, suggesting that they may experience inadequate access to affordable and reliable energy services and have less financial and material resources to buffer harm [3].

Based on the above, previous studies on energy research have mostly discussed the concept of energy literacy [5,6], the relationship between energy knowledge, literacy, and behavior [19–21], scales of energy literacy [26,27], and factors influencing energy literacy [29,30,37], as shown in Table 1. However, few studies have addressed the relationship between energy and other social developments from a transdisciplinary perspective. Tourism, as an integrated industry in modern society, involves the flow and massive consumption of people and services, and the energy behavior of its stakeholders is crucial for environmental sustainability. Therefore, it is useful to focus on energy development from the tourism perspective.

Table 1. The development of energy literacy review (Source: by own study).

Review Items	Research Contents
the concept of energy literacy	the definition of energy literacy [1,6,7]
	relationship between energy knowledge, literacy, and behavior [6,19–21]
	knowledge does not necessarily contribute to energy save behavior [1,22–25]
scales of energy literacy	related to knowledge, related to attitude and behavior [26,27]
	the education and psychological model [28]
factors influencing energy literacy	Gender [29–32]; going away from home to study, and the experience of energy poverty; age [33]
	Different educational backgrounds [38]; different disciplines [39,40]; energy literacy of vulnerable energy users [41]

2.2. *Energy and Tourism*

Recent studies have shown that tourism and energy are interrelated. Energy tourism is often associated with industrial tourism to attract tourists, as some former industrial sites are still open or regenerated for tourism [8,9]. For example, decommissioned coal mining sites or New York's Highline park [41,42]. In addition, agritourism tourism can be considered a type of energy tourism since it is often connected to energy production activities on the farm, such as producing biogas on-site, growing energy crops, or grazing sheep on the meadow of photovoltaic plants [2,43]. Along with tourism destinations, energy tourism could also play an important role in improving the energy literacy of people and changing their energy use behavior, resulting in more sustainable energy citizens [44]. New forms like environmental education, displays of new technologies, interactive science experiments, and various outdoor activities, such as cycling, camping, or hiking, could be used in various types of energy tourism, for example, in ecological education centers, observation towers, and natural trails [2]. The aim is to improve energy knowledge through tourism; however, energy may also impose restrictions on tourism. Energy facilities engaged in the extraction and processing of energy resources can have a negative effect on the character and function of many energy landscapes [2,10]. Environmental pollution and poor landscape vision may discourage visitors from visiting these locations [45,46]. Previous studies also had different views on visitors' perceptions of different energy facilities. Some studies have suggested that visitor perceptions may vary depending on the form, location, and concentration of space; spatial closeness of the energy facility [2]; physical and social values of the local environment [47]; and the type and sociodemographic characteristics of the tourist [48]. In addition, some studies concluded that visually appealing energy facilities, such as large wind farms, may influence the choices of visitors and their intention to revisit. Similarly, studies have reported that there are no significant negative impacts of wind turbines on local tourism [49].

Tourist needs also lead to extensive consumption of energy, for example, fuels burned for traveling, heating, and cooling; chemical products for cleaning; and energy used for cooking. According to a UN report, each tourist produces 1 kg of solid waste per day.

Previous research on the environmental impact of energy consumption in tourism has been broadly discussed. Research showed that new recreational facilities and accommodations had been built to compete with more tourists, leading to more energy consumption [50]. Guests in hotels tend to use more towels, have longer showers and generate more waste than at home; plastic particles such as bags and bottles are discarded in scenic areas [51]. Becken et al. argued that the duration of visitor stay is the main variable affecting total energy use [52], but energy efficiency appears to rise with increasing stay [53]. While these actions are not necessarily caused by malicious intent, they could incur additional management costs, disturb residents, and endanger the ecosystem [54,55]. Furthermore, Becken and Simmons suggested that accommodation in tourism has a low energy consumption of fossil fuels while burning wood is a potential threat to the environment [56]. Fuelwood collecting is one of the prime causes of forest cover loss in remote tourism destinations [57], which may cause landslides and air pollution. Recently some scholars also stated that tourism could have negative externalities such as climate change and air pollution [58]. Similarly, Qureshi et al. confirmed that both environmental variables and air pollution are significantly associated with health services in Malaysia [59]. Some studies also found that economic growth and energy consumption influenced carbon emissions as more tourists arrived [60]. In addition, some studies have also discussed energy saving and pro-environmental behavior in tourism. Given the impact of tourism consumption on the environment, some researchers explored the factors that encourage more sustainable behavior, as these factors may bring about changes in visitor behaviors [61]. Such as pro-environmental behavior are actions that minimize harm or even benefit the environment [62], and pro-environmental tourists refer to those who try to alleviate their negative effects on the environment by adopting energy-saving behaviors in their trip [63]. Studies point to factors such as moral, affective attitudes, and environmental awareness are important to the formation of pro-environmental behavior [64]. Scholars also developed a scale to measure the pro-environmental contextual force that affects urban tourists' PEBs [51].

More recent studies also paid attention to energy literacy in tourism. Studies have shown that tourists are less aware of energy literacy at hotels than at home [65,66]. As previously mentioned, higher energy literacy contributes to energy protection; therefore, it is not surprising that energy literacy has recently gained increasing attention in tourism research. For example, Teng et al. discussed the impact of knowledge and the effect of hotel employees on energy literacy [13]. In addition, some studies have investigated the energy literacy characteristics of peasant households in rural tourism destinations. Zhang also found significant links between the energy knowledge, affect, and behavior of tourism farmers [4]. It has also been suggested that energy feedback is crucial in energy behavior change, and personal values and energy literacy also have an impact on changing energy consumption behavior [67].

Tourism results in a significant increase in energy consumption, which, in turn, leads to an increase in CO_2 emissions and climate change in the long term [68]. Hence, because of the close links between energy consumption and carbon emissions, some studies share significant similarities in carbon and energy literacy in terms of research targets. The reduction in CO_2 emissions depends on the social ethics and responsibility of tourists. Several studies indicate that enhancing carbon literacy is important for reducing carbon emissions and promoting public conversation [69,70]. Juvan and Dolnicar emphasized the crucial role of efficient communication in the low-carbon decisions of tourists [65].

Some researchers have considered energy and carbon literacy. Horng et al. developed a measurement scale for energy-saving literacy and carbon reduction in the tourism and hospitality industries. They also found differences between Taiwanese and Malaysian students in terms of knowledge, ecological concepts, attitude, sensitivity, locus of control, action intention, and action strategy [12]. Similarly, Teng explored the energy and carbon literacy structures in hospitality and tourism practices in Taiwan [71]. Some studies have also suggested that energy-efficient policies are indispensable for tourism. As such, improving energy efficiency and reducing energy waste through energy literacy is important for the sustainable growth of tourism.

Existing energy-related tourism research has primarily discussed energy tourism as a tourist attraction [44], energy facilities' impact on tourists' energy literacy [44], and the constraints that energy imposes on tourism development [10,11,45]. Some studies have also highlighted that tourism is associated with significant energy consumption, while others have discussed the energy literacy of hotel employees [12], types of energy literacy of residents in rural tourism destinations [4], and links between energy and CO_2 emissions [12,13,69,70]. More recently, current research on energy tourism has been expanded to measure the relationship between economic activities and energy consumption in ethnic regions [72]. Some researchers also suggested cycling in ethnic areas is means of low-carbon and fashionable traveling for sustainable tourism [73]. Similarly, the use of renewable energy and locally developed energy-saving technologies is increasing in tourist lodges in Nepal's ethnic region [74]. Luo also found that the absolute total emissions per visitor to one of China's ethnic tourism destinations have reduced slightly [75]. However, few studies have adequately discussed the relationship between tourism development and the energy literacy of residents. Ethnic areas need to break out of poverty and gain knowledge in development to promote local economic and environmental sustainability; these are important issues that require attention in the context of harmonious social development.

3. Research Methods

3.1. Case Introduction

The Longji Terraces are located in Longsheng and Guilin, China. They are among the most beautiful terraces in the world, as shown in Figure 1. According to historical records, they were built in the Qin Dynasty, shaped in the Ming Dynasty, and completed in the early Qing Dynasty nearly 2300 years ago. The terraces are charming and beautiful year-round. Ping'an is the central village of the Longji Terraces Scenic Area, towering over the spine of the Terraces. Since 1993, it has been developed for tourism for nearly 29 years and has attracted amounts of tourists. The village is in a subtropical monsoon climate zone with an average annual temperature of approximately 17.1 °C, with no heat in summer or cold in winter. The rainy season in Ping'an is from April to August, which accounts for almost 72% of the annual rainfall. This area is well watered, and the exposed hills are mostly sandy rocks, which are mostly dark green.

Figure 1. Ping'an village and Longji Terraces (Source: Photo provided by villagers).

The Ping'an village comprises traditional pile-dwelling wooden buildings with typical stilt-style architecture and a "zigzag" stone path running through the entire village. Residents of Ping'an village enjoy glutinous rice, bacon dried fish, and sour bamboo shoots, and they have their own elegant ethnic costumes with strong folklore. In terms of beliefs, there are land gods, thunder gods, frog gods, cows, ancestor worship, and Taoist gods, which are mainly related to the cultivation of terraced agriculture. Presently, the village is still dominated by the original Zhuang ethnic group, and participation in the tourism business has become a major source of income for them, with a total of 108 large and small hotels and dwelling houses, two bars, three Zhuang herbal footbaths, six external operators, and two cafés.

Compared to other ethnic tourist destinations, which rely on folk culture and natural scenery, Ping'an Village is more attractive for its terraced landscape. These terraces come from ancestors who reclaimed the land in front of deep mountains in order to survive. It shows the world the strong will of humanity to survive in nature, the wisdom and strength in understanding nature, and building a homeland.

In addition, the terraced landscape and tourism development in Ping'an village are closely linked to energy sources. As a tourist attraction, the terraces require a high level of water conservation. If construction waste from B&B and tourist waste leads to the contamination of local water sources, this may affect the terraces and ecological sustainability, as well as the sustainability of local water resources and hydropower generation. Traditional pile-dwelling wooden buildings are also posing a safety hazard in terms of fire use, and the development of tourism requires a large amount of energy. Therefore, this village represents a typical case for discussing how ethnic tourism village residents balance tourism development and energy use, which is a concept that needs to be explored in further detail.

3.2. Data Collection and Analysis

The data for this study was mainly obtained from three ongoing field surveys that were conducted from May–July 2022, and the data acquisition methods use an omnibus strategy, including web-based information, participatory observation, and in-depth interviews, the most cited format for qualitative research [76,77]. It is an approach that contains a mixture of information-gathering techniques that include diverse forms of observation [78].

The first stage of this research was conducted in May 2022 through a field pre-survey and web-based information collection [72–78]. The purpose of this stage was to understand the basic situation of the natural environment, tourism resources, infrastructure, and development status of the Longji terraces. The second stage of the research was conducted in June 2022, and a total of 21 in-depth interview samples were obtained through a convenience sampling method due to typical interviewees being more informative and contributing to a deeper understanding [72–79]. The third stage was conducted in July 2022. Based on the collation and analysis of the pre-interview data and reflection, a total of five in-depth interview samples were obtained using a purposive sampling method because typical interviewees with rich information helped to understand the case in greater depth. At the same time, to better understand the research context, the research team followed up on the energy use of tourism in the village using participant observation to understand the perceptions of the residents and their related behaviors towards energy in the field. The semi-structured questions included the knowledge of local residents on energy, the environment, and energy use before tourism development; energy consumption by households after tourism development; energy consumption by tourists related to water, electricity, transport, gas, waste, and sewage; the relationship between energy and resources related to the tourism landscape; residents' perceptions of tourists' energy use; and their own evaluations of energy awareness and opinions about the future of the tourism landscape, environment, and energy use in harmony. The interviews and on-site observations complemented each other during the study until the relevant material was nearly saturated. All Interview recordings were transcribed into text and came to around 174,000 words. In addition, 21,000 words of memo notes were taken, as well as 475 photos that were related to energy literacy and sustainable tourism.

The combined three studies resulted in a sample of 26 in-depth interviews, all of which were within ~0.5–1.5 h and recorded with the consent of the other party. Of these, 15 were male, and 11 were female. There were 23 local people and 3 external local businesses: Two tourism managers in the village, 24 people directly involved in tourism (operating different grades of B&Bs and tourism catering), one person working in tourism transport-related jobs, and one student. As there are 108 households in Ping'an Village, the 26 interviewees involved in the in-depth interviews are all from different households, and they come from different types of businesses such as restaurants, accommodation, bars, souvenir selling, and transportation, etc., and three of them are engaged in the management of both the scenic spot and the village, so they have an in-depth understanding and information of the village. The qualitative research is committed to finding the right interviewees in relation to the theory in the field, and our interviewees are typical, basically representing the different types of business and demographic characteristics in the villages. Also, our third author is of Zhuang ethnicity and knows the ethnic language of this village, enabling us to conduct interviews in greater depth.

Finally, all collected material, including transcribed interview data, web material, field notes, and photos, were stored in memos in NVivo 11 software for analysis [68]. Once the data collection was complete, the authors attempted to obtain the main ideas and key messages by reading transcripts of all observation notes and interview transcripts and then analyzing and qualitatively interpreting all the material using Thematic analysis [79,80]. The data was analyzed and coded based on how the energy literacy of the residents changed before and after tourism development; what contributed to the changes in the energy literacy of the residents; and how the energy literacy of the residents brought about changes in tourism destinations and tourists, thus contributing to the sustainable development of local tourism.

To protect the privacy of the interviewees, the interviewees were coded as "S+ interview No.". The research data were analyzed using a thematic analysis approach, firstly by open coding the original data, labeling and classifying the data sentence by sentence, making initial naming, and marking them as free nodes. Secondly, concepts of similar incidents were gathered to further abstract the free nodes that had previously been marked; then, axial coding aims to put concepts and categories back together by making connections between them. Based on situational understanding, 10 main categories were developed in a long process of continuous breaking up and stitching together of all the material and codes: low level of energy knowledge, unpleasant energy attitude, simple energy behavior before tourism; upward energy knowledge, economic attitude towards energy and positive energy behavior after tourism; external and internal factors; guiding tourist's behavior, change the structure of energy use and improve tourism infrastructure. In the open coding and categorical coding process, the three authors first read all the data materials separately and then open-coded them according to the main research questions; then, the three authors exchanged coding results and made coding decisions after thorough discussion. Finally, based on the coding decisions, all authors discussed the data in context and further analyzed the relationships between the 10 independent clusters to develop the conceptual framework. The overarching concepts and categories that emerged from this process are shown in Figure 2.

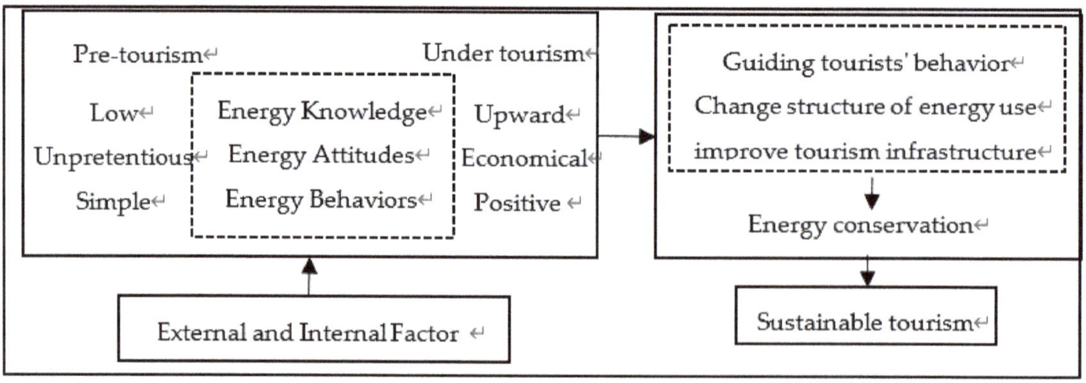

Figure 2. Conceptual Framework of the relationships between energy literacy and sustainable tourism (Source: by own study).

4. Results and Discussion

4.1. Changes in the Energy Literacy of Residents under Tourism Impacts

4.1.1. Energy Literacy of the Ping'an Village Residents in Pre-Tourism

As shown by the field data, based on the perceptions of the residents and their memories of their previous energy awareness during the interviews, energy literacy before tourism development can be described as a low level of energy knowledge, unpretentious energy attitudes, and simple energy behaviors.

Low Level of Energy Knowledge

Before tourism development, knowledge of energy was largely ignored by the residents of the Ping'an village. Their knowledge of energy was still limited to traditional energy sources, such as water, electricity, and fire, which are essential to their lives. In particular, their awareness of new energy sources, the renewable status of traditional energy reserves, and the impact of the use of energy on the environment were still insufficient. For example, before the development of their tourism industry, villagers used to cut down trees in the hills to produce firewood for cooking, and they were not fully aware of the other types of energy for catering; they only prioritized the appearance of the wooden houses in which they lived. Therefore, residents' cognition of energy was at a low level before tourism was undertaken, lacking systematic and profound cognition.

Unpleasant Energy Attitudes

Residents did not give much thought to their attitudes towards energy use nor a sense of responsibility for energy conservation. For example, before the tourism drive, there was no charge for water use in the village. The cost was the main consideration for electricity; an important issue of concern was regarding the use of fire. As the whole village lived in wooden houses, if they were not careful, not only could the wooden houses burn down, but they would also pose a threat to the entire village.

Simple Energy Behavior

For Ping'an village residents, the energy behavior they showed was generally simple before tourism development, particularly with regard to the use of water and electricity. Even though there were no tariffs imposed on water, and they were less aware of the importance of water as an energy source in the overall ecosystem, they do not waste water because of their thrifty habits. Meanwhile, since water is fundamental in rice cultivation in the terraces, village residents took an active interest in the use of water in the terraces during the different seasons. For example, they created simple water storage facilities at the top of the hill to irrigate the farmland during the dry season to prevent damage to the rice harvest. In addition, the village did not have a unified sewage treatment site, and the sewage generated by living organisms could only be treated through septic tanks built by the villagers themselves. As the residents S16 mentioned in the interviews:

"Before there was no sewer pipe, we built a septic tank, divided into two to three septic cells. The sewage flowed straight into the septic tank discharge. When it was full, we took it out to water the vegetables."

4.1.2. Energy Literacy of the Ping'an Village Residents during Tourism Development

Under the influence of tourism, the living standards of the Ping'an village residents have greatly improved, and their awareness and energy consumption are changing. With the development of local tourism, the level of energy literacy of the residents has greatly improved through their daily and continuous tourism practices.

Upward Energy Knowledge

As the data shows, residents are knowledgeable about their energy. For example, regarding water use, residents are familiar with the price of water and know the approximate amount of water they can use in their homes in a month, which can change during the low- and high-tourist seasons and the number of visitors. Regarding electricity consumption, most residents were aware of the price of electricity (RMB 0.56/kWh). Households in the tourist business are also aware that their monthly electricity consumption varies depending on the number of guests. For example, they are clear about the cost of electricity in their homes during the low and high seasons, which is nearly 2000–3000 RMB in the high season and 500 RMB in the low season. Residents are aware of the appliances that consume a large amount of electricity in their homes, such as refrigerators, air conditioners, water heaters, and other appliances that operate continuously (in addition to lighting and cooking). With regards to the use of fire, due to the safety hazards in wooden houses, residents are aware of the dangers that a fire can bring to the village. Each household has someone who knows how to use fire extinguishers, and all households are very careful about the use of fire. Furthermore, 19 out of 26 respondents are aware that the local electricity is generated from hydropower and have a better understanding of the various new energy sources, such as air heating, environmental oil, and other energy-saving products. They are also aware that energy consumption and pollution can affect the natural environment, and they even understand new energy sources that are not suitable for the area. They also mentioned that because of the large number of tourists arriving, if they all drove into the scenic spot, the carrying capacity would exceed the limit, and there would be too many exhaust emissions, which would have a negative impact on the environment.

Economic Attitude towards Energy

Residents can clearly recognize the problems that exist in their villages that require change because of tourism promotion. Twenty-three out of 26 respondents in the interviews mentioned issues that were related to the development of tourism that had occurred, such as excessive consumption of gas for tourist restaurants and excessive household water waste; the excessive use of electricity in their hostel, particularly during the tourist golden week where the village had even experienced power cuts due to the overload of electricity consumption, excess littering of non-biodegradable waste brought by tourists, and lack of road lights in the village, which made it difficult to consume and move around at night. In addition, they can actively seek ways to change, which is important in making decisions to change energy use. For example, when residents realize that the arrival of tourists brings high electricity bills along with economic income, they actively consider whether there are possibilities to reduce their electricity bills and find new energy sources. Water use is a concern for residents in terms of sewage and waste disposal. Concerning the use of water, residents are more sensitive to the disposal of sewage and waste because of the need for clean water for the terraces, the core landscape resource of the area, and the need to maintain the environmental cleanliness of the area. When 15 out of 26 respondents recalled that there was an unpleasant smell resulting from excessive hotel water waste, they mentioned the attitude and sense of responsibility of the scenic residents who took the initiative to push for a solution when faced with the problem. For example, S24 mentioned,

"We asked the government and developer for a long time, hoping to build a sewage treatment station . . . we also take turns to do cleaning for the whole village... and rubbish is transported out daily from the village. It cannot be left in the village because it is not good for the soil and water; it affects the terraces."

Residents are also more aware of the environmental impact of excessive waste disposal and actively address this issue. The positive attitudes of residents towards energy are often more based on awareness of the need for sustainable tourism and more economical energy consumption and less on attitudes towards energy conservation.

Positive Energy Behavior

Residents of the village take positive actions in their energy use to solve the problems mentioned earlier, thus saving and conserving energy. Firstly, new energy sources are used; for example, the widespread use of air heaters is associated with new energy technologies by residents who operate hotels, which largely reduces the cost of electricity. As resident S01 described:

"Compared to electric water heaters, air energy water heaters are more energy efficient; air energy heaters compress air to generate heat. In terms of price, electric water heaters demand more power, reaching three to four thousand watts; if the wattage is too high, the circuit cannot withstand it, and fire safety hazards are also present. However, the wattage of air energy is not very high, it is up to more than thousand watts. Using an electric water heater is equivalent to two or three air energies, and the capacity of air energy is much more affordable than the electric water heater."

The descriptions from the residents further confirm that they are more concerned about energy costs and willing to use new energy products and technologies in their tourism services. With the use of new energy sources, residents have also influenced each other to form a culture of energy conservation, such as the popularity of air energy use in the village. In addition, streetlamps using solar electric panels in scenic areas have solved the problem of lighting streetlamps at night, and villagers have used environmentally friendly oil instead of gas for cooking, reducing gas consumption.

From a water perspective, the village needs to tackle sewage and maintain clean water sources. In terms of sewage disposal, the entire village is built on a unified sewage pipe, which has centralized the treatment. At present, the water in small ditches in the villages is clean after sewage treatment.

Another requiring attention from the data shows that five large cisterns were established for successive classification at the top of the hill for living, terrace irrigation, agricultural production, and fire safety use, as shown in Figure 3. Meanwhile, to prevent fires in wooden houses, the village has built fire hydrants in front of each house, which is connected to the pipes of the cisterns used for fire safety. Finally, during tourism activities, residents are also willing to influence tourists to behave in an energy-saving manner, such as in the use of air conditioning in rooms, raising the temperature to the most energy-efficient level in obvious places, and remembering to turn off lights and air conditioning when going out.

Figure 3. One of the Cistern and Sewage Treatment Centres in Ping'an (Source: by author 1).

4.1.3. Factors Affecting Resident's Energy Literacy Change

Along with the development of local tourism, locals have acquired more energy literacy in their daily tourism practices. According to the field data, the factors influencing residents' change in energy literacy are multiple. Both external and internal factors contribute to the improvement of residents' energy literacy in tourism development. On the external side, economic development, communication with external tourists, diversified access to knowledge, and government support are all factors that have contributed to the change in energy literacy among Ping-an residents. Specifically, economic development has offered more possibilities for residents to choose new energy sources; communication with external tourists, especially pro-environmental visitors, has allowed them to understand that only sustainable environments can attract more tourists; the widespread use of mobile phones has also enabled residents to learn more about energy knowledge from the internet; as for government support, it includes investment in tourism infrastructure, attracts investment from outside, promotion resident's energy knowledge, and other related policy. For example, the local government's continuous promotion of energy knowledge in villages, such as fire and electricity safety, knowledge of energy conservation and environmental protection, etc. All these external factors have contributed to the improvement of residents' energy literacy.

In addition to these extrinsic reasons for promoting energy literacy, the key incentive for local people to become more energy-literate lies in their initiative in tourism development, also referred to as internal factors by residents. First, the goal of achieving better tourism development and poverty alleviation has led them to take the initiative to address the local energy problem. Before tourism, Ping'an Village was extremely isolated and poor. Up until 1992, it was still dependent on relief to survive; the income of residents was mainly from farming and working outside the village [66]. Since the development of tourism from 1993, tourism has gradually become the main source of economic income, including catering, accommodation, shopping, tour guide services, and ticket dividends [66]. By 2002, the villagers' annual per capita income reached about 2000 RMB. Up to 2011, it rose to 13,200 RMB [81]. By 2019 the villagers' annual per capita income had reached 5000 RMB just from the ticket income dividends, with some medium hotel annual income may reach 200,000 RMB. In this process, in order to get out of poverty and achieve wealth through tourism, they made efficient use of a combination of external resources, such as the government, developers, and new sources of energy in technological development, which helped them solve local problems effectively.

Second, the initiatives of residents have a positive impact on their energy literacy. As residents take the initiative to identify various energy problems that exist in local tourism, they actively seek information from the outside to solve local problems. By taking the initiative to learn and communicate with the outside, their energy literacy is enhanced. Such as the resident manager S17 mentioned,

"In the off-season, several of our village committees have gone to other scenic spots around the country to learn, to see how others are doing, to see how people are solving the problems we have, and we often go out to see, which is very helpful to us. Sometimes we also watch Tik Tok and read a lot of relevant knowledge on the internet. We also have two people in the village who specialize in live-streaming terraces, they introduce our beautiful terraces to people outside, and sometime also share and show the balance of our natural ecosystem here."

4.2. Influence of Energy Literacy on Tourism Development in Ethnic Villages

4.2.1. Effect of Residents' Energy Literacy on Tourists' Energy Literacy

The host-tourist relationship has always been central to the tourism development process. Along with tourism growth, energy consumption for restaurants, accommodation, and transportation has shown a growing trend. To properly reduce energy consumption and achieve energy savings and emission reduction in tourism activities, it is necessary for residents and tourists to focus on the input and use of energy products in tourism activities. Ethnic residents are the main actors in the operation of tourism activities and local culture, particularly in ethnic tourism areas. As hosts, the improvement in the energy literacy of residents can have a positive effect on the behavior of tourists.

Guiding Tourists to Focus on Energy and Acquiring Energy Knowledge in Tourism

As residents become more energy-literate, they apply their energy-related knowledge to tourism activities, allowing visitors to gain energy-related knowledge while enjoying the tourist landscape. For example, warning signs stating "please protect water sources" have been installed at prominent locations in the terraced landscape. In the tourist accommodation, there are various warning signs about water and electricity conservation and attention to fire prevention to ensure tourists are aware of the importance of traditional energy sources, such as water, electricity, and fire, to the village during their activities. In addition, residents use their own energy knowledge to guide tourists about energy concerns. For example, they remind tourists about fire in their communication with them; they also caution them about most of their houses being wooden structures and the consequences a fire can have on their village. To reduce costs, residents are actively concerned about the consumption of water, electricity, and other energy sources and, therefore, proactively use new energy-related products to save energy and reduce emissions in tourism. Residents who provide tourist accommodation S05 mentioned,

"It's cooler here at night, so locals don't use air conditioning, but tourists do, and they wouldn't stay without it, so we have to install air conditioning in every room and remind them to turn it off when they go on tours."

In these ways, tourists can also feel the concern of the locals in conserving energy and, thus, understand the importance of energy use for water use in the terraces and for tourism in the villages, which essentially also raises their own energy awareness.

Guiding the Attention of Tourists to the Impact of Energy Consumption Based on Energy Attitudes

Given the ecological fragility of terraces, it is important to understand the impact of energy consumption on terraced-tourism development. Residents have a clear understanding of the environmental impact of excessive energy consumption, such as the impact of cleaning chemicals on water sources and, consequently, food production. Therefore, they try to reduce the use of cleaning products or frequency of cleaning and guide visitors to replace items, such as sheets and towels, as little as possible during continuous stays or send these washed items out of the resort. Because residents recognize that plastic products are harmful to the soil and water supply because they cannot degrade, they guide visitors to reduce the use of plastic products in tourist catering to reduce the impact on the soil on which the terraces depend.

Guiding Tourists to Save Energy and Reduce Carbon Emissions through Tourism Behavior

Residents guide tourist behavior mainly in the provision of tourist accommodation, transport, and other tourist services. In terms of tourist accommodation and catering, air conditioners were installed in rooms provided by the residents. In order to save electricity, the residents provide signage near the air conditioners or on the remote control stating, "To save electricity, it is recommended to turn it on to 26 degrees" to guide tourists. In terms of tourism transport, in the Longji Terraces resort, the tourism management, and residents are aware of the impact of car emissions on the local air and have consciously chosen electric vehicles for their tourist transport services. However, owing to geographical constraints, the lack of motive power of electric vehicles makes them difficult to use in local tourist transport; therefore, tourist transport is mostly available by sightseeing vehicles and buses that burn petrol. Residents involved in tourism management S17 at the resort company mentioned,

"The cars we use now still burn petrol, and the electric cars do not have enough power to go up the mountain, so it's hardly to use them, and burning petrol will definitely have an impact on the air, but we have more trees here, so the impact will not be big."

Nevertheless, the scenic area continues to guide tourists to reduce the use of private cars to drive directly to the scenic area but rather take a scenic bus at the entrance of the scenic area to reach the village, thus, reducing carbon emissions by reducing the amount of vehicle travel.

4.2.2. Transforming the Structure of Energy Use for Sustainable Tourism

Energy consumption relies on outsourcing instead of self-sustaining household energy consumption. Before the onset of tourism, the energy consumption of residents was minimal for basic subsistence use. However, when there was a large influx of tourists, a substantial amount of energy was consumed for food and beverage, accommodation, and transport. For example, leftovers generated by local hotels and restaurants were routinely thrown out by owners into the rubbish collection pond at the entrance of the village at least once a day during the high season and once every two–three days during the low season. The waste was then transported out of the village daily to the rubbish disposal center. The laundry of local hotels was sent to a professional cleaning company in Guilin city during the high season, and used sheets were replaced once or twice a year during the high season.

New energy sources have been used instead of traditional energy sources. Electricity consumption is an important type of energy consumption; before tourists arrived, the villagers primarily used electric lighting. As tourism has grown, the demand for electricity has been constantly increasing with the rapid rise in electricity use for lighting, night landscape creation, air conditioning in tourism lodges, etc., which puts a higher demand on the supply of electricity. Thus, power outages occurred from time to time, which created negative experiences for tourists. To change the excessive demand for electricity, new energy resources, such as solar streetlights and air energy heaters, have been adopted by residents to reduce the reliance on traditional electricity. In particular, the use of air-energy water heaters has effectively improved the problems of long usage time, high replacement frequency, and high-power consumption caused by electric boilers.

A specialized sewage treatment system is used instead of direct discharge. Initially, the sanitary sewage flowed freely into the village ditches; however, as the tourism industry began developing, the ecosystem was no longer able to absorb domestic wastewater. Now, the use of a sewage treatment system has improved the drainage route by separating wastewater and clean water to effectively protect the ecological balance between the terraces for irrigation and the daily use of hostels. As resident S14 said in the interview:

"Water was used to meet the needs of the tourists instead of the irrigation of the terraces earlier, so that many terraces were deserted. Later, the abandoned terraced fields were gradually re-farmed through the construction of the cistern. Before this was done, some of us connected the water pipes randomly, similar to the discharge pipes."

In addition, to avoid damage to vegetation and soil from the use of pesticides, the use of herbicides has been banned instead of manual weeding. In addition, the number of cisterns has increased from one to five, and the function is divided into irrigation for terraces and water for fire protection.

4.2.3. Improved Infrastructure for Sustainable Tourism

Due to the need for tourism development in the village, residents and the management committee, formed by themselves, are constantly appealing to the government and developers for infrastructure changes. For example, traditional streetlights are insufficient, posing safety hazards to tourists at night. The electricity costs were unevenly shared and unmanaged. Improper treatment of sewage affects the irrigation of the terraces, which has environmental and health implications. The improper management of traffic can lead to congestion for visitors as well as excessive exhaust emissions that affect the ecological environment. As owners of the resources, residents are constantly engaging with developers and the government to improve the infrastructure of the village, thus promoting the sustainable development of tourism sites.

The landscape lighting system has been improved in the village. The use of solar-powered streetlights saves electricity while solving the previous situation of no public lighting system in the village at night as well as the apportion of the electricity bill for streetlights. It also creates a beautiful night landscape that enhances the tourism experiences of visitors. As villager S16 said:

"We didn't have enough streetlights before as well as no illumination in the tourist attractions. However, the streetlights are now quite good. Whether it is a rainy or sunny day, solar streetlamps can be bright for a few hours to improve the convenience and safeguard for tourists at night."

The construction of the sewage treatment system, water storage, and firefighting systems has promoted tourism sustainability. The construction of the sewage system in the village was conducted from 2012–2014 near the village. The completion of the sewage system solved the problem of the increasing domestic sewage resulting from the increasing number of tourists, and it ensured irrigation of the terraces and residents' daily use. Its original sewage piping was the first sewage system built in the county. The water storage system was constructed in batches. Before tourism, residents built a cistern at the top of the hill for terrace irrigation. As tourism gradually developed, the villagers then pooled together their money to build two cisterns to meet the need for tourism development and to increase the number of tourists. When the tourism industry grew further, two other cisterns were built with the help of the government and developer near the hiking area for tourists, as well as to increase the firefighting facilities and equipment in the village. More than 300 fire hydrants have been built throughout the village. Regarding safety, as most houses are wooden and brick structures with fire hazards, residents are equipped with fire extinguishers in their tourist accommodation, and the village committee supervises autonomy to ensure the safety of tourist accommodation. Thus, the construction of the water storage system and firefighting system not only ensured seasonal water usage for terrace irrigation to increase tourist attractiveness by maintaining the integrity of the terraced landscape but also ensured the safety of the residents.

Traffic facilities for tourists have been improved and preserved in scenic areas. The construction of car parking at the main entrance of the scenic area and the extension of sightseeing cars and new energy trams has effectively alleviated congestion and excessive energy consumption for tourists. The buses are outsourced to Revitalize Sightseeing Ltd. for their operation. Generally, group visitor cars with more than seven seats must be replaced with scenic buses. The scenery now consists of 30–40 oil-burning vehicles, with six or seven new energy trams added over the years. In addition to car parking at tourist entrances, multi-story car parking has also been built at village entrances. After parking at the gate, tourists are required to hike into the village. Only a few tour buses are available for transporting day trippers to the viewing platform. These measures reduce the energy consumption of the traffic load.

5. Discussions

Rather than the traditional discussion of households [1–5,17] and students [20,23,24], this study extends the understanding of energy literacy from the perspective of ethnic residents and discusses the changes in energy literacy in remote ethnic villages in the context of tourism development. Previous studies of energy literacy have tended to discuss the relationships between knowledge, attitudes, and behavior [6,19–21], focusing on the synchrony perspectives [22–25], with less understanding of energy literacy from diachronic perspectives along with the changing external environment. This study examines how the energy literacy of ethnic minority residents has changed during the development of tourism and the main reason for the changes. In contrast to previous studies that have considered ethnic minorities as a vulnerable group in terms of energy use [3], this study empirically demonstrates that tourism development in ethnic minority areas may also contribute to the transformation of residents from a vulnerable group in terms of energy use to a more energy literate group, thus contributing to the sustainability of tourism.

This study also deepens the understanding of the implications of such changes for local tourism sustainability from a cross-disciplinary perspective and enriches empirical research to better promote energy conservation and sustainable tourism development in ethnic areas. In addition, classical literature on energy and tourism has tended to discuss the impact of energy resources as tourism attractions [8,9,42,43] or aim to improve tourists' energy literacy through tourism [2,10,45], yet how residents influence tourists' energy behavior remains unclear. Our research extends the understanding that an increase in residents' energy literacy can also lead to an increase in visitors' energy behavior. Furthermore, beyond the traditional relationship of host and guest in ethnic tourism research [16], we find that increased energy literacy among residents can also contribute to the structure of energy use and infrastructure development in tourist destinations, which further enhances the understanding of how energy literacy can influence the sustainable development of tourist destinations.

6. Conclusions and Policy Implications

This study proposes a theoretical framework for understanding the relationship between the energy literacy of residents and sustainable tourism development in ethnic areas and highlights the important role of the initiatives of residents in improving their energy literacy in tourism development. The conclusions of this study are as follows:

As the tourism industry in the village has developed, the energy literacy of the residents has changed. Prior to the development of tourism, the energy literacy of Ping'an village residents could be summarized as a low level of energy knowledge, unpleasant energy attitudes, and simple energy behavior. With the development of tourism in Ping'an, the energy literacy of residents has changed to upward energy knowledge, economic attitude towards energy, and positive energy behavior. Both external and internal factors contribute to the improvement of residents' energy literacy in tourism development. External factors contain economic development, communication with external tourists, diversified access to knowledge, and government support. The key reason for this improvement is the internal factors that form the initiative of residents in Ping'an to seek self-change by developing tourism in order to get out of poverty and achieve wealth through tourism. This positive determination to change the backward village has led to a greater increase in the energy literacy of the residents in the village.

The improved energy literacy of the residents affects tourist behavior and sustainable tourism destinations. In Ping'an village, residents guide tourists in three main areas, guiding them to pay attention to energy and gain energy cognition in tourism; in energy attitudes, guiding them to be concerned about the impact of energy consumption; and in tourism behavior, guiding them to save energy and reduce carbon emissions. Moreover, increasing the energy literacy of residents also impacts the sustainable development of the tourist site mainly by changing the structure of energy use and improving the infrastructure of the tourist site.

This study also has implications for policy makers and managers of tourist destinations. When considering how to make public policy on energy literacy in tourist areas, the initiative of residents can be used as a point of regulation. Before providing energy education, in addition to attracting tourism investment, it would be useful to motivate residents to seek their own initiative for tourism development. Only if residents can actively seek tourism development on their own will they be better able to promote their demand for energy literacy improvement. Meanwhile, in the process of training in energy literacy related to tourism, it is also necessary to provide guidance on the energy attitudes of residents until they are internalized in their daily energy behavior, thus encouraging them to optimize their energy use. For the managers of tourist resorts, they should also recognize that improving the energy literacy of residents will contribute to guiding the energy behavior of tourists, which will better protect local tourism resources and promote sustainable tourism development. Therefore, resort managers should take the initiative to cooperate with residents, identify the energy-related problems in the process of tourism development, and guide the resources of various stakeholders to solve the corresponding problems, so that the resort can obtain sustainable development.

Although this study is based on a survey of Ping'an residents, the findings can be extrapolated to other ethnic tourism areas and tourism products that are somewhat dependent on energy and environmental requirements, such as ecotourism, mountain tourism, and rural tourism. Due to the limitations of the research conditions, the research efforts of the researchers, and the overall sample size of the village residents, this study mainly collected data based on qualitative research through field surveys and in-depth interviews to provide a relatively in-depth understanding of the energy literacy of ethnic residents and sustainable tourism. However, it is also worth using quantitative research in the future to measure the energy literacy of the perspectives of tourists to discuss the impact on local, sustainable tourism.

Author Contributions: Conceptualization, S.W., Y.L. and C.F.; Data curation, S.W., C.F. and P.J.; Formal analysis, S.W. and C.F.; Funding acquisition, S.W.; Investigation, S.W., C.F. and P.J.; Methodology, S.W., Y.L., C.F. and P.J.; Project administration, S.W. and C.F.; Supervision, Y.L., C.F. and P.J.; Writing—original draft, S.W. and C.F.; Writing—review & editing, S.W. and Y.L. All authors have read and agreed to the published version of the manuscript.

Funding: This research was funded by the Natural Science Foundation of China Study Foundation of China, grant number 42001165, Natural Science Foundation of Hunan Province of China, grant number 2021JJ40351, the Hunan Education Department Scientific Research Project, grant number 19B336, and the Humanities and Social Sciences Project of the Training Plan for 1000 Young and Middle-aged Key Teachers in Guangxi Universities (2021QGRW061).

Conflicts of Interest: The authors declare no conflict of interest.

References

1. van den Broek, K.L. Household energy literacy: A critical review and a conceptual Typology. *Energy Res Soc. Sci.* **2019**, *57*, 101256. [CrossRef]
2. Frantál, B.; Urbánková, R. Energy tourism: An emerging field of study. *Curr. Issues Tour.* **2014**, *20*, 1395–1412. [CrossRef]
3. Adams, J.; Kenner, A.; Leone, B.; Rosenthal, A.; Sarao, M.; Boi-Doku, T. What is energy literacy? Responding to vulnerability in Philadelphia's energy ecologies. *Energy Res. Soc. Sci.* **2022**, *91*, 102718. [CrossRef]
4. Lee, L.-S.; Chang, L.-T.; Lai, C.-C.; Guu, Y.-H.; Lin, K.-Y. Energy literacy of vocational students in Taiwan. *Environ. Educ. Res.* **2015**, *23*, 855–873. [CrossRef]
5. Zhang, J.; Zhang, Y. Examining the energy literacy of tourism peasant households in rural tourism destinations. *Asia Pac. J. Tour. Res.* **2020**, *25*, 441–456. [CrossRef]
6. DeWaters, J.E.; Powers, S.E. Energy literacy of secondary students in New York State (USA): A measure of knowledge, affect, and behavior. *Energy Policy* **2011**, *39*, 1699–1710. [CrossRef]
7. DeWaters, J.E.; Powers, S. Establishing measurement criteria for an energy literacy questionnaire. *J. Environ. Educ.* **2013**, *44*, 38–55. [CrossRef]
8. Frew, E.A. Industrial tourism theory and implemented strategies. In *Advances in Culture, Tourism and Hospitality Research, Issue 2: Industrial Tourism Theory and Implemented Strategies*; Woodside, A., Ed.; Emerald: Bingley, UK, 2008; pp. 27–42.
9. Otgaar, A. Towards a common agenda for the development of industrial tourism. *Tour Mana Perspect.* **2012**, *4*, 86–91. [CrossRef]
10. Lilley, M.B.; Firestone, J.; Kempton, W. The effect of wind power installations on coastal tourism. *Energies* **2010**, *3*, 1–22. [CrossRef]
11. Ladenburg, J.; Termansen, M.; Hasler, B. Assessing acceptability of two onshore wind power development schemes: A test of viewshed effects and the cumulative effects of wind turbines. *Energy* **2013**, *54*, 45–54. [CrossRef]
12. Horng, J.S.; Hu, M.L.M.; Teng, C.C.C.; Hsiao, H.L.; Liu, C.H.S. Development and validation of the low-carbon literacy scale among practitioners in the Taiwanese tourism industry. *Tour. Manag.* **2013**, *35*, 255–262. [CrossRef]
13. Teng, C.C.; Horng, J.S.; Hu, M.L.M.; Chen, P.C. Exploring the energy and carbon literacy structure for hospitality and tourism practitioners: Evidence from hotel employees in Taiwan. *Asia Pac. J. Tour. Res.* **2014**, *19*, 451–468. [CrossRef]
14. Fraser, R. Cultural heritage, ethnic tourism, and minority-state relations amongst the Orochen in north-east China. *Int. J. Herit. Stud.* **2020**, *26*, 178–200. [CrossRef]
15. Tian, B.; Stoffelen, A.; Vanclay, F. Ethnic tourism in China:tourism-related (dis)empowerment of Miao villages in Hunan province. *Tour. Geogr.* **2021**, 1–20. [CrossRef]
16. Zhang, J.; Xu, H.G.; Xing, W. The host-guest interactions in ethnic tourism, Lijiang, China. *Curr. Issues Tour.* **2017**, *20*, 724–739. [CrossRef]
17. Cowen, L.; Gatersleben, B. Testing for the size heuristic in householders' perceptions of energy consumption. *Environ. Psychol.* **2017**, *54*, 103–115. [CrossRef]
18. Johnson, P.C. Cultural literacy, cosmopolitanism and tourism research. *Ann. Tour. Res.* **2014**, *44*, 255–269. [CrossRef]

19. Hungerford, H.R.; Volk, T.L. Changing Learner Behavior through Environmental Education. *J. Environ. Educ.* **1990**, *21*, 8–21. [CrossRef]
20. Alp, E.; Ertepinar, H.; Tekkaya, C.; Yilmaz, A. A Survey on Turkish Elementary School Students Environmental Friendly Behaviours and Associated Variables. *Environ. Educ. Res.* **2008**, *14*, 129–143. [CrossRef]
21. Rioux, L. Promoting pro–Environmental Behaviour: Collection of Used Batteries by Secondary School Pupils. *Environ. Educ. Res.* **2011**, *17*, 353–373. [CrossRef]
22. Wood, G.; Newborough, M. Dynamic energy-consumption indicators for domestic appliances: Environment, behaviour and design. *Energy Build.* **2003**, *35*, 821–841. [CrossRef]
23. Chen, S.J.; Chou, Y.C.; Yen, H.Y.; Chao, Y.L. Investigating and structural modeling energy literacy of high school students in Taiwan. *Energy Effic.* **2015**, *8*, 791–808. [CrossRef]
24. Davis, P. The attitude and knowledge of Tasmanian secondary students towards energy conservation and the environment. *Res. Sci. Educ.* **1985**, *15*, 68–75. [CrossRef]
25. Gatersleben, B.; Steg, L.; Vlek, C. Measurement and determinants of environmentallysignificant consumer behavior. *Environ. Behav.* **2002**, *34*, 335–362. [CrossRef]
26. Bodzin, A.M.; Fu, Q.; Peffer, T.E.; Kulo, V. Developing energy literacy in US middle-level students using the geospatial curriculum approach. *Int. J. Sci. Educ.* **2013**, *35*, 1561–1589. [CrossRef]
27. Kyriazi, P.; Mavrikaki, E. Development of an Instrument to Measure Environmental Literacy of Post-Secondary Greek Students: Pilot Testing and Preliminary Results. *Accessed Jan.* **2013**, *17*, 2015.
28. Kollmuss, A.; Agyeman, J. Mind the Gap: Why Do People Act Environmentally and What Are the Barriers to pro-Environmental Behavior? *Envrion. Educ. Res.* **2002**, *8*, 239–260. [CrossRef]
29. van den Broek, K.L.; Walker, I. Exploring the perceptions of drivers of energy behavior. *Energy Policy* **2019**, *129*, 1297–1305. [CrossRef]
30. Chaiken, S. Heuristic versus systematic information processing and the use of source versus message cues in persuasion. *Pers. Soc. Psychol.* **1980**, *39*, 752–766. [CrossRef]
31. Kempton, W.; Harris, C.K.; Keith, J.G.; Weihl, J.S. Do consumers know what works in energy conservation? *Marriage Fam. Rev.* **1985**, *9*, 115–133. [CrossRef]
32. Frick, J.; Kaiser, F.G.; Wilson, M. Environmental knowledge and conservation behavior: Exploring prevalence and structure in a representative sample. *Pers. Individ. Differ.* **2004**, *37*, 1597–1613. [CrossRef]
33. Białynicki-Birula, P.; Makieła, K.; Mamica, Ł. Energy Literacy and Its Determinants among Students within the Context of Public Intervention in Poland. *Energies* **2022**, *15*, 5368. [CrossRef]
34. Remund, D.L. Financial literacy explicated: The case for a clearer definition in an increasingly complex economy. *J. Consum. Aff.* **2010**, *44*, 276–295. [CrossRef]
35. Poujade, B.; Hautier, C.; Rouard, A. Determinants of the energy cost of front-crawl swimming in children. *Eur. J. Appl. Physiol.* **2002**, *87*, 102–108. [CrossRef]
36. Lindén, A.L.; Carlsson-Kanyama, A.; Eriksson, B. Efficient and inefficient aspects ofresidential energy behaviour: What are the policy instruments for change? *Energy Policy* **2006**, *3*, 1918–1927. [CrossRef]
37. Urry, J. *The Tourist Gaze: Leisure and Travel in Contemporary Societies*; Sage: London, UK, 1990.
38. Baird, J.C.; Brier, J.M. Perceptual awareness of energy requirements of familiar objects. *Appl. Psychol.* **1981**, *66*, 90–96. [CrossRef]
39. Liu, X.; Mckeough, A. Developmental growth in students' concept of energy: Analysis of selected items from the TIMSS database. *J. Res. Sci. Teach.* **2005**, *42*, 493–517. [CrossRef]
40. Cotton, D.R.; Miller, W.; Winter, J.; Bailey, I.; Sterling, S. Developing Students' Energy Literacy in Higher Education. *Int. J. Sustain. High. Educ.* **2015**, *16*, 456–473. [CrossRef]
41. Edwards, J.A.; Llurde's, J.C. Mines and quarries: Industrial heritage tourism. *Ann. Tour. Res.* **1996**, *23*, 341–363. [CrossRef]
42. Reid, I.R. Adipose tissue and bone. *Clin. Rev. Bone Miner. Metab.* **2009**, *7*, 207–209. [CrossRef]
43. Mbzibain, A.; Hocking, T.J.; Tate, G.; Ali, S. Renewable enterprises on UK farms:Assessing levels of uptake, motivations and constraints to widespread adoption. *Biomass Bioenergy* **2013**, *49*, 28–37. [CrossRef]
44. Devine-Wright, P. Energy citizenship: Psychological aspects of evolution in sustainable energy technologies. In *Governing Technology for Sustainability*; Murphy, J., Ed.; Earthscan: London, UK, 2007; pp. 63–89.
45. Butler, L.; Neuhoff, K. Comparison of feed-in tariff, quota and auction mechanisms to support wind power development. *Renew. Energy* **2008**, *33*, 1854–1867. [CrossRef]
46. Westerberg, V.; Jacobsen, J.B.; Lifran, R. The case for offshore wind farms, artifificial reefs and sustainable tourism in the French Mediterranean. *Tour Manag.* **2013**, *34*, 172–183. [CrossRef]
47. Van der Horst, D. NIMBY or not? Exploring the relevance of location and the politics of voiced opinions in renewable energy siting controversies. *Energy Policy* **2007**, *35*, 2705–2714. [CrossRef]
48. Molnarova, K.; Sklenicka, P.; Stiborek, J.; Svobodova, K.; Salek, M.; Brabec, E. Visual preferences for wind turbines: Location, numbers and respondent characteristics. *Appl. Energy* **2012**, *92*, 269–278. [CrossRef]
49. Franta ́l, B.; Kunc, J. Wind turbines in tourism landscapes: Czech Experience. *Ann. Tour. Res.* **2011**, *38*, 499–519. [CrossRef]
50. Kelly, J.; Williams, P.W. Modelling tourism destination energy consumption and greenhouse gas emissions: Whistler, British Columbia, Canada. *J. Sustain. Tour.* **2007**, *15*, 67–90. [CrossRef]

51. Qin, Q.; Hsu, C. Urban travelers' pro-environmental behaviors: Composition and role of pro-environmental contextual force. *Tour. Manag.* **2022**, *92*, 104561. [CrossRef]
52. Becken, S.; Simmons, D.G.; Frampton, C. Energy use associated with different travel choices. *Tour. Manag.* **2003**, *24*, 267–277. [CrossRef]
53. Gossling, S. National emissions from tourism: An overlooked policy challenge? *Energy Policy* **2013**, *59*, 433–442. [CrossRef]
54. Chen, M.; Tung, P. The moderating effect of perceived lack of facilities on consumers recycling intentions. *Environ. Behav.* **2010**, *42*, 824–844. [CrossRef]
55. Fedyk, W.; Sołtysik, M.; Olearnik, J.; Barwicka, K.; Mucha, A. How overtourism threatens large urban areas: A case study of the city of Wrocław, Poland. *Sustainability* **2020**, *12*, 1783. [CrossRef]
56. Becken, S.; Simmons, D. Understanding energy consumption patterns of tourist attractions and activities in New Zealand. *Tour. Manag.* **2002**, *23*, 343–354. [CrossRef]
57. Cole, V.; Sinclair, A. Measuring the ecological footprint of a Himalayan tourist center. *Mt. Res. Dev.* **2002**, *22*, 132–141. [CrossRef]
58. Saenz-de-Miera, O.; Rossell_o, J. Modeling tourism impacts on air pollution: The case study of PM10 in Mallorca. *Tour. Manag.* **2014**, *40*, 273–281. [CrossRef]
59. Qureshi, M.I.; Rasli, A.; Awan, U.; Ma, J.; Ali, G.; Faridullah, F.; Alam, A.; Sajjad, F.; Zaman, K. Environment and air pollution: Health services bequeath to grotesque menace. *Environ. Sci. Pollut. Res.* **2014**, *22*, 3467–3476. [CrossRef] [PubMed]
60. de Vita, G.; Katircioglu, S.; Altinay, L.; Fethi, S.; Mercan, M. Revisiting the environmental Kuznets curve hypothesis in a tourism development context. *Environ. Sci. Pollut. Res.* **2015**, *22*, 16652–16663. [CrossRef]
61. Han, H. Consumer behavior and environmental sustainability in tourism and hospitality: A review of theories, concepts, and latest research. *J. Sustain. Tour.* **2021**, *29*, 1021–1042. [CrossRef]
62. Esparon, M.; Gyuris, E.; Stoeckl, N. Does ECO certification deliver benefits? An empirical investigation of visitors' perceptions of the importance of ECO certification's attributes and of operators' performance. *J. Sustain. Tour.* **2014**, *22*, 148–169. [CrossRef]
63. Lee, T.H.; Jan, F.-H.; Yang, C.-C. Conceptualizing and measuring environmentally responsible behaviors from the perspective of community-based tourists. *Tour. Manag.* **2013**, *36*, 45–468. [CrossRef]
64. Wu, J.; Font, X.; Liu, J. Tourists' pro-environmental behaviors: Moral obligation or disengagement? *J. Travel Res.* **2020**, *3*, 735–748. [CrossRef]
65. Juvan, E.; Dolnicar, S. The attitude–behaviour gap in sustainable tourism. *Ann. Tour Res.* **2014**, *48*, 76–95. [CrossRef]
66. Baker, M.A.; Davis, E.A.; Weaver, P.A. Eco-friendly attitudes, barriers to participation, and differences in behavior at green hotels. *Cornell Hotel. Restaur. Adm. Q.* **2014**, *55*, 89–99. [CrossRef]
67. Mogles, N.; Padget, J.; Gabe-Thomas, E.; Walker, I.; Lee, J. A computational model for designing energy behaviour change interventions. *User Model. User-Adapt. Interact.* **2018**, *28*, 1–34. [CrossRef] [PubMed]
68. Li, S.; Cheng, Z.; Tong, Y.; He, B. The Interaction Mechanism of Tourism Carbon Emission Efficiency and Tourism Economy High-Quality Development in the Yellow River Basin. *Energies* **2022**, *15*, 69–75. [CrossRef]
69. Alcock, I.; White, M.P.; Taylor, T.; Coldwell, D.F.; Gribble, M.O.; Evans, K.L.; Corner, A.; Vardoulakis, S.; Fleming, L.E. Green' on the ground but not in the air: Pro-environmental attitudes are related to household behaviors but not discretionary air travel. *Glob. Envrion. Change* **2017**, *42*, 136–147. [CrossRef]
70. Howell, R.A. Carbon management at the household level: A defifinition of carbon literacy and three mechanisms that increase it. *Carbon Manag.* **2018**, *9*, 25–35. [CrossRef]
71. Miles, M.B.; Huberman, A.M. *Qualitative Data Analysis: An Expanded Sourcebook*, 2nd ed.; Sage: London, UK, 1994.
72. Hua, H.; Pan, Y.; Yang, X.; Wang, S.; Shi, Y. Dynamic Relations between Energy Carbon Footprint and Economic Growth in Ethnic Minority Autonomous Regions, China. *Energy Procedia* **2012**, *17*, 273–278. [CrossRef]
73. Mou, N.; Liu, Z.; Zheng, Y.; Makkonen, T.; Yang, T.; Zhang, L. Cycling in Tibet: An analysis of tourists' spatiotemporal behavior and infrastructure. *Tour. Manag.* **2022**, *88*, 104418. [CrossRef]
74. Nepal, S.K. Tourism-induced rural energy consumption in the Annapurna region of Nepal. *Tour. Manag.* **2008**, *29*, 89–100. [CrossRef]
75. Luo, F.; Becken, S.; Zhong, Y. Changing travel patterns in China and 'carbon footprint' implications for a domestic tourist destination. *Tour. Manag.* **2018**, *65*, 1–13. [CrossRef]
76. Jaimangal-Jones, D Utilising ethnography and participant observation in festival and event research. *Int. J. Event Festiv. Manag.* **2014**, *5*, 39–55. [CrossRef]
77. Ervin, M. *Applied Anthropology*; Allyn & Bacon: Needham Heights, MA, USA, 2000; p. 142.
78. Kozinets, R.V. *Netnography: Redefined*; Sage: London, UK, 2015.
79. Patton, M. *Qualitative Evaluation and Research Methods*; Sage: Newbury Park, CA, USA, 1990; p. 169.
80. Ritchie, J.; Spencer, L. Qualitative data analysis for applied policy research. In *Analysing Qualitative Data*; Bryman, A., Burgess, R., Eds.; Routledge: London, UK, 1994; pp. 173–194.
81. Wu, Z.; Wu, S. Research on Rural Tourism and Income Increase of Zhuang Farmers. *J. Guangxi Univ. Natl. (Philos. Soc. Sci. Ed.)* **2014**, *03*, 54–61.

MDPI
St. Alban-Anlage 66
4052 Basel
Switzerland
Tel. +41 61 683 77 34
Fax +41 61 302 89 18
www.mdpi.com

Energies Editorial Office
E-mail: energies@mdpi.com
www.mdpi.com/journal/energies

www.ingramcontent.com/pod-product-compliance
Lightning Source LLC
Chambersburg PA
CBHW061826210926
51850CB00006B/820
9783036572802